DYNAMICS OF THE SOLAR SYSTEM

THIS VOLUME IS DEDICATED TO YUSUKE HAGIHARA,
PROFESSOR OF ASTRONOMY AT THE UNIVERSITY OF TOKYO FROM 1935 TO 1957,
DIRECTOR OF THE TOKYO OBSERVATORY FROM 1946 TO 1957,
AUTHOR OF A MONUMENTAL FIVE-VOLUME TREATISE ON CELESTIAL MECHANICS,
AND LONG AN INSPIRING LEADER AND TEACHER.

1897-1979

INTERNATIONAL ASTRONOMICAL UNION

UNION ASTRONOMIQUE INTERNATIONALE

SYMPOSIUM No. 81

PROCEEDINGS OF THE 81ST SYMPOSIUM OF THE INTERNATIONAL ASTRONOMICAL UNION HELD IN TOKYO, JAPAN, 23-26 MAY, 1978

DYNAMICS OF THE SOLAR SYSTEM

EDITED BY

RAYNOR L. DUNCOMBE

The University of Texas at Austin

D. REIDEL PUBLISHING COMPANY

DORDRECHT : HOLLAND / BOSTON : U.S.A. / LONDON : ENGLAND

Library of Congress Cataloging in Publication Data

Main entry under title:

Dynamics of the solar system.

(Symposium - International Astronomical Union; no. 81)
Cosponsored by the Committee on Space Research of the International Council of Scientific Unions (COSPAR) and the International Union of Theoretical and Applied Mechanics (IUTAM).
Includes bibliographical references and index.
1. Solar system–congresses. I. Duncombe, Raynor L. II. International Astronomical Union. III. International Council of Scientific Unions. Committee on Space Research. IV. International Union of Theoretical and Applied Mechanics. V. Series: International Astronomical Union. Symposium; no. 81.
QB500.5.D94 523.2 79-10488
ISBN 90-277-0976-9
ISBN 90-277-0977-7 pbk.

Published on behalf of
the International Astronomical Union
by
D. Reidel Publishing Company, P. O. Box 17, Dordrecht, Holland

Sold and distributed in the U.S.A., Canada, and Mexico
by D. Reidel Publishing Company, Inc.
Lincoln Building, 160 Old Derby Street, Hingham,
Mass. 02043, U.S.A.

Printed in The Netherlands

TABLE OF CONTENTS

PART II. PLANETARY AND LUNAR THEORIES

PART III. EPHEMERIDES, EQUINOX AND OCCULTATIONS

PART IV. SATELLITES AND RINGS

PART VI. COMETS

INTRODUCTION

B. G. Marsden
Harvard-Smithsonian Center for Astrophysics
Cambridge, MA 02138, U.S.A.

IAU Symposium No. 81, "Dynamics of the Solar System", was held at the Hydrographic Office, Tokyo, Japan, during 23-26 May 1978. The Symposium was cosponsored by COSPAR and IUTAM, and generous financial support was also provided by the Japan Society for the Promotion of Science. IAU sponsorship was through Commissions 4, 7 and 20, and the Scientific Organizing Committee consisted of the current Presidents, Vice Presidents and immediate Past Presidents of these Commissions: V. K. Abalakin, R. L. Duncombe, Y. Kozai, L. Kresâk, B. G. Marsden (Chairman), P. J. Message, A. M. Sinzi, G. Sitarski and V. G. Szebehely.

There were 64 participants from 15 countries, and 55 invited and contributed papers were read. The papers covered all branches of research on solar-system dynamics, and the eight sessions (chaired by Y. Kozai, V. G. Szebehely, W. Fricke, A. M. Sinzi, G. Sitarski, B. G. Marsden, R. L. Duncombe and J. Schubart, respectively) basically included papers on the stability of the solar system, the restricted and general three-body problems, the status of the Ephemeris, theories of the motions of the natural satellites, studies on commensurable motions of the minor planets, dynamical and physical investigations on the origin and evolution of comets, planetary theory, and the dynamics of the rings of Saturn and Uranus. The order of the papers has been somewhat rearranged in this proceedings volume (edited by R. L. Duncombe), which presents the scientific contributions in six main sections.

Only twice has an IAU Symposium ever been held in Japan, and it was particularly appropriate that a meeting on the dynamics of the solar system could be held in the homeland of Prof. Y. Hagihara, the most distinguished living celestial mechanician. Prof. Hagihara served as a Vice President of the IAU during 1961-1967 and as President of Commission 7 during 1964-1967, and although now 81 years old, he was able to attend at least part of the sessions on each day of the Symposium. As noted elsewhere, these proceedings are being dedicated to him. During the Symposium Prof. Hagihara was also honored by having his name attached both to a minor planet and to a certain hyperelliptic integral that plays an important role in the theory of the Trojan group of minor planets. The citation for the name of the minor planet appears in full on Minor Planet Circ. Nos. 4419-4420; for information about the Hagihara integral, the reader is referred to the paper by B. Garfinkel in this volume.

The efforts of the local organizers in making IAU Symposium No. 81 a success are greatly appreciated. The Local Organizing Committee consisted of G. Hori, H. Kinoshita (Secretary), Y. Kozai (Chairman) and A. M. Sinzi, but sincere thanks are also due the great number of students and other assistants who ensured that the Symposium was conducted with typical Japanese efficiency.

IN MEMORIAM

The editor notes with sorrow the death of Prof. Y. Hagihara on 29 January 1979.

ACKNOWLEDGEMENT

In the preparation of this volume, the editor gratefully acknowledges the cooperation and assistance of Dr. Y. Kozai, Dr. B. G. Marsden, Mrs. J. S. Duncombe, Chris Wing and Susie Thorn.

---

1. Nakai, H.*
2. Yoshida, H.*
3. Noguchi, M.*
4. Soma, M.*
5. Nakano, S.**
6. Van Flandern, T.C.
7. Sasaki, M.*
8. Inoue, K.
9. Sato, H.*
10. Nagai, R.
11. Yuasa, M.
12. Kubo, Y.
13. Brahic, A.
14. Schubart, J.
15. Sinzi, A.M.
16. Sitarski, G.
17. Garfinkel, B.
18. Szebehely, V.G.
19. Hagihara, Y.
20. Fricke, W.
21. Kozai, Y.
22. Marsden, B.G.
23. Duncombe, R.L.
24. Message, P.J.
25. Kresák, L.
26. Kinoshita, H.
27. Marchal, C.
28. Bretagnon, P.
29. Nacozy, P.E.
30. (Cheon's Son)
31. Cheon, K.J.
32. Goldreich, P.
33. Sugawa, C.
34. Nakamura, T.
35. Froeschlé, C.
36. Scholl, H.
37. Shelus, P.J.
38. Aoki, S.
39. Delsemme, A.H.
40. Mrs. Delsemme**
41. Yabushita, S.
42. Omarov, T.B.
43. Everhart, E.
44. Stellmacher, I.
45. Liu, S.
46. Rickman, H.
47. Kitamura, M.**
48. Hori, G.
49. Carusi, A.
50. Lagerkvist, C.I.
51. Kiang, T.
52. Yoshida, J.
53. Kimura, H.
54. Hatanaka, F.*
55. Matsunami, N.
56. Tomita, K.
57. Greenberg, R.
58. Borderies, N.
59. Hatanaka, Y.
60. Dvorak, R.
61. Bec-Borsenberger, A.
62. Inoue, T.
63. Watanabe, N.
64. Weissman, P.R.
65. Jefferys, W.H.
66. Sekiguchi, N.

* L.O.C. staff. ** registered guest.

IAU Symposium No.81 Dynamics of the Solar System 1978 May 23−26 Tokyo

LIST OF PARTICIPANTS

Aoki, S. : Tokyo Astron. Obs., Tokyo, Japan

Borderies, N. : CNES/GRGS, Toulouse, France

Bec-Borsenberger, A.: Bureau des Longitudes, Paris, France

Brahic, A. : Observatoire de Paris, Meudon, France

Bretagnon, P. : Bureau des Longitudes, Paris, France

Carusi, A. : Laboratorio di Astrofisica Spaziale, Rome, Italy

Chawla, P.D. : Delhi University, Delhi, India

Cheon, K.J. : Korean University, Tokyo, Japan

Delsemme, A.H. : Toledo University, Toledo, U.S.A.

Duncombe, R.L. : Texas University, Austin, Texas, U.S.A.

Dvorak, R. : Universität Graz, Graz, Austria

Everhart, E. : Denver University, Denver, Colorado, U.S.A.

Fricke, W. : Astronomisches Rechen-Institut, Heidelberg, F.R.Germany

Froeschlé, C. : Observatoire de Nice, Nice, France

Garfinkel, B. : Yale University, New Haven, Connecticut, U.S.A.

Goldreich, P. : CALTECH, Pasadena, California, U.S.A.

Greenberg, R. : Planetary Science Institute, Tucson, Arizona, U.S.A.

Hagihara, Y. : Tokyo University, Tokyo, Japan

Hasegawa, I. : Nara, Japan

Hatanaka, Y. : Tokyo Astron. Obs., Tokyo, Japan

Henrard, J. : Namur University, Namur, Belgium

Hirayama, T. : Tokyo Astron. Obs., Tokyo, Japan

Hori, G. : Tokyo University, Tokyo, Japan

Hurukawa, K. : Tokyo Astron. Obs., Tokyo, Japan

Inoue, K. : Hydrographic Department, Tokyo, Japan

Inoue, T. : Kyoto Sangyo University, Kyoto, Japan

Jefferys, W.H. : Texas University, Austin, Texas, U.S.A.

Kiang, T. : Dunsink Observatory, Dublin, Ireland

Kimura, H. : Purple Mountain Obs., Nanking, China

Kinoshita, H. : Tokyo Astron. Obs., Tokyo, Japan

Kozai, Y. : Tokyo Astron. Obs., Tokyo, Japan

Kresák, L. : Astronomical Institute, Bratislava, Czechoslovakia

Kubo, Y. : Hydrographic Department, Tokyo, Japan

Lagerkvist, C.I.: Astron. Obs., Uppsala, Sweden

Marchal, C. : ONERA, Paris, France

Marsden, B.G. : Smithsonian Astrophys. Obs., Cambridge, Mass., U.S.A.

Matsunami, N. : Tokto Astron. Obs., Tokyo, Japan

Message, P.J. : Liverpool University, Liverpool, United Kingdom

Nacozy, P.E. : Texas University, Austin, Texas, U.S.A.

Nagai, R. : Tokyo Astron. Obs., Tokyo, Japan

Nakamura, T. : Tokyo Astron. Obs., Tokyo, Japan

Omarov, T.B. : Astrophys. Institute, Kazakh Academy of Sciences, Almata, U.S.S.R.

Owaki, N. : Tokyo Gakugei University, Tokyo, Japan

Rickman, H. : Observatoire de Nice, Nice, France

Scholl, H. : Astronomisches Rechen-Institut, Heidelberg, F.R.Germany

Schubart, J. : Astronomisches Rechen-Institut, Heidelberg, F.R.Germany

Sekiguchi, N. : Tokyo Astron. Obs., Tokyo, Japan

Shelus, P.J. : Texas University, Austin, Texas, U.S.A.

Shimizu, T. : Bukkyo University, Kyoto, Japan

Sinzi, A.M. : Hydrographic Department, Tokyo, Japan

Sitarski, G. : Space Research Institute, Warszawa, Poland

Stellmacher, I.: Bureau des Longitudes, Paris, France

Sugawa, C. : Mizusawa Latitude Obs., Mizusawa, Japan

Szebehely, V.G.: Texas University, Austin, Texas, U.S.A.

Tanikawa, K. : Mizusawa Latitude Obs., Mizusawa, Japan

Tomita, K. : Tokyo Astron. Obs., Tokyo, Japan

Van Flandern, T.C.: U.S. Naval Obs., Washington, D.C., U.S.A.

Watanabe, N. : Tohoku University, Sendai, Japan

Weissman, P.R. : Jet Propulsion Laboratory, Pasadena, U.S.A

Yabushita, S. : Kyoto University, Kyoto, Japan

Yamazaki, A. : Hydrographic Department, Tokyo, Japan

Yasuda, H. : Tokyo Astron. Obs., Tokyo, Japan

Yoshida, J. : Kyoto Sangyo University, Kyoto, Japan

Yuasa, M. : Tokyo University, Tokyo, Japan

May 25, 1978

INTRODUCTORY ADDRESS

Yusuke HAGIHARA
Department of Astronomy, University of Tokyo, Bunkyo, Tokyo
Japan

The aim of celestial mechanics is to derive all solutions of the differential equations of the three- and *n*-body problems and to reveal the interrelation among all such solutions with various values of the masses and initial conditions, not only to predict the positions of celestial bodies at any instant but to think over the cosmogony of stellar systems and to ascertain the law of universal gravitation in view of the time-space picture of physical theory.

The simplest way to obtain the solution is to find out the necessary number of the first integrals. By Bruns' and Poincaré's theorems there is no further integral, besides the classical integrals, in algebraic form of the momenta or in a uniform function of the orbital elements. Thus we must recourse to transcendental processes for deriving the solution. The perturbation is one of such a kind of successive approximations, such as classical methods of Newcomb, Lindstedt, Delaunay, Bohlin and Whittaker without error estimate and modern method of Krylov-Bogoliubov with error estimates. The formal solution in powers of the disturbing masses and the orbital parameters was shown by Bruns and Poincaré not to be uniformly convergent. All solutions, such that the mean motions — the coefficient of *t* in the arguments of trigonometric functions in the formal series expansions — are modified at each successive approximation, were shown not to be uniformly convergent. In the formal aspect such multiple Fourier series are in the form of quasi-periodic functions of Bohl and Esclangon. The error of truncating at a finite-order term should be estimated like in Krylov-Bogoliubov for the future.

The existence of periodic solutions and asymptotic solutions was proved under suitable choice of the initial conditions according to Poincaré, and also the existence of quasi-periodic solutions under suitable conditions, except the cases when the mean motions are Mahler's *U* numbers, according to Kolmogorov, Arnold and Moser. These theories are based on analytical continuation only for small values of the disturbing masses. For finite values of the masses and for arbitrarily chosen values of the initial conditions we have no means available at present to know in any analytical manner the behaviour of the solution, not to say the

R. L. Duncombe (ed.), Dynamics of the Solar System, 1–4.

mutual relation among the various solutions with varying initial conditions. The condition for representability of the solution in quasi-periodic functions was my problem since 1924 but I could not succeed to solve it yet. It is connected with the stability of motion.

Owing to the failure of investigating the nature and behaviour of the motions in the planetary, satellite and the n-body problems we could study the qualitative characters, that is, the topological theory of the behaviour and interrelation among the solutions of the problems.

In order to apply Cauchy's existence theorem on the solution of a differential equation in an analytic function of the independent variable we must at first regularize the singularities. Painlevé discussed in detail the singularities in the three- and n-body problems. The condition for the singularities must be transcendental. According to the works of Sundman and later of Levi-Civita the singularities for a binary collision in the three-body problem can be regularized and the solution can be analytically continued beyond any binary collision. The solution in powers of the two-thirds' power of the time interval to or from the collision instant was shown to be convergent but the radius of convergence is so small to be used in usual ephemeris computation. For a triple collision the Weierstrass-Sundman condition that the total angular momentum of the three-body system should be zero was derived and the behaviour of the solution in the neighbourhood of the collision was studied by Sundman, Block and Siegel. The motion cannot be analytically continued beyond a triple collision except for very rare cases. Various kinds of imaginary collisions were discussed. The theory can be extended to the n-body problem.

The behaviour at infinity of the orbits was studied by Chazy. He concluded according to Poincaré and Schwarzschild that a distant comet could not be eternally captured by the Sun-Jupiter system. Merman criticized Chazy's work and pointed out that the probability of capture is positive if the total energy is negative. Merman derived complicated conditions for a capture.

To study the behaviour of the solution in the neighbourhood of a singularity the method of Briot-Bouquet was cultivated by Poincaré, Bendixon and Dulac. The nature of the solution-curves, called the characteristics, around a singularity, at the same time as various domains including the domains of ergodic character and with or without limit cycles were distinguished. In particular Poincaré's study on the characteristics on the surface of a torus was developed by Denjoy and they were shown to have different point-set theoretical behaviours according to the initial conditions. As a special case of characteristics the nature of geodesics on different surfaces was studied by Hadamard, Morse and Poincaré. Geodesics on surfaces of any genus were discussed by means of Fuchsian groups of transformations and non-Euclidean geometry.

Poincaré's method of finding out periodic solutions is based on the minimum principle and the analytic continuation. Birkhoff invented the minimax method. Morse defined the type number in contrast to the rotation number of Poincaré and Birkhoff. The conjugate points in the calculus of variations play an important role. Morse went further and developed his calculus of variations in the large and the functional topology and could prove the existence of closed geodesics on the surface of an ellipsoid of any dimension. The analysis situs of Poincaré was sketched with the idea of homology, homotopy, Betti numbers, torsion coefficients, Heegaard diagrams and the fundamental groups. These ideas were applied to the manifolds of motion regarded as topological spaces. The restricted three-body problem was discussed on such analysis situs of the manifolds of motion.

The existence of an invariant point in a surface transformation on the surface of section shows the existence of a periodic solution according to the study of Poincaré and Birkhoff. The behaviour of the transformed points by successive application of the surface transformation reveals the stability character of the motion. The existence of invariant points in a ring transformation conjectured by Poincaré was proved by Birkhoff and extended in several aspects even to a functional space, by Birkhoff, Kerékjartó, Kellog and Nielsen. The theorem was applied to prove the existence of periodic solutions, not by the method of analytic continuation, in the restricted three-body problem by Birkhoff. The invariant points under surface transformations were classified were discussed on the point of view of stability. Birkhoff discovered the ring of instability and remarkable closed curves. He studied the distribution of various types of motion, periodic, asymptotic and recurrent, by means of the surface transformations.

Birkhoff extended the periodic solutions to the recurrent motions on Poincaré's idea of regional recurrence. The theorem was sharpend by Carathéodory, Khintchine and Hopf. The idea was further generalized to the central motions by Birkhoff on point-set theoretical ground and to fleeing points by Hopf. The idea of metrical transitivity was introduced by Birkhoff. Classical forms of the ergodic theorem was discussed by Ehrenfest and others as the foundation of the gas-kinetic theory since Maxwell, Boltzmann and Gibbs. The mean ergodic theorem was proved by von Neumann in a Hilbert space for quantum mechanics. The individual ergodic theorem was proved by Birkhoff. The ideas of mixture and homogeneous chaos were introduced by Hopf and Wiener. The ergodic theorem was recently generalized to functional spaces.

The proof of the existence of mean motions in the theory of secular perturbations was carried out by Bohl and Weyl. It is connected with the distribution of numbers modulo one. Also statistical methods in Diophantine approximation and asymptotic distribution functions were applied to prove the existence of mean motions. The secular constant and the zero-points of an analytical almost periodic functions were studied. Almost periodic motions on a circle and on a surface were studied by Bohr, Jenssen and Fenchel. The relation between the recurrent motion of Birkhoff and the almost periodic motion was studied in connection with the Liapounov stability. Strongly and weakly stable motions were defined in relation to Liapounov stability and Poisson stability.

These topological theories, in combination with the numerical computation of the trajectories with concrete numerical values of the parameters, such as the works of Hénon, Rabe, Bartlett and Szebehely, if a sufficient number of numerical solutions are derived, will throw light on a rough idea on the behaviour and interrelation of the motions of the proposed dynamical problem. Hénon, Deprit and Contopoulos have embarked in such research of their individual particular problems. With the accumulation of the results of such numerical computations the topological theories are expected to reveal the key to know the interrelation and interlacement of various kinds of trajectories as the solutions of the problem. Hénon discovered by numerical computer analysis the domains of periodicity and ergodicity as islands and seas on the surface of section. This result reveals the existence of a non-uniform integral in the three-body problem in contrast to Poincaré's theorem on the non-existence of uniform integrals other than the classical integrals. The problem is to derive the explicit expressions for the boundaries of those domains.

Now the study of topological theories and the development of powerful computers are showing the way to lead to the main problem of celestial mechanics, long-cherished in the heart of human culture.

PART I

STABILITY, N- AND 3-BODY PROBLEMS, VARIABLE MASS

STABILITY OF THE SOLAR SYSTEM

Victor Szebehely
The University of Texas at Austin

ABSTRACT

This paper reviews the present status of research on the problem of stability of satellite and planetary systems in general. In addition new results concerning the stability of the solar system are described. Hill's method is generalized and related to bifurcation (or catastrophe) theory. The general and the restricted problems of three bodies are used as dynamical models. A quantitative measure of stability is introduced by establishing the differences between the actual behavior of the dynamical system as given today and its critical state. The marginal stability of the lunar orbit is discussed as well as the behavior of the Sun-Jupiter-Saturn system. Numerical values representing the measure of stability of several components of the solar system are given, indicating in the majority of cases bounded behavior.

INTRODUCTION

No more appropriate subject than the unsolved problem of the stability of the solar system may be contributed to a volume honoring Professor Y. Hagihara. His fundamental contributions to celestial mechanics underline not only the importance of this subject but point out the basic difficulties encountered such as the non-uniform convergence of series solutions and the non-existence of sufficient number of uniform integrals. Indeed, the semi-convergent series as shown by Poincaré (1892-1899), may be truncated and give solutions for certain intervals of time with excellent accuracy, but these series are of no use when solutions in the domain $-\infty \leq t \leq \infty$ are to be studied. Numerical integrations give meaningful results for limited periods of time, but, once again, their applicability to stability problems raises serious questions. Modern, powerful topological theories by Kolmogorov, Arnold and Moser, while representing significant advances in the theory of dynamical systems are not applicable to the stability of the solar system because the actual perturbations are above the limits set by the theory.

R. L. Duncombe (ed.), Dynamics of the Solar System, 7–15.

With Hagihara (1970-1976), we may state that if the trigonometric series used in celestial mechanics are not uniformly convergent then they do not represent the solution of the differential equations describing the solar system. On the other hand, if these series were convergent and would contain no secular terms, the solar system's stability would be established by the classical methods of general perturbation theory.

A complete solution would be available if sufficient number of uniform integrals would exist. Once again, the non-existence of such integrals was shown by Poincaré (1892-1899) and by Bruns (1887). The use of existing integrals allows the reduction of the order of the problem as well as allows the establishment of certain topological properties of the manifold of the motion. This approach originally proposed by Hill (1878) is one of the few methods which, with all its disadvantages, may be used in a precise, mathematically well formulated and exact manner to establish - if not the stability - at least certain boundaries of the motion.

The basic problem of the stability of the solar system may be stated in a variety of ways. Hagihara's (1957) unambiguous and clear mathematical formulation is: "What is the interval of time at the end of which the solar system deviates from the present configuration by a previously assigned small amount?"

Professor Hagihara's basic question has only partial answers today. An introductory, modern, non-mathematical treatment of the subject of the stability of the solar system is available elsewhere (Jefferys and Szebehely, 1978).

THREE APPROACHES TO THE PROBLEM OF STABILITY

In this section three fundamental approaches to stability are shortly described and their advantages and disadvantages are listed.

(A) The method applicable to the solar system and offering mathematically precise results is the previously mentioned method of Hill (1878). It uses integrals of the motion and was employed by Hill to study the stability of the lunar orbit. The method is also applicable to the model of the restricted problem (Szebehely, 1967) and it was generalized to other dynamical systems such as the general problem of three bodies. Between 1968 and 1977, a series of papers appeared in which the topology of the general problem of three bodies are discussed by Golubev (1968), Smale (1970), Marchal and Saari (1975), Bozis (1976), Zare (1977) and Szebehely (1977).

The ideal in its simplest form may be expressed as:

$$V^2 = V(x,y)-C \ ,$$

which for a two-degrees of freedom dynamical system represents the energy integral with $V(x,y)$ the potential, v the velocity and C the constant of integration, determined by the initial conditions. For a given value of C motion is possible in those regions of the plane (x,y) for which $V(x,y) > C$. In this way boundaries of possible motions may be established by a study of the topology of the curves $C = V(x,y)$. Values of C associated with topological changes are bifurcation of critical values.

The greatest advantage of the method is that it is an exact, non-linear analysis. Consequently, it's free of the criticisms raised against the method of small disturbances, periodic or quasi-periodic motions have no importance and the fact that it is also devoid of any reference to commensurabilities may be considered an advantage because of the ambiguous role of resonances in celestial mechanics. Another great advantage of the method is its applicability to non-integrable dynamical systems, such as the gravitational problem of n bodies. On the other hand, its greatest disadvantage is that it gives partial results only and at best it establishes upper or lower boundaries of the motion. The importance of these limitations calls for an explanation by two simple examples. The "stability" of the lunar orbit proved by Hill may be formulated precisely as follows: if the dynamical model adopted by Hill is accepted to describe the motion of the Moon, then the Moon may not leave the Earth and may not become an (independent) planet of the Sun. Note that this result does not mean that the Moon cannot collide with the Earth and that all Hill's assumptions must be dynamically correct. In this example Hill meant by "stability" the existence of an upper bound of the Moon's distance from the Earth.

The second example is the application of the method to the stability of Saturn in the Sun-Jupiter-Saturn system. The meaning of "stability" now becomes quite different. If the assumptions of the (circular) restricted problem are accepted then Saturn can not penetrate the region occupied by the Sun and by Jupiter, i.e., it can not become an interior planet or it can not form a binary system with Jupiter. But, Saturn may escape the system. Therefore, by "stability" we mean a lower bound for Saturn's distance from the Sun. Both examples lack two bounds. The lunar distance has no lower bound and the orbit of Saturn has no upper bound. Both examples ignore other influences such as tidal effects, resonance conditions and perturbations by other planets. Nevertheless, the limited results of these examples, accepting the validity of the dynamical models are exact and no mathematical approximations are involved.

As the degree of freedom of the dynamical system increases the results remain exact but their significance becomes more limited. If the model of the restricted problem is not applicable and the general three-body problem must be used, the topology of the boundary surfaces becomes rather complicated and in few cases may we offer statements concerning boundedness or stability of general validity. For instance, with our present day knowledge regarding the existence of uniform integrals of the three-body problem we find that <u>all</u> boundary surfaces are open to infinity, making escape always possible. Direct exchange, on the other hand, between the roles played by the three bodies, is not always possible.

A distinct advantage of Hill's method is that it may be used to define a measure of stability via bifurcation theory. When the controlling parameters of the system are such that in its present state the system is far removed from the nearest bifurcation then a strong stability exists. On the other hand, when slight changes in the dynamic parameters describing the system throw it into bifurcation, its stability is low. A measure of stability $S = (s_{ac} - s_{cr})/s_{cr}$ was introduced by Szebehely (1977) indicating the differences between the present (actual) system and its nearest bifurcation value. This parameter is the Jacobian constant s=C for dynamical systems described by the restricted problem and it is the quantity $s = -(c^2h)/G^2m^5$ for the general three-body problem. (Here c is the angular momentum, h the total energy, G the constant of gravity and m the average mass). The subscripts "ac" and "cr" refer to the actual or present value and to the critical or bifurcation value respectively. It may be shown for the restricted problem that $C_{cr} \geq 3$ and that for the general problem $s_{cr} \neq 0$ as long as two of the three bodies have non-zero mass. The normalization is quite arbitrary as is the definition used for C and h. Nevertheless, accepting the same definition for all examples offers meaningful comparisons regarding the measures of stability. Note that if $S > 0$ the dynamical system is removed from its nearest possible bifurcation. When $S < 0$ the possibility of changing to a different type of motion exists but will not necessarily occur.

(B) Stability investigations by numerical methods is the second basic approach. A great variety of possibilities exist along these lines and with present day high speed electronic computers, interesting experimental results became available in the past few years. The stability of numerically established periodic and quasi-periodic planetary and satellite type orbits give indication of the long-range behavior of the system. Because of the numerical nature of these investigations, they are approximate and their precise mathematical validity is always questionable. Using the restricted problem Hénon (1970) studied the linear stability of periodic and quasi-periodic orbits as well as their global, non-linear behavior using the method of the surfaces of section (Hénon and Heiles, 1964). These results reveal the existence of stable retrograde orbits, far removed from the primaries. Similar numerical investigations for planetary type three and many-body periodic orbits, performed by Hadjidemetriou (1976), indicate linear stability of the solar system using the model of the general problem of three bodies.

It is known that when the participating masses in the general problem of three bodies are of the same order of magnitude, random initial conditions result, most of the time, in escapes even for negative total energy (Szebehely, 1971). Such unstable motions contradict the generally found stable behavior of planetary type motions when the mass of one body dominates. Kuiper (1973), therefore, recommended that increased masses for the planets be used in numerical experiments and an artificial instability be created. With increased masses a shorter numerical integration time is expected to determine the behavior of the system. Numerical integrations have shown (Nacozy, 1976) no secular trends in the motion of the Sun-Jupiter-Saturn system until the masses of Jupiter and Saturn were increased 25 times of their actual value. The sudden appearance of instability with increased masses is an indication of the stability of the system with its present (actual) masses. This method is known as the Kuiper-Nacozy-Szebehely (K-N-S) theory. Note that Hill's method gives a factor of 15 instead of 25 for the onset of possible instability, (Szebehely and McKenzie, 1977).

Other long-time numerical integrations seem to show stability for over 10^7 years. These integrations do not include the inside planets, and do not offer analytical proof of stability inspite of their unquestionably ingenious techniques and reliable results (Birn (1973), Cohen, Hubbard and Oesterwinter (1968)).

(C) The third approach to stability investigations is the classical general perturbation method. Once again no mathematical results exist which would show long-time behavior because of the previously mentioned divergence of the series. Until new, uniformly convergent series solutions become available, the various averaging techniques, truncations, secular perturbations, etc., will furnish results useful for finite time only. Sundman (1912), has shown the existence of the solution of the pertinent differential equations. The radius of convergence of his series is so small that not even a general pattern of dynamical behavior emerges, not to mention stability properties. Message's (1978) recent proof of the non-existence of secular terms to any order in the formal solution for the semi-major axis is an important result considering formal solutions. An analytical approach along these lines is the K-A-M (Kolmogorov (1954), Arnol'd (1963), Moser (1973)) theory which states that if the perturbations to a soluble (two-body) problem are small enough, the commensurabilities are of high enough order and certain continuity conditions are satisfied, then quasi-periodic solutions exist. Unfortunately, for the solar system the first condition is not satisfied and, therefore, the KAM theory is not applicable.

STABILITY MEASURES FOR THE SOLAR SYSTEM

In this section, the results of calculations are given according to the first approach (A). The satellite systems of the solar system are discussed first, followed by remarks on the planetary system itself. The numerical values given are based on the most recent astronomical constant accepted by the International Astronomical Union in 1976. Table I. lists values of S for the satellite systems.

PLANET	SATELLITE	STABILITY	PLANET	SATELLITE	STABILITY
Earth	Moon	0.00015	Saturn	Mimas	0.72023
Mars	Phobos	0.00254		Enceladus	0.56154
	Deimos	0.00098		Tethys	0.45187
Jupiter	V	1.33570		Dione	0.35133
	Io	0.56521		Rhea	0.25030
	Europa	0.35058		Titania	0.10458
	Ganymede	0.21478		Hyperion	0.08531
	Callisto	0.11663		Iapetus	0.03221
	XIII	0.0106		Phoebe	0.00313
	VI	0.00988	Uranus	Miranda	0.33493
	VII	0.00944		Ariel	0.21569
	X	0.00925		Umbriel	0.15462
	XII R	-0.00536		Titania	0.09358
	XI R	-0.00612		Oberon	0.06939
	VIII R	-0.00656	Neptune	Triton	0.21629
	IX R	-0.00666		Nereid	0.01206

Table I. Stability of Satellites in the Solar System

The model of the restricted problem is used with the three bodies being the Sun, the planet and its satellite. The assumptions of the restricted problem are acceptable for this case since the satellites' effects of the primaries are less than approximately 1% when compared to the effects of the primaries on each other. (The only exception is of Neptune's Triton). Attention is directed to the low stability of the Earth's moon. This small positive number ($S = 1.5 \times 10^{-4}$) becomes negative ($S = -2.75 \times 10^{-4}$) if the model of the general problem of three bodies is used (Szebehely and McKenzie, 1977), indicating a borderline case of bifurcation. The four outer retrograde satellites of Jupiter are below the bifurcation value ($S < 0$) suggesting the possibility of capture-origin.

The discussion of satellite stability in the solar system may be closed by mentioning several approaches to the problem of finding the limiting orbital radii of natural or artificial satellites around the planets of the solar system. Such formulae were given among others by Hagihara (1952), Kuiper (1953), and Szebehely (1978).

Regarding planetary stability, the dynamical situation is considerably more complicated. Table II. shows the results of the bifurcation computations, assuming the model of the restricted problem and using the Sun and Jupiter as primaries. This Table needs careful interpretation. The numbers decrease as the planetary orbit approaches Jupiter's orbit and the minimum is shown for Saturn. Uranus, Neptune and Pluto show increasing numbers since they are farther removed from Jupiter's orbit.

MERCURY	VENUS	EARTH	MARS	SATURN	URANUS	NEPTUNE	PLUTO
3.60	1.61	1.00	0.48	0.07	0.35	0.64	0.85

Table II. Stability of Sun-Jupiter-Planet Systems.

Table III. shows more appropriate combinations from a dynamical point of view. Since the principal effect in the secular variation in the mean motion on Mercury is by Venus, on Venus is by the Earth, on the Earth by Jupiter and on Mars by the Earth, Table III. shows also the corresponding stability measures.

PRIMARIES	SUN-MERCURY	SUN-VENUS	SUN-EARTH
PLANET	VENUS	EARTH	MARS
STABILITY	0.110	0.027	0.041

Table III. Planetary Stabilities in the Solar System.

Studies performed regarding the effect of Jupiter's orbital eccentricity e_J on the stability of planetary orbits shows that the increase of e_J has a destabilizing effect. The same is true if the values used for the planetary masses are increased and the general problem is used as the underlying dynamical model.

REFERENCES

Arnol'd, V.I., (1963), Usp. Math. Nauk., **18**, 5, 6, 13, 91.

Birn, J., (1973), Astronaut. Astrophys., **24**, 283.

Bozis, G., (1976), Astrophys. Sci., **43**, 355.

Bruns, H., (1887), Acta Math., **11**, 25.

Cohen, C.J., E.C. Hubbard and C. Oesterwinter, (1968), Astron. J., **72**, 973.

Golubev, V.G., (1968), Doklady Akad. Nauk, USSR, **180**, 308.

Hadjidemetriou, J., (1976), Proc. NATO Adv. Study Inst., Reidel Publ.

Hagihara, Y., (1952), Proc. Japan Acad., **28**, 182.

Hagihara, Y., (1957), "Stability in Celestial Mechanics," Kasai Publ., Tokyo.

Hagihara, Y., (1970-1976), "Celestial Mechanics," Vol. 1-5, MIT Press and Japan Soc. Promotion of Science, Tokyo.

Hénon, M., (1970), Astron. and Astrophys., 9, 24.

Hénon, M. and C. Heiles, (1964), Astron. J., 69, 73.

Hill, G.W., (1878), Am. J. Math., 1, 5, 129, 245.

Jefferys, W.H. and V. Szebehely, (1978), "Dynamics and Stability of the Solar System," Cosmic Physics, in print.

Kolmogorov, A.N., (1954), Doklady Akad. Nauk, 98, 4.

Kuiper, G.P., (1953), Proc. Natl. Acad. Sci., U.S., 39, 1154, 1159.

Kuiper, G.P., (1973), Celest. Mech., 9, 359.

Marchal, C., (1975), ONERA Rept., No. 77.

Marchal, C. and D. Saari, (1975), Celest. Mech., 12, 115.

Message, J., (1978), Proc. IAU Symp. 41, Reidel Publ.

Moser, J., (1973), "Stable and Random Motions," Princeton Univ. Press, Princeton, N.J.

Nacozy, P., (1976), Astron. J., 81, 787.

Poincaré, H., (1892-1899), "Les Méthodes Nouvelles," Gauthier-Villars, Paris.

Smale, S., (1970), Invent. Math., 11, 45.

Sundman, K.F., (1912), Acta Math., 36, 105.

Szebehely, V., (1967), "Theory of Orbits," Acad. Press, N.Y.

Szebehely, V., (1971), Celest. Mech., 4, 116.

Szebehely, V., (1974), Celest. Mech., 9, 359.

Szebehely, V., (1977), Celest. Mech., 15, 107.

Szebehely, V., (1978), Celest. Mech., 17, in print.

Szebehely, V. and R. McKenzie, (1977), Astron. J., 82, 79.

Szebehely, V. and R. McKenzie, (1978), Celest. Mech., 17, in print.

Szebehely, V. and K. Zare, (1977), Astron. and Astrophys., 58, 145.

Zare, K., (1977), Celest. Mech., 16, 35.

ACKNOWLEDGEMENTS

No greater honor may be awarded to a worker in celestial mechanics, than to be invited to deliver a paper celebrating Professor Hagihara's birthday. The preparation of this paper was one of the greatest challenges of this author. Its actual delivery, his greatest pleasure. For this opportunity, I wish to acknowledge the partial support of The University of Texas and of the U.S. National Science Foundation, Division of Astronomy. My students and colleagues contributed significantly by their assistance and critiques. Mr. R. McKenzie, graduate student at The University of Texas deserves special mention for contributing valuable computational assistance.

DISCUSSION

Kiang: I would like to suggest a new approach to questions of stability, less sophisticated than the topological method. As soon as we write Hill's equation for a closed orbit (in phase space, in general), we can define its stability in terms of the solution of Hill's equation. In this way, I was able to show that a typical Hilda asteroid (at the 3/2 commensurability) is stable while a Hecuba asteroid (at the 2/1 point) with the same eccentricity is unstable. In this particular problem it is possible to write down Hill's equation only after radically departing from the classical treatment of the virtual displacement.

Szebehely: Thank you.

NUMERICAL STUDIES ON THE STABILITY OF THE SOLAR SYSTEM

Paul Nacozy
The University of Texas at Austin

In this paper, recent numerical and analytical results of the author and others, concerning the stability of the solar system are presented. The results indicate that the actual planetary system could possibly have stable and bounded motion.

In order to clearly distinguish the type of stability that will be referred to here, we will first discuss and define various types of stability of solutions of dynamical systems and, in particular, of gravitational systems. Perhaps one of the most stringent types of stability is asymptotic stability. This type of stability requires that the distance between two particles on neighboring solutions or orbits approaches zero as time approaches infinity. A weaker type of stability is Liapunoff stability. This type of stability requires that the distance between two particles on neighboring solutions can be made as small as desired as time goes to infinity, by adjustment of the initial conditions of the two solutions. An even weaker type of stability, often referred to as orbital stability, merely requires that the distance between two neighboring solutions or orbits can be made to remain as small as desired as time approaches infinity, by adjustment of the initial conditions.

The first two of these types of stability are generally not encountered in gravitational systems. The third type is encountered in some gravitational systems, but in discussions on the stability of the major planetary system we usually do not need to require stability as stringent as orbital stability.

Three somewhat weaker and less stringent types of stability may be defined as follows. The first type is often referred to as Laplacian stability. This type of stability requires that a solution for a system of particles have mutual distances bounded both from below and from above and that the particles have no close approaches or collisions and that no particle escapes to infinity. A second type of stability is referred to as the Komolgoroff-Arnol'd-Moser (KAM) stability. The KAM type of stability requires that a solution be represented by a quasi-periodic function, possessing a finite number of non-commensurable basic frequencies. (If the frequencies were commensurable, then the

R. L. Duncombe (ed.), Dynamics of the Solar System, 17–21.

solution would be periodic.) A third type of stability is referred to as Poisson stability. This type of stability requires certain restrictions be placed on three of the osculating orbital elements that represent the orbits of the particles of the system. These three orbital elements are the semi-major axis, the eccentricity, and the inclination. One definition of Poisson stability places the restriction on these three elements that they have no secular trends. Another closely related deformation is that the functions of time that represent these three osculating elements be bounded functions both below and above. Some definitions of Poisson stability restrict their discussion to the semi-major axes, assuming that the eccentricity and inclination behave in a similar fashion to the semi-major axes.

Laplacian stability, KAM stability, and Poisson stability, as they are defined above, are closely related in that they all insure bounded motion, and it is this property of bounded motion that is usually referred to in discussions of the stability of the solar system. The KAM stability is more stringent than the other two for it requires that the motion be represented by purely quasi-periodic functions. We will restrict our discussion here to bounded motion in either the Laplacian or Poisson sense.

Many recent numerical results appear to show that for orbits of the planetary type there exists initial conditions for which the orbits of the planets have bounded motion with a lack of secular trends in the planetary semi-major axes, eccentricities and inclinations, at least for very long times. Cohen, Hubbard and Oesterwinter (1972) have recently provided a solution for the five outer planets by numerical integration over a period of one million years. Their solution shows no noticeable secular trends in these orbital elements, indicating that the motion of the outer planets is apparently bounded, exhibiting Laplacian and Poisson stability, at least for the one million year period. If secular trends do in fact exist, the one million year solution shows that they are so small that their detection would require a numerical integration over times very much longer than the one million year period.

Using the method of surface of sections, numerical calculations have recently been made by Henon and Heiles (1964), Jefferys (1966) and Contopolous (1967). Their results confirm that the KAM type of stability exists for certain gravitational systems, and their results indicate that quasiperiodic motion is possible with masses that are at least as large as the planetary masses and perhaps much larger.

The recent works of Harrington, Szebehely, Ovenden, Birn, Lecar and Franklin are related and all show that for systems of the planetary type, Laplacian stability can occur. Harrington (1972) and Szebehely (1972) show, separately, that if the planetary masses are small enough and their mutual distances are large enough, Laplacian stability can exist. Ovenden, et al (1974) shows that for small enough planetary masses and with mutual distances satisfying certain properties related to near-commensurabilities, Laplacian stability is possible. Birn (1973) and Lecar and Franklin (1974) show that for planetary masses small enough there are certain regions in the phase space which allow stable motion in the sense of Laplace. Within these regions planetary orbits appear stable and outside these regions they appear unstable.

The study that we have undertaken extends the solution of Cohen, Hubbard and Oesterwinter by increasing the values of the planetary masses (Nacozy, 1976 and 1977). It is hoped that a system having increased planetary masses would partially and in some sense simulate the solutions with the actual masses over much longer intervals of time. For the solution with increased masses, the amplitudes of some of the periodic perturbations and the trends of the secular perturbations, if they exist, will be increased as long as the general character of the motion remains unaltered. In particular, third-order secular terms in the semi-major axes, factored by the third power of the masses, if they exist, should be apparent sooner if the planetary masses are increased. If third-order secular terms do exist they would occur in 1/1000 of the time if the planetary masses were increased by ten times. Since, as Kuiper (1973) has pointed out, the stability of the Jupiter-Saturn system is obviously the dominant criterion for the continued existence of the planetary system, we undertook a study in a general three-body problem consisting of Jupiter, Saturn and the Sun. The system utilized the actual mutual inclination of Jupiter and Saturn and the actual osculating initial conditions. The masses of Jupiter and Saturn were increased by a factor γ so that the ratio of the mass of Jupiter to the mass of Saturn remained constant. The factor γ was increased from $\gamma = 1$ (the actual planetary masses) to $\gamma = 1000$ (which gives Jupiter the mass of the Sun). Solutions were obtained by a numerical integration wherein regularization was employed for the closest pair and two different types of integration methods were used; one being an eighth-order Runge-Kutta and the second being a variable-order recurrent power series method. In the integration, various stepsizes were used in order to obtain an estimate of the global truncation error. In addition the energy integral of the system was monitored and its constancy was held to one part in 10^9 (Nacozy, 1977).

For the mass parameter γ less than about 29, our results show that the motions of Jupiter and Saturn are qualitatively similar to the one million year solution of the outer planets by Cohen, Hubbard, and Oesterwinter. As γ is increased beyond 29, the system is altered significantly and unbounded motion is immediately apparent. For γ between 29 and 100, Saturn is ejected from the system after about one or two thousand years, on a highly elliptical orbit. For γ greater than about 100, Saturn is ejected from the system much quicker on a hyperbolic orbit. The crucial result here is that there is apparently a range of values of γ ($\gamma < 29$) that provide only bounded and stable motion and a range of large values ($\gamma > 29$) that provide only unbounded and unstable motion with a very sharp transition from the stable character to the unstable character for the value of $\gamma = 29$. These results are related to the results of Henon and Heiles, Jefferys, Contopoulos and others using the method of surface of section where they show that quasi-periodic and bounded motion exist for small values of the mass parameter and as the mass parameter is increased past a certain value they also show a sudden disruption of the system and sudden transition from quasi-periodic motion to non-quasi-periodic motion.

All of the above mentioned numerical results indicate that the actual Jupiter-Saturn-Sun system has masses of Jupiter and Saturn that are much smaller than those causing breakup of the system. The results might imply that the actual Jupiter-Saturn-Sun system is stable in the sense of Laplace (and Poisson) probably at least for many millions of years and perhaps for very much longer times.

Our results also may be placed in context with a result given by Szebehely (1977) and an additional analytical result that has been obtained recently by Nacozy and Kwok (1978) and Hadjidimetriou (1978). Szebehely has found that the Hill curve corresponding to the general three-body problem considering the Jupiter-Saturn-Sun system is closed around Jupiter and the Sun, excluding Saturn for the mass parameter $\gamma < 14$. As the mass parameter is increased beyond 14 the Hill curve opens allowing Jupiter, Saturn and the Sun to have the possibility of an interchange and hence a subsequent Saturn ejection. Our result, showing that Saturn is ejected when the mass parameter γ is increased beyond 29 shows that even after the Hill curve opens, bounded motion can still persist for larger values of the mass parameter. This result is analogous and possibly an extension of similar results in the circular restricted problem of three bodies where a zero-velocity curve around one of the primaries can open as the mass parameter μ increases while bounded motion persists for the infinitesimal particle until the mass parameter is increased further. The additional result recently found by Nacozy and Kwok, and separately by Hadjidemetriou, shows that the periodic orbit that is close to the actual Jupiter-Saturn-Sun system having the commensurability ratio of 5:2 and having the actual masses of Jupiter, Saturn and the Sun, is stable in the linear sense. As we increase the mass parameter and obtain the family of periodic orbits with the mass parameter as the parameter of the family, we obtain stability up to the mass parameter $\gamma = 39$. For $\gamma > 39$, instability occurs. This result shows that a periodic orbit can persist and remain stable, for larger values of the mass parameter than that which causes breakup for a non-periodic (but possibly quasi-periodic) orbit.

ACKNOWLEDGEMENTS

This study was made possible by the Department of Aerospace Engineering and Engineering Mechanics, and by the Bureau of Engineering Research, both of the University of Texas at Austin, and by the National Science Foundation, Astronomy Section.

REFERENCES

Birn, J., 1973: Astron. Astrophys., Vol. 24, p. 283.

Cohen, C.J., Hubbard, F.C., Oesterwinter, C., 1972: Astron. Pap. Amer. Eph., Vol. 22, Pt. 1, p. 9.

Contopolous, G., 1967: Bull. Astron., 3rd Ser., Vol. 2, Fasc. 1, p. 223.

Hadjidimetriou, J., 1978: Appears in this volume.

Harrington, R.S., 1972: Celest. Mech., Vol. 6, p. 322.

Henon, M., and Heiles, C., 1964: Astron. Jour., Vol. 69, p. 73.

Jefferys, W., 1966: Astron. Jour., Vol. 71, p. 306.

Kuiper, G., 1973: Celest. Mech., Vol. 9, p. 321.

Lecar, M., Franklin, F., 1974: Appears in The Stability of the Solar System, ed. Y. Kozai, D. Reidel Publ., p. 25.

Nacozy, P., 1976: Astron. Jour., Vol. 81, p. 787.

Nacozy, P., 1977: Celest. Mech., Vol. 16, p. 77.

Nacozy, P., Kwok, J., 1978: to be published.

Ovenden, M., Feagin, T., Graf, O., 1974: Celest. Mech., Vol. 8, p. 455.

Szebehely, V., 1972: Celest. Mech., Vol. 6, p. 84.

Szebehely, V., 1977: Astron. Jour., Vol. 82, p. 79.

STABILITY OF PERIODIC PLANETARY-TYPE ORBITS OF THE GENERAL PLANAR N-BODY PROBLEM

John D. Hadjidemetriou
University of Thessaloniki, Thessaloniki, Greece

1. INTRODUCTION

It is known that families of periodic orbits in the general N-body problem ($N>3$) exist, in a rotating frame of reference (Hadjidemetriou 1975, 1977). A special case of the above families of periodic orbits are the periodic orbits of the planetary type. In this latter case only one body, which we shall call sun, is the more massive one and the rest N-1 bodies, which we shall call planets, have small but not negligible masses. The aim of this paper is to study the properties of the families of periodic planetary-type orbits, with particular attention to stability. To make the presentation clearer, we shall start first with the case N=3 and we shall extend the results to N>3. We shall discuss planar orbits only.

2. PERIODIC PLANETARY-TYPE ORBITS INVOLVING THE SUN AND TWO PLANETS

(a) Families of Periodic orbits

In order to compute periodic orbits of the planetary-type involving 3 bodies, we may start from a degenerate family of periodic orbits where the massless bodies P_1 and P_3 revolve around P_2, whose mass is finite, in circular orbits in the same plane. We shall also assume that they revolve in the same direction. This motion is periodic with respect to a rotating frame of reference Oxy whose x axis contains always P_2 and P_1 (the positive direction being from P_2 to P_1). In fact, if we normalize the units so that the gravitational constant is equal to 1, the total mass is equal to 1 and the radius of P_1 around P_2 is equal to 1, the motion is periodic in the Oxy frame for any value R of the radius of P_3 around P_2, with a period equal to

$$T = 2\pi/(1-T_1/T_3), \tag{1}$$

where T_1, T_3 are the periods of the two planets, respectively, in the inertial frame and $T_1/T_3=R^{-3/2}$, according to the normalization mentioned

R. L. Duncombe (ed.), Dynamics of the Solar System, 23–28.

above. Thus, we have a monoparametric family of periodic orbits, with R as the parameter, which can be considered as a particular case of the restricted circular 3-body problem. The continuation of the orbits of this family to periodic orbits of the general 3-body problem ($m_1,m_3>0$) is possible for all values of R except for those where $T=2\pi n$, $n=1,2,3,...$ In this latter case the continuation theorem is not applicable (Hadjidemetriou 1975). These orbits correspond to the resonances for the periods of the two planets, given by

$$T_1/T_3 = n/(n+1), \tag{2}$$

as obtained from (1). The same result was obtained by Griffin (1920), by a different proceedure.

The numerical computations have revealed that the orbits of the above degenerate family, for $m_1=m_3=0$, can be extended to the general case (i.e. $m_1,m_3>0$) even for large values of m_1 and m_3 (Hadjidemetriou 1976). The motion obtained by this continuation process in a symmetric periodic motion of the general 3-body problem in a rotating frame Oxy whose origin coincides with the center of mass of P_2 and P_1 and the x axis contains always these two bodies. The two planets P_1 and P_3 describe (in the inertial frame) nearly circular or elliptic orbits around the Sun (P_2), as we shall describe below.

Let us see now how we can represent the above planetary-type periodic orbits. The position of the three bodies, in the rotating frame Oxy can be determined by the coordinates x_1 of P_1 and (x_3,y_3) of P_3. Consequently, a symmetric periodic motion is specified by the initial conditions $x_{10},x_{30},\dot{y}_{30}$ (for the rest variables we have $\dot{x}_{10}=y_{30}=\dot{x}_{30}=0$). Thus, a monoparametric family of periodic orbits is represented by a smooth curve in the space $x_{10}\ x_{30}\ \dot{y}_{30}$. In this paper, we shall use the projection of this curve to the plane $x_{10}\ x_{30}$, which suffices for illustration purposes.

We consider the space of initial conditions $x_{10}\ x_{30}$ (Fig.1). Evidently, the family of degenerate orbits, where P_1 and P_3 have zero masses and describe circular orbits around P_2, is given by the straight line $x_{10}=1$.

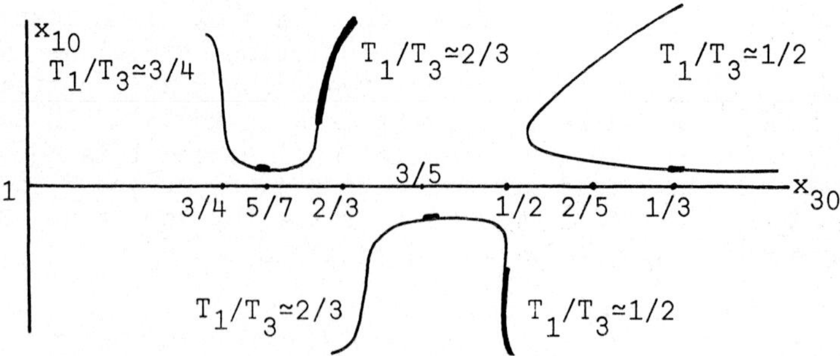

Fig.1: The space $x_{10}x_{30}$ of the initial conditions and the families of planetary-type periodic orbits (graphically).

On this line we mark the resonant orbits of the form $T_1/T_3=n/(n+1)$ for which the continuation theorem is not applicable. These orbits have an accumulation point at $x_{30}=1$. When the degenerate family $x_{10}=1$ is extended to $m_1,m_3>0$, we obtain the set of families of periodic orbits shown in Fig.1 (Hadjidemetriou 1976, Delibaltas 1977). The numerical computations were made for the cases $m_1=m_3=0.001$ and also for the cases m_1=mass of Jupiter, m_3=mass of Saturn and vice-versa. (The total mass was normalized to 1). The results are qualitatively the same for all cases. We note that the degenerate family breaks to (presumably) an infinite number of families of periodic orbits, for $m_1,m_3>0$, which are "separated" by the resonant orbits $T_1/T_3=1/2$, 2/3, 3/4,... The part of the families for $m_1,m_3>0$ which is near the line $x_{10}=1$ corresponds to nearly circular orbits of the two planets. The rest part corresponds to nearly elliptic orbits of the two planets, with eccentricities which increase as we go outwards. Along each elliptic branch the ratio T_1/T_3 of the osculating periods (at t=0) is almost constant for all members of the family. Note that there are two separate branches of nearly elliptic orbits of the two planets for each value of $T_1/T_3=1/2$, 2/3... The difference between them is that at t=0 the two planets are in a different initial situation (at pericenter or apocenter). These elliptic branches can be associated with families of asteroids with the corresponding ratio T_1/T_3, with Jupiter as one of the planets and the asteroid as the other. Also, one member of the first family, in the circular part corresponds to the resonance 2/5 and approximates the Sun-Jupiter-Saturn case.

(b) Stability

We shall study first the stability of the periodic planetary-type orbits where the two planets describe nearly circular orbits. We note that the degenerate periodic orbits of the family $x_{10}=1$ (Fig.1) are, evidently, all stable. It can be proved (e.g. Hadjidemetriou 1978) that there are for each degenerate periodic orbit two stability indices, corresponding to the nonzero characteristic exponents, which are equal to each other and are given, for the normalization used here, by

$$K=-2\cos T, \tag{3}$$

where T is given by (1). These stability indices are not, in general, critical (equal to $K=\pm 2$) and consequently the orbits obtained by the continuation procces, by increasing the masses m_1 and m_3 are stable, for continuation reasons. (The two zero characteristic exponents of the degenerate orbit in the rotating frame are preserved when m_1, $m_3>0$, due to the existence of the energy integral). However, there are degenerate periodic orbits with critical stability indices, (K=2), corresponding to $T=(2\nu+1)\pi$, i.e. to the resonant orbits $T_1/T_3=(2\nu-1)/(2\nu+1)$. The resonant orbits $T_1/T_3=1/3$, 3/5, 5/7, of this kind are shown in Fig.1. These orbits can become unstable when extended to $m_1,m_3>0$, i.e. $|K|>2$. And in fact, the numerical computations have revealed that in each of the families for $m_1,m_3>0$, in their part of circular orbits, there is a small unstable region, generated from the above critical degenerate periodic orbits.

We come now to the branches of the elliptic orbits of the families for $m_1, m_3 > 0$. Such an orbit can be considered to be obtained by the continuation of a degenerate periodic orbit where one planet describes an elliptic orbit. The stability indices in this latter case are all critical and may become stable or unstable when the masses are increased. The numerical computations have revealed that some of the branches of elliptic orbits are stable and others are unstable (Hadjidemetriou 1976, Delibaltas 1977). The unstable parts of the families are shown by bold lines in Fig.1. It was also revealed by the numerical computations that the stability of the same branch of elliptic orbits depends on the ratio m_1/m_3. For example, the resonant branch 1/2 of the first family is stable when the inner planet P_1 has the mass of Saturn and the outer planet P_3 has the mass of Jupiter, but it is unstable when the inner planet P_1 has the mass of Jupiter and the outer planet P_3 the mass of Saturn.

Another question concerning the stability is how the stability evolves when the masses of the planets increase. A general result is that the unstable regions in the families of periodic planetary-type orbits extend when the masses of the planets increase. For example, the unstable region in the first family, at the resonance 1/3, extends and when the masses of the planets become about 38 times larger than the masses of Jupiter and Saturn, (the ratio m_1/m_3 being kept fixed) the unstable region covers the resonant orbit 2/5 which represents the Jupiter-Saturn system (Hadjidemetriou and Michalodimitrakis, 1978a).

3. PERIODIC PLANETARY-TYPE ORBITS INVOLVING THE SUN AND THREE PLANETS

The method used to compute periodic orbits of the planetary type involving two planets can be extended to any number of planets. We shall study here the case N=4, i.e. the Sun (or planet) and three planets (or satellites). We call P_1, P_3 and P_4 the three planets, respectively and P_2 the Sun and consider the degenerate system with $m_1=m_3=m_4=0$, $m_2=1$, where the three planets revolve around the sun in circular orbits in the same plane. We shall also assume that they revolve in the same direction. The motion is periodic with respect to a rotating frame Oxy whose x axis contains the bodies P_2 and P_1, and the origin is at P_2, if

$$(\omega_3-\omega_1)/(\omega_4-\omega_1) = p/q, \tag{4}$$

where p,q are integers and $\omega_i=2\pi/T_i$, i=1,3,4, T_i being the periods of the planets in the inertial frame. Using the same normalization as in the case N=3 (i.e. G=1, $m_1+m_2+m_3+m_4=1$, radius of P_1=1) we have for the period of the above degenerate periodic orbit in the rotating frame,

$$T = 2\pi q/(1-T_1/T_4). \tag{5}$$

Note that when p/q is fixed, for each radius R_3 of P_3 there corresponds a certain radius R_4 of P_4, as can be seen from (4). Thus, for fixed p,q, we have a monoparametric family of degenerate periodic orbits with R_3 as the parameter, in which the three planets describe circular orbits. All these orbits can be continued to the general case m_1, m_3, $m_4>0$ with the exception of those orbits where $T=2\pi n$, $n=1,2,3,...$ (Hadjidemetriou, 1977). As a consequence, the degenerate family of periodic orbits mentioned above breaks to an infinite number of families of periodic orbits when the masses are increased, in the same way as in the case N=3. These families, for m_1, m_3, $m_4>0$, contain symmetric periodic orbits with respect to a rotating frame Oxy whose origin is at the center of mass of P_2,P_1 and its x axis contains always these bodies. As in the N=3 case, these families contain branches along which the ratios T_1/T_3 and T_1/T_4 are nearly constant, but the eccentricities of the osculating orbits of P_1, P_3, P_4 around P_2 vary. A detailed analysis of this type of planetary-type orbits, corresponding to the case p/q=2/3 will be given elsewhere (Hadjidemetriou and Michalodimitrakis, 1978b). We describe here four distinct branches, A,B,C,D, of the above monoparametric families, corresponding to the ratios $T_1/T_3=1/2$, $T_1/T_4=1/4$. This case represents the motion of the three inner Galilean satellites of Jupiter. The computations were made for the actual masses of Jupiter and its three satellites. Branch A corresponds to the actual case. The main characteristics of these branches are shown below. a stands for apojove and p for perijove at t=0.

Branch	I	II	III	Initial situation	Stability
A	p	p	p	I, III in conjunction, II in opposition	stable
B	a	a	a	I, III " " , II " "	unstable
C	p	p	p	I, II " " , III " "	unstable
D	a	a	a	I, II " " , III " "	unstable

De Sitter (1908, 1909, 1918, 1928) (see Hagihara 1961) has obtained one periodic orbit with the properties of the branch A and one with the properties of B. It is clear however from this analysis that there exists an inifinite number of such orbits, with the same resonance but different eccentricities.

4. THE RESTRICTED PLANETARY 4-BODY PROBLEM.

As an approximation to the above families of periodic orbits we consider now the case where the mass of one planet is negligible. This is justified in the Solar System where the main planets are Jupiter and Saturn. Thus, we may consider a four-body system with P_2 as the Sun, P_1 as Jupiter, P_3 as Saturn and P_4 as a massless planet, and take the case $T_1/T_3=2/5$, as in the actual motion of the Sun-Jupiter-Saturn system. Since the motion of P_4 does not affect the motion of P_1,P_2 and P_3, its motion is determined from a system with two degrees of freedom (for planar motion), in the rotating frame of reference Oxy defined in section 3. This case can be considered as the generalization of the restricted 3-body problem. The conditions under which the motion of P_4 is periodic are similar to the general case N=4, discussed in section 3. A complete description of this problem will be given elsewhere (Hadjidemetriou 1978).

We shall present here one example of a periodic motion, corresponding to the case $p/q=-1/9$. This motion approximates the system Sun-Mars-Jupiter-Saturn, as the "radius" of the nearly circular orbit of P_4 is equal to 1.51 A.U. There are two such periodic orbits with the same resonance, differing in phase. One of them corresponds to $x_{10}>0$, $x_{30}<0$, $x_{40}>0$, and the other to $x_{10}>0$, $x_{30}<0$, $x_{40}<0$, at $t=0$. The numerical computations revealed that the first periodic orbit is stable and the second is unstable.

Let us draw now our attention to the unstable case. The orbits of all the planets are nearly circular and this periodic motion can be considered to be obtained from a degenerate periodic motion where all the planets have zero masses and describe circular orbits, by continuing through the masses. Evidently, if only the mass of Jupiter is increased, and its motion is kept circular, the motion of P_4 is determined from the well-known restricted circular 3-body problem. The numerical integrations reveal in this case that the motion of P_4 is stable. Let us introduce now the planet Saturn with its actual mass. The above results show that the orbit of P_4 may become unstable due to the effect of Saturn! This indicates that the stability analysis of planetary systems based on the study of the restricted 3-body problem may not be always correct, because the effect of other planets than Jupiter is not taken into account.

REFERENCES

1. Delibaltas, P.: 1976, Astrophys. Space Science 45, 207.
2. Griffin, F.L.: 1920, in F.R. Moulton "Periodic Orbits", Ch. 14, Garnegie Inst.
3. Hadjidemetriou, J.D.: 1975, Celes. Mech. 12, 155.
4. Hadjidemetriou, J.D.: 1976, Astrophys. Space Science 40, 201.
5. Hadjidemetriou, J.D.: 1977, Celes. Mech. 16, 61.
6. Hadjidemetriou, J.D.: 1978 (in preparation).
7. Hadjidemetriou, J.D. and Michalodimitrakis, M.: 1978a in V. Szebehely (ed). Proceeding of I.A.U. Colloqium No 41 (in press).
8. Hadjidemetriou, J.D. and Michalodimitrakis, M.: 1978b (in preparation).
9. Hagihara, Y.: 1961, in Kuiper, G.P. and Middlehurst, B.M. (eds), "The Solar System", Univ. of Chicago Press, p.123.

PERIODIC ORBITS OF ARBITRARY INCLINATIONS AND ECCENTRICITIES IN THE GENERAL 3-BODY PROBLEM.

C. Marchal
O.N.E.R.A. - 92320 - Châtillon - France

ABSTRACT

Usual periodic orbits have periods of the order of magnitude of a few revolutions. However if we consider much longer periods it is possible to find, for three given masses, periodic orbits in any neighbourhood of arbitrary given initial eccentricities and inclinations provided that the distance of the outer body is sufficiently large with respect to the mutual distance of the two inner bodies.

INTRODUCTION

Euler and Lagrange found the first periodic orbits of the 3-body problem. Later, using the symmetries of the problem, Poincaré described a very general way for the construction of periodic orbits (Poincaré 1892 - 1893 - 1899) and considered them as our essential key for the understanding of that problem. He formulated this conjecture : "In the phase space of the 3-body problem the set of periodic orbits is dense in the set of bounded orbits". Since then many families of periodic orbits have been found both in the restricted case and in the general case and a method is proposed here for the research of periodic orbits of arbitrary inclinations and eccentricities, but of very long period.

NOTATIONS

Let us use the ordinary Jacobi decomposition of the 3-body motion into the "inner motion" (i.e. the relative motion of the two nearest point-mass m_1 and m_2) and the slower outer motion (i.e. the motion of the third point-mass m_3 with respect to $O_{1.2}$ the center of mass of m_1 and m_2). The usual osculating elements of the inner and outer orbits will be:

$$a, e, i, \Omega, \omega, M, \qquad a_3, e_3, i_3, \Omega_3, \omega_3, M_3 \qquad (1)$$

(semi-major axis, eccentricity, inclination, longitude of the node, argument of the pericenter, mean anomaly; with subscripts 3 for the orbit of m_3).

R. L. Duncombe (ed.), Dynamics of the Solar System, 29–32.

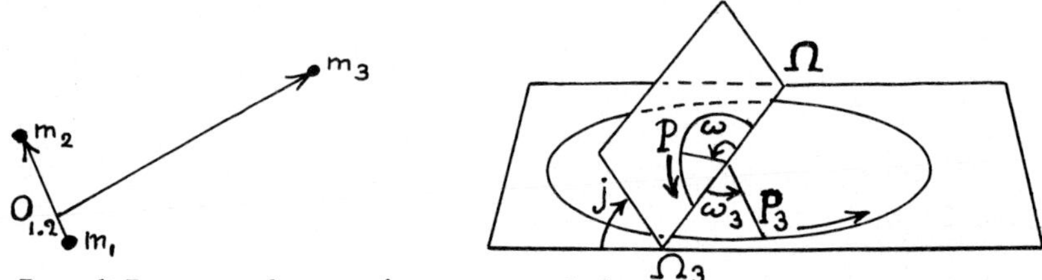

Figure 1 - P and P_3 are the pericenters of inner and outer orbits.

We will choose the polar reference direction in the direction of angular momentum, the line of nodes is then in the equatorial plane, the angles ω and ω_3 are also the angles between the line of nodes and the pericenters directions (fig. 1), the mutual inclination j is equal to i + i_3 and the longitudes Ω and Ω_3 are related by $\Omega_3 = \Omega + \pi$

1. PERIODIC ORBITS

The Poincaré conditions of symmetry are :

$$0 = \sin M = \sin M_3 = \sin 2\omega = \sin(\omega_3 - \omega) \tag{2}$$

If they are satisfied at some instant t_1 there is a past-future symmetry : the three mutual distances are even function of time with respect to t_1 , if they are also satisfied at some other instant t_2 the two symmetries imply the periodicity with the period 2 ($t_2 - t_1$). Hence starting from initial conditions satisfying (2) we shall look for a second passage at these conditions and thus we shall look for initial a, e, i, a_3 , e_3 , i_3 leading to such a second passage.

2. FIRST ORDER SECULAR APPROXIMATION

If the ratio $(m_1 + m_2 + m_3)\, a^3 / (m_1 + m_2)\, a_3^3 (1 - e_3^2)^3$ is small the use of Delaunay's variables and a proper Von Zeipel transformation (Marchal 1977) lead to "secular elements" a_S , e_S , i_S , Ω_S , ω_S M_S and a_T , e_T , i_T , Ω_T, ω_T, M_T very near to the corresponding osculating elements a, --, M, a_3 , -- M_3 but with much smaller short period variations. The "secular mutual inclination" j_S is equal to $i_S + i_T$ and is near $j = i + i_3$

The symmetries of the transformation lead to symmetry conditions :

$$0 = \sin M_S = \sin M_T = \sin 2\omega_S = \sin(\omega_T - \omega_S) \tag{3}$$

and in the first order approximation the 3-body problem is integrable : a_S , a_T , e_T , dM_S/dt, dM_T/dt are constant.

Ω_S, Ω_T $(= \Omega_S + \pi)$ and ω_T are ignorable and given by final quadratures.

i_S and i_T are given interms of a_S , e_S , a_T , e_T , j_S by :

$$0 \leq i_S \leq j_S \leq \pi \quad ; \quad 0 \leq i_T \leq j_S \leq \pi \quad ; \quad i_S + i_T = j_S \tag{4}$$

$$\sin i_T / \sin i_S = m_1 m_2 m_3^{-1} (m_1+m_2)^{-3/2} (m_1+m_2+m_3)^{1/2} \left[a_S(1-e_S^2)/a_T(1-e_T^2)\right]^{1/2} \tag{5}$$

Finally the three remaining elements e_S , ω_S, j_S are related by :

$$(1 - e_S^2)(1 + \sin^2 j_S) + 5 e_S^2 \sin^2 j_S \sin^2 \omega_S = Z = \text{constant} \tag{6}$$

$$\cos j_S = A(1 - e_S^2)^{-1/2} - B(1 - e_S^2)^{1/2} \tag{7}$$

(A and B being two constants : $B = 0.5\, m_1 m_2 m_3^{-1} (m_1+m_2)^{-3/2} (m_1+m_2+m_3)^{1/2} \left[a_S/a_T(1-e_T^2)\right]^{1/2}$

and the variations of e_S, ω_S, j_S are given by the final quadrature :

$$de_S/dt = K\, e_S (1-e_S^2)^{1/2} \sin^2 j_S \sin 2\omega_S \qquad (8)$$

with, G being the constant of the law of universal attraction :

$$K = 1.875\, G^{1/2} m_3 (m_1+m_2)^{-1/2} \left[a_S/a_T^2(1-e_T^2)\right]^{3/2} = \text{constant} \qquad (9)$$

It leads to motions in which e_S, j_S, ω_S, $d\Omega_S/dt$, $d\omega_T/dt$ are generally periodic (with the same period much larger than that of the mean anomalies M_S and M_T) but sometimes the secular motion is asymptotic (either to $e_S = 0$ or to $j_S = \pi$ or to $\sin\omega_S = 0$ or to $\cos\omega_S = 0$ for proper values of the integrals A, B, Z; for instance the motion is asymptotic to $e_S = 0$ if $Z = 2-(A-B)^2$; $1+AB \geqslant B$ and $3 \geqslant (A-B)(5A-3B)$). With that secular motion it is easy to look for periodic orbits by the Poincaré method, i.e. to determine initial conditions satisfying the Poincaré symmetry conditions (3) and leading to a second passage at these conditions, these "approximated periodic orbits" are even dense everywhere in the region of interest. However we have only studied a first order approximation, let us consider now the upper order effects.

3. ANALYSIS OF THE UPPER ORDER EFFECTS

Let us use M_T as parameter of description instead of the time, it gives:
A) dM_S/dM_T is very near to the ratio of the mean angular motions n_S/n_T, that is $\left[(m_1+m_2)a_T^3/(m_1+m_2+m_3)a_S^3\right]^{1/2}$ or, in terms of the "quasi integrals" B and e_T, we can write :

$$dM_S/dM_T = \left[m_1^3 m_2^3 (m_1+m_2+m_3)/8 m_3^3 (m_1+m_2)^4 B^3 (1-e_T^2)^{3/2}\right].\left(1+O(\varepsilon_1)\right) \qquad (10)$$

with : $\varepsilon_1 = \left[m_1 m_2 a_S^2 a_T + m_3(m_1+m_2) a_S^3\right] / \left[a_T^3 (1-e_T^2)^3 . \inf\{e_S, e_T\}.(m_1+m_2)^2\right]$ (11)

Note that, since by hypothesis $(m_1 + m_2 + m_3)\, a^3 / (m_1 + m_2)$. $a_3^3 (1 - e_3^2)^3$ is small, the ratio dM_S/dM_T is large.

B) With $x = 1 - e_S^2$ we obtain similarly :

$$dx/dM_T = \pm 12 B^3 m_1^{-3} m_2^{-3} m_3^4 (m_1+m_2)^4 (m_1+m_2+m_3)^{-2} \left[\{Z-2x+(A-Bx)^2\}\{(5-Z)x-3x^2+(4x-5)(A-Bx)^2\}+O(\varepsilon_2)\right]^{1/2} \qquad (12)$$

with the $\pm$ sign $= -\text{sign}(e_S') = \text{sign}\,(-\sin 2\omega_S + O(\varepsilon_2))$ and :

$$\varepsilon_2 = \left[a_S/a_T e_T (1-e_T^2)^{5/2}\right].\sup\left\{\left[|m_2-m_1|/(m_1+m_2)\right];\left[(m_1+m_2+m_3)a_S/(m_1+m_2)a_T(1-e_T^2)\right]^{1/2}\right\} \qquad (13)$$

On the other hand, with $\varepsilon_3 = \varepsilon_2 (1-e_T^2)^{-1/2} \geqslant \varepsilon_2$:

$$d\omega_T/dM_T = 6B^3 m_1^{-3} m_2^{-3} m_3^4 (m_1+m_2)^4 (m_1+m_2+m_3)^{-2}\left[B(5-5Z+2x)+(Z-x)(A+Bx)\left[x-(A-Bx)^2\right]^{-1} + O(\varepsilon_3)\right] \qquad (14)$$

And finally, with $\varepsilon_4 = \varepsilon_2\, e_T (1-e_S^2)^{-1/2} / \inf\{e_S ; e_T\}$:

$$\frac{d\omega_S}{dM_T} = \frac{6B^3 m_3^4 (m_1+m_2)^4 x^{1/2}}{m_1^3 m_2^3 (m_1+m_2+m_3)^2}\left[2+2AB-2B^2x+\frac{[Z-2x+(A-Bx)^2][(A-Bx)(A+Bx-2Bx^2)-x^2]}{x(1-x)[x-(A-Bx)^2]}+O(\varepsilon_4)\right] \qquad (15)$$

Note that : A) dx/dM_T, $d\omega_T/dM_T$, $d\omega_S/dM_T$ are small, of the order of dM_T/dM_S or even smaller. The three-body motion of interest is the composition of two slowly perturbed Keplerian motions.

B) If we neglect the error terms and the small variations of the "quasi integrals" e_T, A, B, Z, the equation (12) is a quadrature and we obtain the integrable system of the previous section.

C) e_T only appears in the error terms and in dM_S/dM_T. Hence, taking account of the continuity of the problem and of the small variations and the independance of e_T, A, B, Z, if we consider an "approximate periodic orbit" of the previous section we can obtain a true periodic orbit by a modification of the initial values of e_T^2, A, B, Z of the order of sup (ε_1, ε_3, ε_4) provided that the three modifications δA, δB and δZ imply independant final modifications $\delta\omega_T$ and $\delta\omega_S$ in the integration of (14) and (15) without error terms

(as it is everywhere the case except on some submanifolds).
More precisely let us put $\varepsilon = \sup(\varepsilon_1, \varepsilon_3, \varepsilon_4)$ in the interval (t_1, t_2):

$$\varepsilon = \frac{a_S}{a_T(1-e_T^2)^{5/2}} \cdot \sup\left\{\frac{|m_2-m_1|}{m_1+m_2} ; \left[\frac{(m_1+m_2+m_3)a_S}{(m_1+m_2)a_T(1-e_T^2)}\right]^{1/2}\right\} \cdot \sup\left\{\frac{1}{e_T\sqrt{1-e_T^2}} ; \frac{1}{e_S\sqrt{1-e_T^2}} ; \frac{1}{e_S\sqrt{1-e_S^2}}\right\} \quad (16)$$

(t_1 and t_2 are the Poincaré instants of symmetry; e_S varies in (t_1, t_2)).
Then, if ε is small with respect to 1 and if the above condition of independance of $\delta\omega_S$ and $\delta\omega_T$ is respected, we obtain in the vicinity of the "approximate periodic orbit" of interest a one non trivial parameter family of true periodic orbits (of the same $\Delta M_S, \Delta M_T, \Delta\omega_S, \Delta\omega_T$ per period) inside the region defined by:

$$\delta Z \; ; \; \delta(A - Bx) \; ; \; \delta B/B \; ; \; \delta(e_T^2) = O(\varepsilon) \quad (17)$$

That is, with the ordinary osculating parameters a, e, a_3, e_3 and j (mutual inclination) in the region defined by:

$$\delta(Ln\,(a_3/a)) \; ; \; \delta(e^2) \; ; \; \delta(e_3^2) \; ; \; \delta(\cos j) = O(\varepsilon) \quad (18)$$

Note 1 - The true periodic orbits, obtained by this method are sometimes called "relative periodic orbits" or "periodic orbits in a rotating frame of reference". The absolute periodicity must take account of the motion of the elements Ω and Ω_3 (with $\Omega_3 = \Omega + \pi$).
In general the absolute periodic orbits are dense along the one-parameter families of relative periodic orbits, they correspond to rational values of $\Delta\Omega/2\pi$ during one period of the relative motion.
Note 2 - ε is infinite when e_S goes to one, hence the method doesn't work for oscillating orbits of the second kind (Marchal 1977) in which the bodies m_1 and m_2 have an infinite number of approaches as close as desired (but they don't have a strict collision). These orbits fill a set of positive measure of phase space, one of them has been integrated by Hadjidemetriou (Hadjidemetriou 1977).
Note 3 - The relation (18) doesn't imply the denseness of periodic orbits even in the regions of small ε and the Poincaré conjecture remain open. However, if we consider arbitrary orientations, we can select any small open set of the 11-dimension space of elements (a, e, i, Ω, ω, M, e_3, i_3, Ω_3, ω_3, M_3) : that set is crossed by periodic orbits for all sufficiently large values of a_3.

CONCLUSION

The research of periodic orbits of very long period has led to many new families of periodic orbits of arbitrary inclinations, eccentricities and orientations (for any 3 given masses, both in the restricted and in the general case), the Poincaré conjecture on periodic orbits remain open but it is likely true.

REFERENCES

Hadjidemetriou. J.D. (1977) "Integration of an oscillating orbit of the second kind". Private correspondance.
Marchal. C. (1977) "Collisions of stars by oscillating orbits of the second kind" ONERA - T.P. N° 1977-97E
Poincaré. H. (1892 - 1893 - 1899) "Les méthodes nouvelles de la mécanique céleste" - Gauthier Villars - Paris.

A NOTE AS TO A PERTURBATION OF HILL'S CURVES

V.R. Matas
Astronomical Institute, Czechoslovak Academy of Sciences,
Praha, Czechoslovakia

Let us consider motion of an infinitesimal body in the gravitational field of two homogeneous spheroids with coincident equatorial planes. Let mass of the infinitesimal body M - a material point - be negligible if compared with masses of the spheroids S_1, S_2 respectively. We will study the motion of M in the common equatorial plane provided that the spheroids move along circular orbits (in an inertial frame of reference) - see Kondurar', 1952. System of the equations of motion of M posseses Jacobi's integral and Hill's generalized curves of zero relative velocity of M are consequently described by the equation

$$H(r_1) = \left(1 + \frac{3}{10}(a_1^2e_1^2 + a_2^2e_2^2)\right)\frac{1 - m}{2}r_1^2 + \frac{1 - m}{r_1} + \frac{1 - m}{10}a_1^2e_1^2r_1^{-3} = \tag{1}$$
$$C - \left(1 + \frac{3}{10}(a_1^2e_1^2 + a_2^2e_2^2)\right)\frac{m}{2}r_2^2 - \frac{m}{r_2} - \frac{m}{10}a_2^2e_2^2r_2^{-3} = C - B(r_2) = A$$

where C is a "Jacobian" constant of integration, r_1, r_2 are the dimensionless distances between M and the centers of mass of the spheroids S_1, S_2; $1 - m$, m are the dimensionless masses of the spheroids; a_i, e_i are their respective semimajor axes and eccentricities ($i = 1,2$).

Let r_2 be given. Then solving (1) is equivalent to seeking positive roots of the function $G(x) = H(x) - A$, where A is a constant parameter. If $A \leqq 0$ then there does not exist any solution of the equation $G(x) = 0$. In the case $A > 0$ let us study the functions $G(x)$, $H(x)$ in the interval $(0,\infty)$. It follows that the functions G, H are convex in $(0,\infty)$ and have there an absolute minimum at a point $x_o \leqq 1$ (1 is the dimensionless distance of the centers of mass of the spheroids) which is the only root of the derivative G' (or H') - an increasing and concave function in $(0,\infty)$. If the minimum $G(x_o)$ is positive, equation $G(x) = 0$ evidently has no real solution. Provided that $G(x_o) = 0$ we put $r_1 = x_o$ and consequently pair r_1, r_2 (r_2 was given) satisfies Eq. (1) of the zero relative

R. L. Duncombe (ed.), Dynamics of the Solar System, 33–36.

velocity curve. Eventually in the case $G(x_o) < 0$ there are two different positive roots r_{11}, r_{12}, $(r_{11} < x_o < r_{12})$ of the equation $G(x) = 0$. If moreover

$$r_{1k} + r_2 \geqq 1, \; r_{1k} + 1 \geqq r_2, \; r_2 + 1 \geqq r_{1k}, \tag{2}$$

where k equals either 1 or 2 (r_2 was given), then the pair of distances r_{1k} and r_2 (from S_1 and S_2 respectively) defines two points (or one point) on the zero relative velocity curve (1).

An analogical examination of the function $B(x)$ in the interval $(0,\infty)$ immediately gives that B is concave in $(0,\infty)$ and has there an absolute maximum at a point $x_1 \leqq 1$ which is the only root of the derivative B' (that is decreasing and convex in $(0,\infty)$). The above considerations as to the functions H, B imply that the curves of the zero relative velocity of M are real if, and only if, $H(x_o) \leqq C + B(x_1)$. In this case the Jacobian constants C have a lower bound, viz. $C \geqq H(x_o) - B(x_1) = C_o$. Consequently if $C = C_o$ equation (1) posseses only one solution $r_1 = x_o$, $r_2 = x_1$ which defines, if moreover $x_o + x_1 > 1$, two equilibrium solutions L_T, L_T' of the equations of motion of M.

Let $C \geqq C_o$ and $x_o + x_1 > 1$ be valid. It follows from the preceding study of the function B that the equation $H(x_o) = C + B(r_2)$ has two solutions $r_{2min} \leqq r_{2max}$ in the most. The pair x_o, r_{2min} defines actual points on the zero relative velocity curve in case $r_{2min} \geqq 1 - x_o$. The pair x_o, r_{2max} represents actual points on the curve if $r_{2max} \leqq 1 + x_o$. Analogically the equation $H(r_1) = C + B(x_1)$ posseses two solutions $r_{1min} \leqq r_{1max}$ in the most. The pair r_{1min}, x_1 characterizes actual points on the curve examined when $r_{1min} \geqq 1 - x_1$. The pair r_{1max}, x_1 determines actual points on the curve if $r_{1max} \leqq 1 + x_1$. r_{1min}, r_{2min} are decreasing functions of C and r_{1max}, r_{2max} are increasing functions of C, all of them in $\langle C_o,\infty)$. Let C_1, C_2, D_1, D_2 be values of C for which

$$\begin{aligned} r_{1min}(C_1) &= 1 - x_1, \; r_{1max}(D_1) = 1 + x_1, \\ r_{2min}(C_2) &= 1 - x_o, \; r_{2max}(D_2) = 1 + x_o. \end{aligned} \tag{3}$$

C_1, C_2, D_1, D_2 are to be determined uniquely (if $x_1 < 1$, $x_o < 1$). We easily find

$$\begin{aligned} C_1 &= H(1 - x_1) - B(x_1), \; C_2 = H(x_o) - B(1 - x_o), \\ D_1 &= H(1 + x_1) - B(x_1), \; D_2 = H(x_o) - B(1 + x_o). \end{aligned} \tag{4}$$

Let us first consider the zero relative velocity curves determined by $C \in \langle C_o,C_1\rangle$ (or $C \in \langle C_o,C_2\rangle$). Then the properties found as to the functions H, B justify us to state that the locus of all the points least distant from the spheroid S_1 (or S_2) on the concerned zero relative velocity curves is the minor arc $L_T L'_T$ (intersecting the line segment S_1S_2 at a point P_1 (or P_2)) of a circle center of which is the center of mass of the other spheroid. As the Jacobian constant C approaches C_o, the points of the determined locus approach L_T, L'_T ($C = C_o$ implies that L_T, L'_T are on the locus). As C approaches C_1 (or C_2), the points of the locus approach the intersection of the locus and the line segment S_1S_2. If $C \geqq C_1$ (or $C \geqq C_2$) then, with respect to (3), the points – on the corresponding generalized Hill's curves – which are nearest to the spheroid S_1 (or S_2) fill the line segment S_1P_1, excluding S_1 (or S_2P_2, excluding S_2); the finite dimensions of the spheroids are not considered now. As C becomes equal to C_1 (or C_2) the concerned points approach the intersection P_1 (or P_2) – these points approach the center of mass of S_1 (or S_2) if the constant C becomes infinite.

If, for instance, S_2 approaches a sphere with a spherical density distribution then $x_o \to 1$ and C_2 becomes infinite. If $x_o = 1$ there does not exist any real C_2 so that $r_{2min}(C_2) = 1 - x_o = 0$. Instead of this we have $r_{2min}(C) \to 0$ for $C \to \infty$. In this case the concerned minor arc $L_T L'_T$ of the circle centered at S_1 goes through S_2 (provided that S_2 is considered to be a material point) but S_2 is not a point of the locus. The points of this minor arc merely approach S_2 if C becomes infinite. The same considerations may be made for the body S_1. It is to be seen that the locuses found represent a generalization of a similar geometric property of the Hill's curves in the restricted three-body problem (see Matas, 1978).

Notice now the zero relative velocity curves defined by $C \in \langle C_o, D_1 \rangle$ (or $C \in \langle C_o, D_2 \rangle$). The derived characteristics of the functions H, B imply that the locus of all the points most distant from the spheroid S_1 (or S_2) on the considered zero relative velocity curves is the major arc $L_T L'_T$ of the circle centered at the remaining spheroid. As the Jacobian constant C approaches C_o, the points of this locus approach L_T, L'_T ($C = C_o$ corresponds to the equilibrium points L_T, L'_T). As C approaches D_1^o (or D_2), the points of the locus approach an intersection R_1 (or R_2) of the locus and the ray S_1S_2 (or S_2S_1) extended from S_1 (or S_2). In the case when $C \geqq D_1$ (or $C \geqq D_2$) obviously (see (3)) the locus of all the most distant points – on the given generalized Hill's curves – from the spheroid S_1 (or S_2) is a ray which: (i) is situated on the straight line S_1S_2, (ii) has endpoint R_1 (or R_2) and (iii) does not go through the spheroids. As the Jacobian constant C approaches D_1 (or D_2) points of the ray approach R_1 (or R_2); the points of the ray have unlimited extent in the opposite direction if the constant C becomes infinite.

An analytical approach gives (see the precision adopted in (1))

$$x_o = 1 - \frac{a_2^2 e_2^2}{10}, \quad x_1 = 1 - \frac{a_1^2 e_1^2}{10}, \tag{5}$$

$$
\begin{aligned}
&C_o = \frac{3}{2} + \frac{1}{4} a_1^2 e_1^2 + \frac{3}{20} a_2^2 e_2^2 + \frac{m}{10}(a_2^2 e_2^2 - a_1^2 e_1^2) \geqq \frac{3}{2} , \\
&C_o \rightarrow \frac{3}{2} \quad \text{for} \quad e_1, e_2 \rightarrow 0 ; \\
&C_1 \approx \frac{100(1 - m)}{a_1^2 e_1^2} \rightarrow \infty \quad \text{for} \quad e_1 \rightarrow 0 ; \\
&C_2 \approx \frac{100m}{a_2^2 e_2^2} \rightarrow \infty \quad \text{for} \quad e_2 \rightarrow 0 ; \\
&D_1 = \frac{5}{2} - m + \frac{7}{16} a_1^2 e_1^2 + \frac{3}{5} a_2^2 e_2^2 - \frac{m}{20}\left(\frac{23}{4} a_1^2 e_1^2 + 7a_2^2 e_2^2\right) , \\
&D_1 \rightarrow \frac{5}{2} - m \quad \text{for} \quad e_1, e_2 \rightarrow 0 ; \\
&D_2 = \frac{3}{2} + m + \frac{1}{4} a_1^2 e_1^2 + \frac{3}{20} a_2^2 e_2^2 + \frac{m}{20}\left(7a_1^2 e_1^2 + \frac{23}{4} a_2^2 e_2^2\right) , \\
&D_2 \rightarrow \frac{3}{2} + m \quad \text{for} \quad e_1, e_2 \rightarrow 0 .
\end{aligned}
\tag{6}
$$

REFERENCES

Kondurar', V.T.: 1952, "Trudy Gos. Astron. Inst. Shternberga" 21, pp.135-158.

Matas, V.: 1978, Celes. Mech. (in press).

DISCUSSION

Garfinkel: Are the two spheroids of zero obliquity with respect to the orbital plane?

Hori: Yes.

Garfinkel: Can the restriction that the spheroids are of constant density be removed by introducing the moments of interia in the place of a_1, a_2, e_1, e_2?

Hori: Very likely.

LES ORBITES DE COLLISION DU PROBLEME RESTREINT

Fernand S. Nahon
Universite P. et M. Curie, Paris, France

ABSTRACT: The article studies collisions with the smaller primary, m_2, in the planar restricted problem of three bodies. The formulation is in polar coordinates and the corresponding momenta. The results are based on the fact that the collisions correspond to a solution of a partial differential equation.

Considérons pour fixer les idées les orbites de collision avec le primaire M_2 de faible masse.

Soit : $F(r,\theta,R,\Theta) = \frac{1}{2}(R^2 + \frac{\Theta^2}{r^2}) - \Theta - (U + \nu r \cos\theta)$

où : $U = \frac{\mu}{r} + \frac{\nu}{\Delta}$, Δ distance de M au primaire M_1 le hamiltonien du problème restreint en coordonnées "jovicentriques", c'est-à-dire d'origine M_2. Levi-Civita a montré que les orbites de collision vérifient sur la surface $F = -C$ une relation invariante :

$$\Theta = \varphi(r,\theta;C)$$

où φ est développable en série entière de $\sqrt{r} = \rho$. Les coefficients a_n de φ^n sont des fonctions périodiques de période 2π de la variable θ. Nous nous proposons de simplifier la procédure de Levi-Civita et de mettre en évidence, pour la valeur critique de la constante de Jacobi, la famille qui généralise les orbites de collision du problème képlérien.

Nous nous appuierons sur le théorème suivant que nous donnons sans démonstration :

Proposition 1 - Aux orbites de collision avec M_2 correspond une solution $S(r,\theta,C)$ de l'équation aux dérivées partielles :

$$\frac{1}{2}\left[\left(\frac{\partial S}{\partial r}\right)^2 + \frac{1}{r^2}\left(\frac{\partial S}{\partial \theta}\right)^2\right] - \frac{\partial S}{\partial \theta} = U + \nu r \cos\theta - C$$

C'est la solution unique de cette équation qui est développable en série entière de $\rho = \sqrt{r}$ au voisinage de $\rho = 0$.

Cas où μ est petit :

Si $\mu \longrightarrow 0$, on peut montrer que les orbites de collision

R. L. Duncombe (ed.), Dynamics of the Solar System, 37–39.

a) n'existent que pour $C \leqslant 3/2$

b) pour $C < 3/2$, elles peuvent être représentées en première approximation par un assemblage d'orbites képlériennes qui se raccordent en M_2.

Le cas intéressant est le cas $C = 3/2$; pour l'étudier on peut poser

$$r = \mu^{1/3}\,\bar{r}.$$

Cela revient :

1) à remplacer F par l'approximation de Hill :

$$H = \tfrac{1}{2}\left(\bar{R}^2 + \frac{\bar{\Theta}^2}{\bar{r}^2}\right) - \bar{\Theta} - \left[\frac{1}{\bar{r}} + \frac{\bar{r}^2}{2}\,(3\cos^2\bar{\theta} - 1)\right]$$

2) à considérer l'intégrale $H = -h$ qui correspond à $C = 3/2$.

On peut montrer que :

$$C = \frac{3}{2} + \mu^{2/3}\,h + o(\mu^{2/3})$$

On est donc ramené (en supprimant les barres)

1°) à étudier les orbites de collision avec l'origine du hamiltonien :

$$H = \tfrac{1}{2}\left(R^2 + \frac{\Theta^2}{r^2}\right) - \Theta - \left[\frac{1}{r} + \frac{r^2}{r}\,(3\cos^2\theta - 1)\right] \qquad (1)$$

2°) et ceci est équivalent d'après la proposition (1) au problème de chercher la solution de l'équation aux dérivées partielles

$$\frac{1}{2}\left[\left(\frac{\partial S}{\partial r}\right)^2 + \frac{1}{r^2}\left(\frac{\partial S}{\partial \theta}\right)^2\right] - \frac{\partial S}{\partial \theta} = \frac{1}{r} + \frac{r^2}{2}\,(3\cos^2\theta - 1) \qquad (2)$$

développable en série entière de $r = \rho^2$ autour de l'origine.

Le cas des orbites de collision képlériennes suggère de poser

$$r = a\,\tau^{2/3}$$

c'est-à-dire de faire le changement de variables

$$\left.\begin{aligned} a^3 &= \frac{9}{2} \\ \rho &= \sqrt{a}\;\tau^{1/3} \\ S &= 2\sqrt{2a}\;\tau^{1/3} + a^2\,\tau^{7/3}\,\varphi(\tau,\theta) \end{aligned}\right\} \qquad (3)$$

Passons sur le détail des calculs est donnons le théorème récapitulatif

Proposition 2 - Soit la famille des orbites de collision avec le primaire de masse μ pour la valeur $C = 3/2$ de la constante de Jacobi relative au barycentre.

1°) Ces orbites généralisent les orbites de collision képlériennes en ce sens que le paramètre est le temps fictif τ défini par la relation képlérienne :

$$r^3 = \frac{9}{2}\,\mu\,\tau^2 \qquad (4)$$

2°) Soit θ_o la limite de θ lorsque $r \longrightarrow 0$. Les orbites dépendent de θ_o suivant la formule :

$$\theta = \theta_o - \tau - \frac{9}{14} \sin 2\,\theta_o \frac{\tau^2}{2} + o(\tau^2) \qquad (5)$$

où les termes non écrits représentent une série entière en τ à coefficients périodiques de période 2π en θ_o.

3°) Elles vérifient sur la variété $F = -3/2$ une relation invariante :

$$\Theta = \left(\frac{9}{2}\right)^{2/3} \tau^{7/3} f(\theta, \tau) \qquad (6)$$

où f est la solution unique développable en série entière de τ, à coefficients périodiques de période 2π en θ, de l'équation :

$$7\varphi + 3\tau(\varphi'_\tau - \varphi'_\theta) + o(\tau^2) = \frac{3}{2}(3 \cos^2\theta - 1) \qquad (7)$$

Les premiers termes du développement sont donnés par

$$f(\theta, \tau) = -\frac{9}{14} \sin 2\,\theta - \left(\frac{27}{70}\right) \cos 2\,\theta\,\tau + o(\tau^2) \qquad (8)$$

où τ est positif pour les orbites d'éjection et négatif pour les orbites de collision.

Référence :

Levi-Civita, T.: 1903, "C.R. Acad. Sc. Paris", t. 136, p. 221.

LIBRATION OF RETROGRADE SATELLITE ORBITS IN THE CIRCULAR PLANE RESTRICTED THREE-BODY PROBLEM

Daniel Benest
Observatoire de Nice, Nice, France

INTRODUCTION

In the circular plane restricted three-body problem, we study the stable large retrograde non-periodic satellite orbits. We use rotating axes with the origin in the body around which turns the satellite, called its primary. We choose the initial conditions such as $Y_0=0$ and $U_0=0$, so that an orbit can be represented by a point in the (X_0, V_0) plane. In this plane, the set of stable orbits is represented by a limited region, which we call the stability zone. This zone is composed in general by a large continental region, approximately limited by Lagrange points, and a peninsula more or less elongated. Inside, takes place the characteristic of the single-periodic symmetrical family f which can be called the backbone of the zone (figure 1).

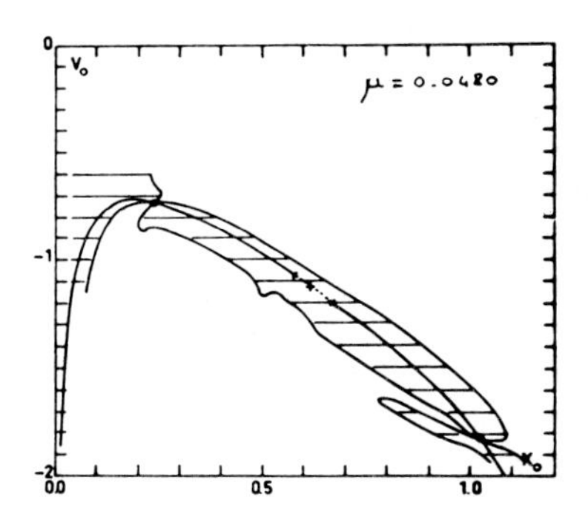

Figure 1. An example of the stability zone.

The numerical explorations have shown that the non-periodic orbits can be approximately decomposed into a fast "reference motion" and a slow libration of its centre around the primary of the satellite. Moreover, the amplitude of the libration, which is zero for the periodic

R. L. Duncombe (ed.), Dynamics of the Solar System, 41–44.

orbits, increases when the initial conditions of the non-periobic orbits move off from the initial conditions of the periodic orbits.

HILL'S CASE

In Hill's case, the reference motion is elliptic, with a period of the order of 2π, and the centre of this ellipse librates on a very elongated oval with a period much larger than 2π. The analysis of this libration is developped in a paper published in 1976. In this paper, we establish the equations of motion for the coordinates of the centre of the ellipse and we found two integrals of motion: the first is the semi-major axis of the ellipse; the second is essentially Jacobi's integral, translated into the new coordinates. A numerical verification gives very good agreement for all these results.

GENERAL CASE

We turn now to the general case, i.e. $\mu \neq 0$. In this case, neither the reference motion, nor the trajectory of its centre cannot be described by simple curves (figure 2). The trajectory of the centre of the reference motion can be considered as a very narrow "bean", elongated along the circle of centre B_1 and radius 1.

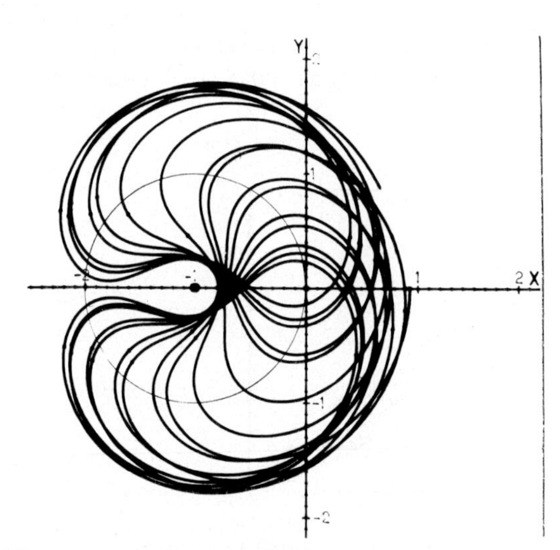

Figure 2. An example of libration.

To use the same analysis as in Hill's case, we must first "rectify the bean". This lead us to use the transformation defined by:

$$X = (\xi + 1)\cos\theta - 1 \quad \text{and} \quad Y = (\xi + 1)\sin\theta \,. \tag{1}$$

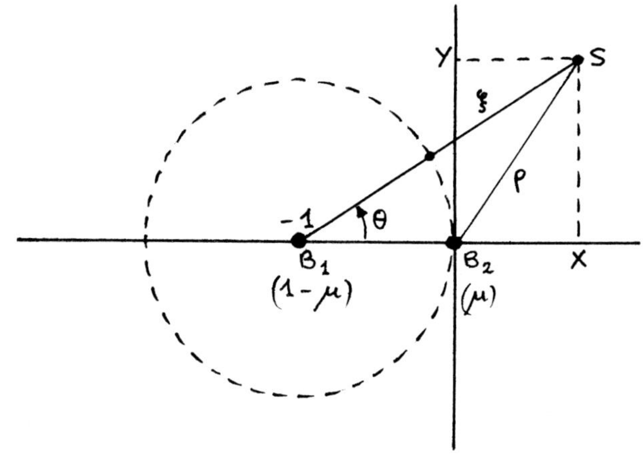

Figure 3. Representation of the coordinate systems (X,Y) and (ξ,θ).

Then, the classical equations of motion for the satellite become:

$$\left.\begin{aligned} &\dot{\xi} = \eta \;, \quad \dot{\theta} = \psi \;, \\ &\dot{\eta} = (\xi+1)(\psi+1)^2 - (\xi+1)^{-2} + \mu\, f(\xi,\theta) \;, \\ &\dot{\psi} = -2\eta(\psi+1)/(\xi+1) + \mu\, g(\xi,\theta) \,. \end{aligned}\right\} \quad (2)$$

In the limiting case $\mu=0$ (which corresponds to the two-body problem), the family f reduces (in fixed axes) to a family $\mathcal{E}$ of ellipses of focus B_1 and semi-major axis 1. In this case, the terms in μ vanish in equations (2) and the reduced equations have a well-known solution. Rather than the constants of integration, which are the elements of $\mathcal{E}$, the physically interesting quantities are the coordinates (x,y) of the centre of $\mathcal{E}$, the amplitude A in ξ and the phase difference φ between the motions of S and B_2, given by the equations:

$$\left.\begin{aligned} &x=a-1, \qquad y=(1-a^{3/2})\alpha+\omega+t_o, \\ &A=-ae, \qquad \varphi=t_o, \end{aligned}\right\} \quad (3)$$

where a, e, ω and t_o are the elements of $\mathcal{E}$ and α is the eccentric anomaly of S on $\mathcal{E}$.

Then the solution of the reduced equations can be written:

$$\begin{aligned} &t=\sqrt{x+1}\,(A\sin\alpha+(x+1)\alpha)+\varphi, \\ &\xi=A\cos\alpha+x+1, \\ &\theta=-A\sqrt{x+1}\sin\alpha-\alpha+2\,\mathrm{arc\,tg}\left(\sqrt{\frac{x+1-A}{x+1+A}}\,\mathrm{tg}\,\alpha/2\right)+Y. \end{aligned} \quad (4)$$

In the approximation $\mu=0$ (considering x, y, A and φ as constant), we differentiate equations (4) to obtain η and ψ:

$$\begin{aligned} &\eta=-A\sin\alpha/\sqrt{x+1}(A\cos\alpha+x+1), \\ &\psi=\sqrt{(x+1)^2-A^2}\,/\sqrt{x+1}(A\cos\alpha+x+1)^2-1. \end{aligned} \quad (5)$$

We return now to exact equations (2) and we effect a change of variable, replacing (ξ,θ,η,ψ) by (x,y,A,φ). Therefore, we differentiate equations (4) and (5), considering now x, y, A and φ as variables, to obtain the values of $(\dot\xi,\dot\theta,\dot\eta,\dot\psi)$; then, substituting in equations (2) and solving for $(\dot x,\dot y,\dot A,\dot\varphi)$, we obtain the differential equations for the new variables:

$$\left.\begin{aligned} &\dot x=\mu F(x,y,A,\varphi,\alpha),\\ &\dot y=(1-(x+1)^{3/2})/\sqrt{x+1}(A\cos\alpha+x+1)+\mu\, G(x,y,A,\varphi,\alpha),\\ &\dot A=\mu\, H(x,y,A,\varphi,\alpha),\qquad \dot\varphi=\mu\, K(x,y,A,\varphi,\alpha),\end{aligned}\right\}\quad(6)$$

where F, G, H and K are rather complicated expressions, which integration does not seem to be feasible analytically in general.

Fortunately, we may put some approximations. The period T of the libration is much greater than the period 2π of the reference motion; therefore we can say that $\dot x$ is of the order of x/T and $\dot y$ of the order of y/T. On the other hand, the numerical results show that $x \ll 1$, while A and y are of the order of 1. And from the equations (6), we deduce that $\dot x$ is of the order of μ, so that finally μ must be much smaller than x. This means particularly that we can neglect in $\dot y$ the term in μ. Moreover, we can average the equations (6) over 2π with the assumption that x, y, A and φ stay constant over this period. Finally, we obtain the following equations:

$$\dot x=\mu\frac{\partial I}{\partial y},\qquad \dot y=-3x/2,\qquad \dot A=0,$$

$$\text{where}\quad I=\frac{1}{\pi}\int_0^{2\pi}\left((A\cos\alpha+1)/\rho-(A\cos\alpha+1)^2\cos\theta\right)d\alpha \qquad (7)$$

$$\text{with}\quad \rho^2=(A\cos\alpha+1)^2-2(A\cos\alpha+1)\cos\theta+1$$

$$\text{and}\quad \theta=-A\sin\alpha-\alpha+2\,\mathrm{arc\,tg}\left(\sqrt{\frac{1-A}{1+A}}\,\mathrm{tg}\,\alpha/2\right)+y.$$

The integration of the integral I does not seem to be feasible analytically in general; nevertheless, the numerical integration for a set of values of A and y is in progress.

As in Hill's case, we have two integrals of motion. The first is:

$$A=c^{st}. \qquad (8)$$

The other:

$$B(x,y,A)=3Ax^2/4+\mu\,AI=c^{st}, \qquad (9)$$

is essentially Jacobi's integral, translated into the new variables and averaged over 2π under the same assumptions which lead us to the final equations (7).

Now we have to compute numerically some examples of curves B=cst and verify all these results for some actual orbits.

REFERENCE

Benest,D., 1976, "Cel. Mech." 13, pp 203-215

TOPOLOGY OF THE NEGATIVE ENERGY-MANIFOLD OF THE KEPLER MOTION

Junzo Yoshida
Department of Physics, Kyoto Sangyo University, Kita-ku,
Kyoto 603, Japan

1. Introduction

The motion of a particle of mass m according to the central force of Newton, is denoted by

$$m\frac{dx}{dt} = y, \qquad \frac{dy}{dt} = -\frac{K}{|x|^3}x, \tag{1}$$

where K is a constant. x=0 corresponds to singular points of this equations. The domain of (1), denoted by $\mathfrak{X}=(R^3-\{0\})\times R^3$, is called the phase space of the Kepler motion. In the sequel we set m=K=1 for simplicity and also transform the independent variable from t to s by $dt = |x|\,ds$ $(x\neq 0)$, then the Kepler motion in the phase space $\mathfrak{X}$ is written as

$$\frac{dx}{ds} = |x|\,y, \qquad \frac{dy}{ds} = -\frac{x}{|x|^2}. \tag{2}$$

Further, we shall confine the following discussion to the case of the negative energy value, except the preliminary discussion.

Moser [1] investigated the Kepler motion, describing the problem in an n-dimensional space, and showed that: (i) The energy surface with a negative constant is homeomorphic to the unit tangent bundle of the n-sphere S^n, punctured at one point which corresponds to the collision states. (In particular for n=2, the unit tangent bundle of the sphere S^2 is homeomorphic to the 3-dimensional projective space P^3.) (ii) The orbit space of the Kepler flow on an energy surface is homeomorphic to S^2 for n=2, and to $S^2\times S^2$ for n=3.

In the investigation of Moser, however, the problem of the reconstruction of the energy-manifold on the basis of the orbit space was not treated. The goal of this note is to fill up this gap. To this end, we consider first the Kepler flow on the Kustaanheimo-Stiefel's parametric space, by making use of theKustaanheimo-Stiefel's transformation (abbreviated as KS-map in the sequel) (Kustaanheimo-Stiefel [2]), and discuss the topology of the relevant energy-manifold, and then pull back the results to the phase space.

R. L. Duncombe (ed.), Dynamics of the Solar System, 45–48.

2. Preliminaries

We collect here the important properties with respect to the KS-map (cf. Stiefel-Scheifele [3]), and define the notions convenient in what follows.

Def. 1. If we set: $i=\left(\begin{array}{cc|cc} & -1 & & \\ 1 & & & \\ \hline & & & 1 \\ & & -1 & \end{array}\right)$, $j=\left(\begin{array}{cc|cc} & & -1 & \\ & & & -1 \\ \hline 1 & & & \\ & 1 & & \end{array}\right)$, $k=\left(\begin{array}{cc|cc} & & & 1 \\ & & -1 & \\ \hline & 1 & & \\ -1 & & & \end{array}\right)$,

and construct iu, ju, ku from a 4-vector $u=(u_1, u_2, u_3, u_4)'$, where prime denotes "transposed", and if any point of R^3 is automatically supplemented to a point of R^4 with fourth-component of value zero, then the KS-map $T_1: R^4 \to R^3$ is defined by $u \mapsto x=L(u)u$, where $L(u)=(u, iu, ju, ku)$.

Def. 2. For any $u \in R^4$ $(u \neq 0)$, $\mathcal{L}_u$ is defined by $\mathcal{L}_u=\{v \in R^4 \mid L(u)v= =L(v)u\}$.

Def. 3. For any $u \in R^4$ $(u \neq 0)$, a map $T_u: \mathcal{L}_u \to R^3$ is defined by $v \mapsto y=(2/|u|^2)L(u)v$.

Def. 4. In $R^4 \times R^4$ we call a set $\mathcal{P}=\{(u,v) \mid u \in (R^4-\{0\}),\ v \in \mathcal{L}_u\}$ the KS-space of Kustaanheimo-Stiefel.

Def. 5. We define a C^∞-surjection $T: \mathcal{P} \to \mathfrak{X}$; $(u,v) \mapsto (x,y)$ by $x=T_1(u)= =L(u)u$, $y=T_u(v)=(2/|u|^2)L(u)v$. We call T the enlarged KS-map.

Lemma 1. For any $u, \bar{u} \in R^4$ $(u, \bar{u} \neq 0)$ and $v \in \mathcal{L}_u$, $\bar{v} \in \mathcal{L}_{\bar{u}}$, a necessary and sufficient condition for $L(u)u=L(\bar{u})\bar{u}$, and $T_u(v)=T_{\bar{u}}(\bar{v})$ is that there exists $\chi \pmod{2\pi}$ such that $\bar{u}=\exp(-\chi k)u$, $\bar{v}=\exp(-\chi k)v$.

Lemma 2. The system of differential equations on $\mathcal{P}$:

$$\frac{du}{ds} = v, \qquad \frac{dv}{ds} = \frac{1}{|u|^2}\left[|v|^2 - \frac{1}{2}\right]u \tag{3}$$

transforms to the Kepler motion (2) on the phase space $\mathfrak{X}$ by the enlarged KS-map T.

Lemma 3. The system of differential equations (3) on the KS-space has the following integrals:

(i) $L(u)kv-L(v)ku=C$, $\quad L(u)v-L(v)u=C_0$, $\quad L(u)iv-L(v)iu=C_1$, $L(u)jv-L(v)ju=C_2$.

The first one corresponds to the integral of angular momentum and the constant C_0 of the second one must be zero for the consistency of the theory (cf. Def. 4 and 2).

(ii) $\dfrac{2}{|u|^2}\left[|v|^2 - \dfrac{1}{2}\right] = h$; this corresponds to the integral of energy.

Taking the energy-integral into account, the system of differential equations (3) reduces to the following system:

$$\frac{du}{ds} = v, \qquad \frac{dv}{ds} = \frac{h}{2}\,u, \tag{4}$$

on each energy-manifold P_h^* in KS-space, where

$$P_h^* =\{(u,v)\ u \neq 0,\ v \in \mathcal{L}_u,\ \frac{2}{|u|^2}\left[|v|^2 - \frac{1}{2}\right] = h\}. \tag{5}$$

Since, for a given constant h, the functions on the right-hand sides of (4) have no singular points on $R^4 \times R^4$, we shall extend its domain from

(5) to

$$P_h = \{(u,v)\,|\,2|v|^2 - h|u|^2 = 1, \text{ and } v\varepsilon\mathcal{L}_u \text{ if } u\neq 0\}. \tag{6}$$

Hence P_h is a union of P_h^* and P_{col}, which is defined as:

$$P_{col} = \{(u,v)\,|\,u=0,\ |v|^2 = \tfrac{1}{2}\}. \tag{7}$$

Def. 6. Any two points (u,v) and $(\bar{u},\bar{v})$ on P_h are call orbit-equivalent each other, if they are on a same solution curve of the flow (4) on P_h, and are called KS-equivalent each other, if there exists such χ(mod. 2π) that $\bar{u}=\exp(-\chi k)u$, $\bar{v}=\exp(-\chi k)v$.

3. Results

In what follows we shall restrict ourselves to a discussion of a negative energy-manifold with an energy constant $h = -2\omega^2$. And we here introduce an important diffeomorphism H on $R^4\times R^4$ after Stiefel-Scheifele ([3], § 45) by

$$u = \tfrac{1}{2}(\xi+\eta), \qquad v = -\tfrac{1}{2}\omega k(\xi-\eta). \tag{8}$$

By the transformation (8), the Kepler flow (4) on P_h is transformed into

$$\frac{d\xi}{ds} = -\omega k\xi, \qquad \frac{d\eta}{ds} = \omega k\eta. \tag{9}$$

In the (ξ,η)-space we can easily consider the topological structure of the corresponding energy-manifold induced by the diffeomorphism H. In the sequel we will state the main results without proof.

We define a C^∞-surjection Π by $(\xi,\eta)\to\Pi(\xi,\eta)=(L(\xi)\xi,L(\eta)\eta)$ in (ξ,η)-space, which defines a C^∞-projection of $S^3\times S^3$ onto $S^2\times S^2$. With this Π we obtain the following:

Prop. 1. P_h is diffeomorphic to $S^3\times S^3$.

Prop. 2. $S^3\times S^3$ is a bundle space of a fibre bundle with base space $S^2\times S^2$, fibre $S^1\times S^1$ and structure group $SO(2)\times SO(2)$.

Let us consider a quotient space of $S^3\times S^3$ by the KS-equivalence relation, induced in (ξ,η)-space by the same way as § 2, Def. 6. We denote it by $(S^3\times S^3)/KS\sim$, and let T be a canonical projection of $S^3\times S^3$ onto $(S^3\times S^3)/KS\sim$. Since Π is compatible with the KS-equivalence relation, there exists uniquely a C^∞-surjection Π^* of $(S^3\times S^3)/KS\sim$ onto $S^2\times S^2$ such that $\Pi = \Pi^*\circ T$. By the projection Π any points which are orbit- or KS-equivalent each other, are mapped onto a same point. The set of equivalence classes $S^2\times S^2$ corresponds to the orbit space of the Kepler motion.

Prop. 3. $((S^3\times S^3)/KS\sim,\ \Pi^*,\ S^2\times S^2,\ S^1,\ SO(2))$ is a fibre bundle.

On the basis of Prop. 2 and 3 we obtain the following theorems.

Theo. 1. P_h is diffeomorphic to a bundle space $S^3\times S^3$ of a fibre bundle: $(S^3\times S^3,\ \Pi,\ S^2\times S^2,\ S^1\times S^1,\ SO(2)\times SO(2))$.

Theo. 2. $P_h/KS\sim$ is diffeomorphic to a bundle space $(S^3 \times S^3)/KS\sim$ of a fibre bundle: $((S^3 \times S^3)/KS\sim, \Pi^*, S^2 \times S^2, S^1, SO(2))$.

Let $P_{h,c}$ be a set of all (u,v) with energy $h=-2\omega^2$ and with angular momentum C:

$$P_{h,c} = \{(u,v) \,|\, L(u)kv - L(v)ku = C,\ 2|v|^2 + 2\omega^2 |u|^2 = 1, \text{ and } v \in \mathcal{L}_u \text{ if } u \neq 0\}.$$

Theo. 3. If $|C| \neq 1/2\omega$, $\neq 0$, $P_{h,c}/KS\sim$ is diffeomorphic to $S^1 \times S^1$, if $|C| = 1/2\omega$, $P_{h,c}/KS\sim$ is diffeomorphic to S^1, and if $C=0$, $P_{h,c}/KS\sim$ is diffeomorphic to $S^2 \times S^1$.

Theo. 4. The energy-manifold $X_h \subset \mathfrak{X}$ of the Kepler motion is homeomorphic to a subset $(P_h \setminus P_{col})/KS\sim$ of $P_h/KS\sim$.

In a planar case, $P_h/KS\sim$ is diffeomorphic to a projective space P^3 and Theo. 2 should be rewritten as:

Theo. 2'. In the planar case $P_h/KS\sim$ is diffeomorphic to a bundle space P^3 of a fibre bundle with base space S^2, fibre S^1 and structure group SO(2).

References

1. Moser, J.: Regularization of Kepler's Problem and the Averaging Method on a Manifold. Comm. Pure Apple. Math. 23 (1970), 609-636.
2. Kustaanheimo, P. and E. Stiefel: Perturbation Theory of Kepler Motion Based on Spinor Regularization. J. Reine Angew. Math. 218 (1965), 204-219.
3. Stiefel, E.L. and G. Scheifele: Linear and Regular Celestial Mechanics, Springer-Verlag, Berlin 1971, Chap. II, III, and XI.

ON NON-STATIONARY PROBLEMS OF CELESTIAL MECHANICS

T.B.Omarov
Astrophysics Institute, Kazakh Academy of Sciences,
Alma-Ata, USSR

Some non-stationary problems of celestial mechanics can be described in an inertial system of right-angled coordinates $O x_1 x_2 x_3$ with gravitational potential of the form:

$$\mathcal{U}(x_1, x_2, x_3, t) = \frac{\gamma(t)}{\gamma_0} \gamma_0 \tilde{\mathcal{U}}(x_1, x_2, x_3) \quad , \qquad (I)$$

where $\gamma(t)$ is a sufficiently arbitrary function of time and γ_0 is the meaning of $\gamma(t)$ in the initial epoch t_0. For example, in a two-body problem of variable mass $M(t)$ we have:

$$\mathcal{U} = -\frac{G M(t)}{r} , \qquad r^2 = x_1^2 + x_2^2 + x_3^2 . \qquad (2)$$

We can also remember a generalized problem of two immovable centres with the variable gravitational constant $G(t)$, when

$$\mathcal{U} = -\frac{G(t) m}{2}\left[\frac{1+zi}{r_1} + \frac{1-zi}{r_2}\right] ,$$
$$i = \sqrt{-1} , \qquad (3)$$
$$r_1^2 = x_1^2 + x_2^2 + [x_3 - c(z+i)]^2 ,$$
$$r_2^2 = x_1^2 + x_2^2 + [x_3 - c(z-i)]^2 ,$$

where m, z, c - some constants. It is of interest for analysis of effects of variable gravitation in an orbital motion of earth artificial satellites [I]. By a transformation of the time constant we can easily ascertain the following result: if the stationary problem with the potential $\gamma_0 \tilde{\mathcal{U}}(x_1, x_2, x_3)$ is integrated, then the following system of equations is also integrated:

$$\frac{d^2 x_i}{dt^2} = -\frac{\partial \mathcal{U}}{\partial x_i} + \frac{1}{2\gamma} \cdot \frac{d\gamma}{dt} \frac{dx_i}{dt} , \qquad (i = 1, 2, 3) . \qquad (4)$$

R. L. Duncombe (ed.), Dynamics of the Solar System, 49–52.

When in the gravitational potential (I) the value $\gamma(t)$ is changed sufficiently slowly, the solution of this system can be considered as an unperturbed motion in a corresponding non-stationary problem. Specifically, for the gravitational potential (2) we have the aperiodic motion along a conic section, described by equation [2]

$$\frac{d^2 x_i}{dt^2} = -\frac{G\mathcal{M}(t)}{r^3} x_i + \frac{1}{2\mathcal{M}} \cdot \frac{d\mathcal{M}}{dt} \frac{dx_i}{dt}, \quad (i = 1,2,3). \tag{5}$$

With the purpose to use the well worked out canonical theory of perturbations we can try to construct Lagrangian of the non-conservative system (4). Multiply both parts of equations (4) by some function $R(x_1, x_2, x_3, t)$:

$$R\ddot{x}_i = -R\left(\frac{\partial \mathcal{U}}{\partial x_i} - \frac{\dot{\gamma}}{2\gamma} \dot{x}_i\right), \qquad (i = 1,2,3) \quad , \tag{6}$$

where the point above means a differentiation in time. Take R so as the equalities to be identically fulfilled:

$$R\left(\ddot{x}_i + \frac{\partial \mathcal{U}}{\partial x_i} - \frac{\dot{\gamma}}{2\gamma} \dot{x}_i\right) = \frac{d}{dt}\left(\frac{\partial L}{\partial \dot{x}_i}\right) - \frac{\partial L}{\partial x_i} = \frac{\partial^2 L}{\partial \dot{x}_i \partial t} + \sum_{j=1}^{3}\left(\frac{\partial^2 L}{\partial \dot{x}_i \partial \dot{x}_j} \ddot{x}_j + \frac{\partial^2 L}{\partial \dot{x}_i \partial \dot{x}_j} \dot{x}_j\right) - \frac{\partial L}{\partial x_i}; \quad (i = 1,2,3), \tag{7}$$

where

$$L = R\left(\frac{1}{2}\sum \dot{x}_i^2 - \mathcal{U}\right). \tag{8}$$

Comparing in the expression (7) members, contaning the arbitraries of the same order, we shall receive:

$$R \equiv m(t) = \mu_0 \gamma^{-\frac{1}{2}}, \qquad \mu_0 = const. \tag{9}$$

With due regard for expression (9) in the formula (6), we have the following system of equations:

$$m\frac{d^2 x_i}{dt^2} = -m\frac{\partial \mathcal{U}}{\partial x_i} - \frac{dm}{dt} \cdot \frac{dx_i}{dt}, \quad (i = 1,2,3), \tag{IO}$$

which is equivalent to the system (4) and has Lagrangian (8).

Describe the Lagrangian, which we have found, in a spherical coordinates:

$$L = m\left[\frac{1}{2}\left(\dot{r}^2 + r^2\dot{\varphi}^2 + r^2 \cos^2\varphi\, \dot{\lambda}^2\right) - U\right] \quad . \tag{11}$$

Turn to Humilton's function:

$$H = \frac{1}{2m}\left(P_1^2 + P_2^2/r^2 + P_3^2/r^2\cos^2\varphi + 2m^2 U\right) , \tag{12}$$

where P_1, P_2, P_3 - are generalized impulses:

$$P_1 = \frac{\partial L}{\partial \dot{r}} , \quad P_2 = \frac{\partial L}{\partial \dot{\varphi}} , \quad P_3 = \frac{\partial L}{\partial \dot{\lambda}} \quad . \tag{13}$$

The corresponding Humilton-Jakoby's equation has the form:

$$\frac{\partial \Psi}{\partial t} + \frac{1}{2m}\left[\left(\frac{\partial \Psi}{\partial t}\right)^2 + \frac{1}{r^2}\left(\frac{\partial \Psi}{\partial \varphi}\right)^2 + \frac{1}{r^2\cos^2\varphi}\left(\frac{\partial \Psi}{\partial \lambda}\right)^2 + 2m^2 U\right] = 0 . \tag{14}$$

For the case of the system of equations (5) variable values t, r, φ, λ in equation (14) can be divided [3]. Suppose

$$\Psi = \Psi_0(t) + \Psi_1(r) + \Psi_2(\varphi) + \Psi_3(\lambda) \quad , \tag{15}$$

we can find:

$$\begin{aligned}
\Psi_3 &= \alpha_3 \lambda \quad , \\
\Psi_2 &= \int_0^{\varphi} \sqrt{\alpha_2^2 - \frac{\alpha_3^2}{\cos^2\varphi}}\, d\varphi \quad , \\
\Psi_1 &= \int_{r_0}^{r} \sqrt{\frac{2G\mu_0}{r} - \frac{\alpha_2^2}{r^2} - 2\alpha}\, dr \quad , \\
\Psi_0 &= \alpha_1 \int_{t_0}^{t} \frac{dt}{m(t)} \equiv \alpha_1 F(t) \quad ,
\end{aligned} \tag{16}$$

where $\alpha_1, \alpha_2, \alpha_3$ are constants of integration. The general solution of the system of equations (5), which have been received by this method, is anologous to the solution of classical two-body problem, but with one exclusion: instead of time t in the corresponding expression the function of time $F(t)$ takes part. This circumstance permits us to write at once the analogy of different systems of canonical elements [3].

R E F E R E N C E S

Bekov, A.A., Omarov, T.B.: 1978, "Pis'ma v Astronomicheskij zhurnal". Akademija Nauk SSSR. Izdatel'stvo "Nauka", Moskva, 4, pp.30-33.

Omarov, T.B.: 1975, Dynamics of gravitating systems of metagalaxy, Izdatel'stvo "Nauka", Kaz.SSR, Alma-Ata.
Demchenko, B.I., Omarov, T.B.: 1977, "Trudy Astrofizicheskogo Instituta Akademii Nauk Kaz.SSR", 29, pp.16-21.

DISCUSSION

Szebehely: Is the unsteadiness always represented by the mass as a function of time?

Omarov: The mass and/or the constant of gravity may be functions of the time.

CONSTRUCTION ET PRECISION DE NOUVELLES THEORIES PLANETAIRES

P. Bretagnon and J. Chapront
Bureau des Longitudes, Paris, France

Au Bureau des Longitudes, la construction de théories planétaires a été développée dans trois directions : Une théorie générale du mouvement des quatre grosses planètes du système solaire est en cours d'élaboration à la Faculté des Sciences de Lille, avec L. Duriez (1977), à la suite des travaux de V.A. Brumberg et J. Chapront (1973). Des théories de type classique, à variations séculaires, du mouvement de l'ensemble des planètes de Mercure à Neptune, sont en voie d'achèvement au Bureau des Longitudes. Elles sont construites par P. Bretagnon et J.L. Simon (1975, 1978). Des compléments numériques à l'ensemble de ces recherches, intégrations numériques, représentation des solutions en séries de Tchebychev, sont apportés par P. Rocher, en ce qui concerne le mouvement des petites planètes, et par J. Piraux pour les actions de Pluton sur Uranus et Neptune, dans le cadre des théories à variations séculaires.

1- LES THEORIES PLANETAIRES DU BUREAU DES LONGITUDES.

Les théories à variations séculaires forment le noyau fondamental de l'ensemble de ces travaux qui ont pour but d'aboutir à la construction de nouvelles éphémérides.

On donne la définition suivante à l'expression "variations séculaires" : les arguments de longues périodes (mouvement des périhélies et des noeuds) sont développés par rapport au temps. Les variables sont les éléments elliptiques :

$$\sigma = (a , \lambda , h , k , p , q)$$

a : demi-grand axe,
λ : longitude moyenne,
$z = k + \sqrt{-1}\, h = e.\exp\sqrt{-1}\,\tilde{\omega}$; e : excentricité, $\tilde{\omega}$: longitude du périhélie
$\zeta = q + \sqrt{-1}\, p = \sin i/2.\exp\sqrt{-1}\,\Omega$; i : inclinaison, Ω : longitude du noeud

Elles sont représentées par des développements de Taylor, en puissance du temps, sous la forme de séries de Fourier, en multiples des longitu-

R. L. Duncombe (ed.), Dynamics of the Solar System, 61–67.

des. Elles sont calculées avec $\sigma = \sigma^{(0)}$ où $\sigma^{(0)}$ est une constante d'intégration qui doit être ajustée à l'Observation.

Deux méthodes ont été envisagées pour le calcul des solutions :
- Une méthode itérative qui substitue de proche en proche les solutions à partir des $\sigma^{(0)}$, et des solutions du problème des deux corps, dans un formulaire fermé des équations différentielles pour les variables elliptiques.
- Une méthode d'accroissement où les seconds membres des équations sont développés en puissance des masses, à partir de la même solution de départ que ci-dessus. Cette méthode nécessite, pour tous les ensembles de planètes, le calcul de dérivées partielles, en très grand nombre.

Les solutions, une fois obtenues à partir d'un jeu de constantes initiales $\sigma^{(0)}$, il reste un problème fondamental à traiter : il faut ajuster la théorie avec l'Observation, après que l'on se soit assuré de la bonne précision interne des fonctions $\sigma(t,\sigma^{(0)})$. Avant d'aborder la comparaison avec les observations elles-mêmes, ce qui nécessite, sur un long intervalle de temps, un travail de compilation considérable, nous avons choisi d'effectuer un ajustement sur un modèle qui représente au mieux l'Observation. Dans ce but, nous avons utilisé l'intégration numérique d'Oesterwinter et Cohen (1972), qui couvre 55 années d'observations méridiennes de l'U.S. Naval Observatory. Un nouveau jeu d'éléments corrigés $\sigma^{(0)}$ est alors déduit.

Par ailleurs, une théorie générale planétaire en variables elliptiques héliocentriques est élaborée. Les éléments σ sont identiques à ceux choisis plus haut. A la place de a et λ on pose :

$\beta = (\frac{a}{A})^{-3/2} - 1$, où A est la valeur numérique calculée à partir d'un moyen mouvement N vérifiant la relation : $N^2A^3 = n^2a^3 = GM_{\odot}$

et, $\varepsilon = \lambda - Nt$

Les seconds membres des équations de Lagrange relatives à ces variables sont développés analytiquement sur ordinateur, selon les puissances croissantes de β , z , $\bar{z}$, ζ , $\bar{\zeta}$ et de $\exp \pm\sqrt{-1}(Nt+\varepsilon)$. On en recherche une solution par approximations successives, ordonnées selon les puissances des masses des planètes.

On a choisi de prendre pour chaque planète n^2a^3 égal à la même valeur $GM_{\odot}$ pour éviter l'apparition de perturbations séculaires dans le demi-grand axe à la deuxième approximation, qui est alors rigoureusement d'ordre 2 des masses. L. Duriez (1978) donne une démonstration du théorème de Poisson en variables héliocentriques.

σ désignant l'une quelconque des variables, on obtient une solution sans termes séculaires de la forme : $\sigma = \sigma^{(0)} + S_\sigma(\sigma_i^{(0)}, t)$.

$\sigma^{(0)}$ est une série trigonométrique quasipériodique de t, à longues périodes, dont les coefficients sont numériques, et qui est la solution générale de la partie autonome des équations; c'est dans cette solution qu'on introduit les constantes d'intégration. Elle est d'ordre zéro des masses au moins.

$S_\sigma(\sigma_i^{(0)}, t)$ est une série trigonométrique quasipériodique de t, à courtes périodes, dont les coefficients sont des polynômes des fonctions $\sigma_i^{(0)}$, et qui est une solution particulière (sans termes séculaires) de la partie non autonome des équations. Cette solution est d'ordre 1 des masses au moins. Pratiquement, on construit $S_\sigma(\sigma_i^{(0)}, t)$ ordre par ordre en masses, en développant au voisinage des fonctions $\sigma_i^{(0)}$ les seconds membres des équations initialement exprimés en variables σ_i. On y retrouve toutes les inégalités des théories classiques, du type de Le Verrier, y compris celles de petit diviseur, comme la grande inégalité : $(2N_J - 5N_S)t$.

La solution complète est finalement conditionnée par la résolution du système autonome, ce qu'on réalise par approximations successives, à partir de la solution du système linéaire de Laplace-Lagrange. Le calcul du système autonome est poussé aux termes de degré 1, 3 et 5 en variables $z_i^{(0)}$ et $\zeta_i^{(0)}$, pour les ordres 1 et 2 en masses, et pour les 4 grosses planètes.

En ce qui concerne le développement des équations et des solutions, un programme ordinateur permet d'atteindre tous les termes de degré global inférieur ou égal à 5 dans les variables elliptiques : β, z, $\bar{z}$, ζ et $\bar{\zeta}$, et les multiples de $\pm Nt$ inférieurs à 20. L'extension aux degrés supérieurs est envisagée.

Actuellement une tentative de raccordement des solutions du problème général avec les développements des théories à variations séculaires décrites plus haut est envisagée. Elle devrait permettre, dans l'avenir d'étendre la durée de validité des théories de type classique qui restent très précises sur un intervalle de temps d'environ 1000 ans mais qui se dégradent au delà.

Il faut enfin rattacher à ce travail l'autre extrémité de l'étude qui consiste à compléter les solutions par des méthodes numériques, en vue d'un ajustement complet à l'Observation. C'est le cas, en particulier, des perturbations de Pluton, essentielement sur Neptune et Uranus. Une intégration, en séries de polynômes de Tchebychev, des écarts entre la solution de P. Bretagnon et le problème complet est entreprise. A la suite d'un ajustement au modèle d'observation d'Oesterwinter, des corrections significatives ont été déterminées.

2- RESULTATS DES THEORIES A VARIATIONS SECULAIRES.

Nous avons développé nos solutions dans les variables a, λ, $z = k + \sqrt{-1}h$, $\zeta = q + \sqrt{-1}p$ sous forme semi-numérique où seules les longitudes moyennes apparaissent explicitement, les autres éléments étant substitués numériquement. Il est, en effet, impossible de conserver une forme entièrement analytique qui conduirait à une fonction perturbatrice de plusieurs centaines de milliers d'arguments : les développements doivent, en effet, être poussés jusqu'à l'ordre 15 en excentricités-inclinaisons pour le couple Terre-Mars et jusqu'à l'ordre 70 en α (rapport des demi-grands axes) pour le couple Vénus-Terre afin d'atteindre la précision recherchée. Nous avons par ailleurs construit les dérivées de nos solutions par rapport à toutes les constantes d'intégration ce qui constitue, avec les solutions semi-numériques, une théorie entièrement analytique dans un petit domaine autour des constantes que nous avons choisies. L'ensemble des expressions intervenant dans les solutions et dans les dérivées est donné sous forme fermée ce qui facilite la détermination de la précision des résultats.

Au départ nous avons cherché à construire les solutions par la méthode itérative mais il n'a pas été possible, pour les planètes inférieures, de déterminer les arguments quasi-résonants à une précision suffisante. Par contre, cette méthode a donné de bons résultats pour les grosses planètes avec toutefois des difficultés de convergence dans les longitudes d'Uranus et de Neptune.

Pour les besoins des planètes inférieures, nous avons utilisé la méthode d'accroissement par rapport aux masses et actuellement nous avons construit une solution au premier et au deuxième ordre à la précision de $10^{-4''}$ pour les planètes inférieures et $10^{-3''}$ pour les grosses planètes. Nous avons de plus obtenu par la méthode itérative une solution développée jusqu'à l'ordre 6 par rapport aux masses (jusqu'à la puissance cinquième du temps dans les développements de Taylor) pour les grosses planètes. C'est cette solution de la méthode itérative que nous avons retenue pour les grosses planètes en l'améliorant pour certains arguments quasi-résonants à l'aide des résultats de la théorie au deuxième ordre. Nous avons ainsi réduit les difficultés de convergence des longitudes d'Uranus et de Neptune.

Au cours de ce travail, nous avons trouvé des termes séculaires dans les demi-grands axes et cela dès la deuxième approximation dans l'ordre des masses. Ces résultats ont confirmé les termes séculaires du demi-grand axe déjà trouvés par J.L. Simon. Ces termes, négligeables pour les planètes inférieures, ne prennent une importance numérique que pour Jupiter et Saturne. La méthode itérative, développée à un ordre élevé par rapport aux masses, a donné des résultats très différents pour ces termes. Par la troisième loi de Kepler, le terme séculaire du demi-grand axe donne, dans la longitude, un terme en t^2 qui atteint au bout de 1000 ans $-30''.2$ pour Jupiter et $+75''.4$ pour Saturne. Nous donnons dans le tableau ci-dessous, pour Jupiter et Saturne, le

AN EXTENSION OF NEWTON'S EQUATION OF MOTION

Ki Jae Cheon
Faculty of Science Korea University,
Ogawa, Kodaira, Tokyo 187, Japan

Classical mechanics failed to solve two problems in own defensive area, namely the motion of Mercury's perihelion, and the high-velcity motion of a charged particle. Today it is generally believed that the concepts of classical mechanics are completely invalid in a treatment of these problems. In this paper, however, we discuss these problems throughly with the concepts of classical mechanics — Euclidean space-time, point of mass and central force. Thereat we introduce a new concept "absolute mass variation", with which we extend the Newton's second law of motion. In following chapters we show that this extended equation explains the motion of Mercury's perihelion, and that it throws new light on the atomic physics. We also make a study of reconstruction of internal structure of mechanics. We discuss the possibility of the revival of the principal frame of Newton's mechanics.

1. BASIC EQUATIONS AND ANGULAR MOMENTUM At first we write "new" equations of motion of a mass-point m.

$$\dot{\bar{P}} \equiv \dot{m}\bar{v} + m\dot{\bar{v}} = \bar{F} \qquad (1)$$

$$\dot{m}c^2 = s\bar{F}\cdot\bar{v} \quad , \qquad (2)$$

where $s \equiv 3/2$, c=light velocity and

$$\bar{F} = \begin{cases} -GMm\bar{r}/r^3 & (3a) \\ \mp e^2\bar{r}/r^3 \quad . & (3b) \end{cases}$$

The introduction of $\dot{m}$-term prescribed by (2) is the fundamental point of our theory. It is evident that the law of conservation of momentum and angular momentum still holds independently of $\dot{m}$-term. Nevertheless areal velocity is no more conservative, but varies in proportion to mass as

$$\dot{m}\bar{h} + m\dot{\bar{h}} = 0 \quad , \quad \text{or} \quad \dot{m}/m = -\dot{h}/h \quad . \qquad (4)$$

Generally speaking, the law of conservation exists where dynamical quanti-

R. L. Duncombe (ed.), Dynamics of the Solar System, 53–56.

ties are concerned, but corresponding kinematical quantities vary with mass. In view of this conclusion, the "curved space-time" in general relativity can be interpreted as the result of forcing the conservation theorem to the areal velocity.

2. MERCURY'S PERIHELION We apply the new equations of motion to one body problem. Consider motion of a mass-point m around the sun M. From (1), (2) and (3a), the equations of two components of polar coordinates are written as $\dot{m}v_r+ma_r=F_r$, $\dot{m}v_\phi+ma_\phi=F_\phi$. From these, we easily obtain (4) and the orbit equation

$$d^2/d\phi^2(1/r)+1/r=GM/h(r)^2 \simeq GM(1+3GM/c^2r)/h_0{}^2 \quad , \tag{5}$$

where $h(r)=h_0\exp(-3GM/2c^2r)\simeq h_0(1-3GM/2c^2r)$ and $h_0=(\infty)$. This is exactly the Einstein's orbit equation which explains the motion of Mercury's perihelion. In case of hydrogen atom, taking the Coulomb force, R.H.S. of (5) is $e^2/mh^2=e^2m/A^2\simeq e^2m_0(1+3e^2/2m_0c^2r)/A^2$. This result differs in the perturbation term by numerical factor 3 from the Sommerfeld's (1916) which was deduced from the special relativity.

3. THE ANNUAL CHANGE OF EARTH'S MASS From (2) and (3a) we get a following expression of earth's mass variation

$$m(r)=m_0\exp(3GM/2c^2r) \quad . \tag{6}$$

The difference of earth's mass between perihelion and aphelion is evaluated as

$$\Delta m \simeq 3m_0\varepsilon v^2/c^2 \simeq 5\times10^{-10}m_0 \quad , \tag{7}$$

where ε is orbital eccentricity and v is orbital velocity. The measurement of the annual change of gravity caused by this Δm is expected to prove the existence of $\dot{m}$-effect. The annual variation of earth's rotational velocity is taken for another corroboration of our $\dot{m}$-effect. The moment of inertia I, angular velocity ω and length of one day T also change in proportion to the change of mass. Relation among these quantities are given by $\dot{m}/m=\dot{I}/I=-\dot{\omega}/\omega=\dot{T}/T$. Amplitude of annual variation of length of one day ΔT is evaluated as

$$\Delta T=(\Delta m/m)\times(\text{half a year})\simeq 8\times10^{-3} \quad (\text{sec}) \quad . \tag{8}$$

The observational value is reported to be 20∿25 ms (Rochester, 1973).

4. INERTIAL SYSTEM For simplicity we consider two mass-points m_1, m_2 whose masses equal to each other. If there is not external force, then the total momentum $(m_1+m_2)\bar{V}_0$ is constant. But in this case, the velocity of center of gravity $\bar{V}_0$ is not always constant, for the total mass m_1+m_2 may vary owing to internal forces. This conclusion can be extended to the general case. Therefore it is concluded that a kinematical definition of inertial system is impossible.

5. EXISTENCE OF CRITICAL VELOCITY v_c We consider the direct collision of two mass-points. By eliminating $\bar{F}\cdot\bar{v}$ from $\dot{\bar{P}}\cdot\bar{v}=\bar{F}\cdot\bar{v}$ and (2), and noticing that $\bar{v}$ is parallel to $\dot{\bar{v}}$, we obtain an integral

$$m(v)=m_0/\surd(1-sv^2/c^2) \quad . \tag{9}$$

Since mass cannot be infinite, velocity of the mass-point cannot exceed the upper bound $v_c=c/\surd s$. Then it is concluded that the relative velocity of two mass-points cannot exceed $2v_c$.

6. INTEGRALS From (2) and (3b) we have

$$m(r)=m_0+3e^2/2c^2r \quad , \tag{10}$$

where $m_0=m(\infty)$. In case of the Coulomb force, by taking scalar product of $\bar{P}$ and the expression (1), and using (10) we obtain the integral

$$D_1=m^2v^2-(m+m_0)e^2/r \quad . \tag{11}$$

Similarly from $\bar{P}\cdot\dot{\bar{P}}=\bar{F}\cdot\bar{P}$, (2) and (3a) or (3b), we have another integral

$$D_2=m^2v^2-m^2c^2/\surd s \quad . \tag{12}$$

7. SUNDMAN'S RESULT Let us apply our $\dot{m}$-effect to hydrogen atom. From (10) we get the increment of an electron mass at Bohr radius a_0 as $\Delta m \equiv m(a_0)-m_0=(3/2)m_0\alpha^2$, where α is Sommerfeld's fine structure constant. In case of free fall, D_1 in (11) is zero. Therefore, considering the process of free fall at which m is nearly equal to m_0, we get $r(dr/dt)^2=2e^2/m_0$ from (11). By integrating this equation we finally have

$$r(t)=(9e^2/2m_0)^{1/3}(t-t_c)^{2/3} \quad . \tag{13}$$

It is surprising that this Sundman's result holds not at $r\simeq 0$, but at $r\simeq a_0$.

8. ELECTRON RADIUS r_0 We consider the direct collision of an electron and a positron with the coordinates of center of gravity. From (9) and (10), replacing r by 4r, we obtain an integral

$$4mrv^2=e^2(1+m_0/m) \quad . \tag{14}$$

In this integral, let us notice the moment when r goes to zero. At this moment v goes to critical velocity $c/\surd s$ and m tends to infinity, but R.H.S. stays at the finite value e^2, so L.H.S. also must be finite. On the other hand we know the experimental fact that two photons radiate after pair annihilation of an electron and a positron. On the view point of the conservation of total energy we may assume that at the moment of the pair annihilation m(t) becomes equal to m_0. From this assumption and from our discussion about (14), we can define the electron radius r_0 uniquely:

$$r_0 = 3e^2/8m_0c^2 \quad . \tag{15}$$

Velocity and mass of electron at r_0 are given as $m(r_0)=2m_0$, $v(r_0)=c/\sqrt{2}$.

9. A DYNAMICAL INTERPRETATION OF MICROSCOPIC MAGNETIC FIELD Our non-relativistic modification of Newton's mechanics naturally suggest the critique of electromagnetic field theory. Depending on the Bohr's semi-classical model of hydrogen atom we show that the magnetic field has its origin in our $\dot{m}$-effect. From (1), (2) and (3b) we get the equation of motion of an electron

$$m d\bar{v}/dt = \bar{F} - s(\bar{F}\cdot\bar{v})\bar{v}/c^2 \quad . \tag{16}$$

On the other hand, that of special relativity is

$$m d\bar{v}/dt = \bar{f} - (\bar{f}\cdot\bar{v})\bar{v}/c^2 \quad , \tag{17}$$

where $m=m_0/\sqrt{(1-v^2/c^2)}$ and $\bar{f}$ is the Lorentz's force. Both equations consist of corresponding three terms, but substances of each term are different. Let us compare our equation (16) to relativistic equation (17) in v^2/c^2 approximation. At first, it is noticed that the absolute mass variation (10) can be rewritten as $m=m_0/(1-v^2/c^2)^{3/2}+O(v^4/c^4)$, where $m_0=m(r=\infty)$ has the same physical meaning as proper mass in special relativity. This expression corresponds to longitudinal mass in electrodynamics. We artificially devide the increment of mass Δm given by (10) into two parts $\Delta m/3$ and $2\Delta m/3$. Then $m_0+\Delta m/3=m_0/\sqrt{(1-v^2/c^2)}+O(v^4/c^4)$. This part is substantially equal to the relativistic mass in (17). To analyze the second part $2\Delta m/3$, we use a kinematical relation of relative motion $d\bar{v}/dt=\bar{\omega}\times\bar{v}$, where $\bar{\omega}=\bar{r}\times\bar{v}/r^2$ is the angular velocity of the electron. Then $(2/3)\Delta m d\bar{v}/dt =e^2\bar{\omega}\times\bar{v}/c^2r=(e/c)\bar{v}\times\bar{H}$, where $\bar{H}(\bar{r})=-(e/c)\bar{r}\times\bar{v}/r^3$. Usually, this $\bar{H}$ is regarded as the microscopic magnetic field at the position of the electron. Consequently, we can express our equation (16) as

$$m_0/\sqrt{(1-v^2/c^2)}\, d\bar{v}/dt = -e(\bar{E}+(1/c)\bar{v}\times\bar{H}) - s(\bar{F}\cdot\bar{v})\bar{v}/c^2 \quad . \tag{18}$$

Thus we traced the dynamical origin of the magnetic field. The difference between the results of two theories, namely factor $(\bar{F}\cdot\bar{v})\bar{v}/2c^2 \sim O(v^2/c^2)$ gives us the hint for the experimental veritification of our $\dot{m}$-effect.

REFERENCE

Rochester, M. G. : 1973, Trans. Amer. Geophys. Union, 54, 769.

EFFECT OF PERTURBED POTENTIALS ON THE STABILITY OF LIBRATION POINTS IN THE RESTRICTED PROBLEM

K. B. Bhatnagar and P. P. Hallan
Zakir Husain College, Delhi University, India

ABSTRACT

The location and the stability of the libration points in the restricted problem have been studied when there are perturbations in the potentials between the bodies. It is seen that if the perturbing functions involving the parameters α,α_1,α_2 satisfy certain conditions, there are five libration points, two triangular and three collinear. It is further observed that the collinear points are unstable and for the triangular points, the range of stability increases or decreases depending upon whether the perturbation point $(\alpha,\alpha_1,\alpha_2)$ lies on one or the other side of the plane $A\alpha + B\alpha_1 + C\alpha_2 = 0$, and it remains the same if the point lies on the plane, where A,B,C depend on the perturbations. The theory is verified in the following four cases: (1) there are no perturbations in the potentials (classical problem), (2) only the bigger primary is an oblate spheroid, (3) both the primaries are oblate spheroids, and (4) the primaries are spherical in shape and the bigger is a source of radiation.

R. J. Duncombe (ed.), Dynamics of the Solar System, 57.

PART II

PLANETARY AND LUNAR THEORIES

terme séculaire du demi-grand axe obtenu à la deuxième approximation dans la méthode d'accroissement, sa valeur obtenue par la méthode itérative et le terme en t^2, qui résulte de cette dernière valeur dans la longitude.

Terme séculaire du demi-grand axe.

Planète	Terme séculaire de a en UA/an		Terme en t^2 de la longitude en $''/an^2$
	Ordre 2	méthode itérative	
Jupiter	$+ 0,624 \times 10^{-9}$	$+ 1,886 \times 10^{-9}$	$- 30,2 \times 10^{-6}$
Saturne	$+ 0,733 \times 10^{-9}$	$- 21,268 \times 10^{-9}$	$+ 75,4 \times 10^{-6}$

Des comparaisons de nos solutions à des intégrations numériques internes ont donné pour les grosses planètes des écarts ne dépassant pas, sur un intervalle de 1000 ans, 0",3 en longitude et 0",1 pour les autres éléments (excepté dans le cas des longitudes d'Uranus et de Neptune pour lesquelles les écarts atteignent plusieurs secondes dans l'état actuel de nos solutions).

Nous avons, par ailleurs, comparé la théorie de Le Verrier-Gaillot à un prolongement sur 1000 ans de l'intégration numérique des Astronomical Papers of American Ephemeris (1951). Les écarts entre ces deux solutions atteignent 60" en longitude et sont dus en partie à l'absence dans la théorie de Le Verrier-Gaillot des termes en t^2 provenant des termes séculaires du demi-grand axe. Notons toutefois que la solution de Gaillot a très bien représenté le mouvement de Jupiter entre 1750 et 1907 et que cette solution lui a permis de déterminer une remarquable valeur de la masse de Saturne : 1/ 3 499,8. L'UAI a proposé en 1976 à Grenoble 1/ 3 498,5.

Pour Mercure, Vénus et la Terre les comparaisons entre la théorie au deuxième ordre et une intégration numérique interne donnent, sur un siècle, des écarts ne dépassant pas 0",007. Pour Mars, ces écarts atteignent 0",035. Nous avons examiné la théorie de Mars que Clemence (1949) a publiée entre 1949 et 1961 et qui est développée jusqu'au troisième ordre des masses. Clemence s'est étonné d'avoir des écarts en longitude atteignant 0",042 par comparaison à l'intégration numérique de Herget qui couvre une période de 35 ans. Il considère que l'explication la plus probable est une erreur dans les théories de Vénus, la Terre, Jupiter et Saturne de Newcomb et de Hill qu'il a utilisées. Mais il envisage également une erreur de l'intégration numérique ou de sa théorie. Nous n'avons pas effectué sous forme analytique, une comparaison rendue difficile par le choix de variables différentes entre notre solution et la théorie de Clemence. Toutefois, nous avons trouvé quelques termes des perturbations de la longitude moyenne au deuxième ordre des masses qui dépendent d'arguments que Clemence n'a pas retenu, par exemple :

$$+ 0''\!,2405 \sin(10\lambda_T - 19\lambda_M + 3\lambda_S) + 0''\!,5124 \cos(10\lambda_T - 19\lambda_M + 3\lambda_S)$$
$$- 0''\!,0059 \sin(8\lambda_T - 16\lambda_M + 6\lambda_J) - 0''\!,0362 \cos(8\lambda_T - 16\lambda_M + 6\lambda_J)$$
$$- 0''\!,0190 \sin(\lambda_{Me} - 10\lambda_T + 11\lambda_M) + 0''\!,0236 \cos(\lambda_{Me} - 10\lambda_T + 11\lambda_M)$$

Ceci explique probablement que les écarts, qui atteignent 0",042 en longitude entre la théorie de Clemence et l'intégration numérique sur une période de 35 ans, soient comparables à ceux que nous avons sur un siècle.

Nous avons amélioré nos constantes d'intégration par comparaison de notre théorie des huit planètes à l'intégration numérique d'Oesterwinter et Cohen. Les éléments moyens issus de cette comparaison sont plus proches de ceux de Newcomb que de ceux de Le Verrier pour Vénus et la Terre. C'est le contraire pour Mars. La théorie de Le Verrier donne, par exemple, la valeur moyenne de l'excentricité pour 1950.0 : 0,093 356 41; la valeur de Newcomb est 0,093 358 91; notre théorie donne 0,093 355 39. La valeur de Clemence : 0,093 359 71 est voisine de celle de Newcomb.

Après avoir repris nos calculs avec les nouvelles constantes d'intégration, nous allons entreprendre la construction des perturbations du troisième ordre par rapport aux masses afin d'améliorer les théories des planètes inférieures, en particulier la théorie de la Terre. Une bonne connaissance du mouvement de la Terre est, en effet, indispensable car ce mouvement intervient dans toute comparaison à l'observation de théories planétaires. Par ailleurs, une bonne théorie de la Terre est nécessaire à l'amélioration des perturbations planétaires indirectes de la Lune. De plus, une amélioration des constantes d'intégration de la Terre agit directement sur le problème central de la Lune par modification des constantes de ce problème.

Nous entreprendrons ensuite la nécessaire comparaison de nos théories à l'observation et il sera intéressant de déterminer de nouvelles valeurs des masses des grosses planètes.

REFERENCES

Brumberg, V.A., Chapront, J. 1973, Celes. Mech. 8, 335
Clemence, G.M. 1949, APAE Vol. 11, p. 225
Duriez, L. 1977, Astron. & Astrophys. 54, 93-112
Duriez, L. 1978, Astron. & Astrophys. A paraître
Eckert, W.J., Brouwer, D., Clemence, G.M. 1951, APAE Vol. 12, p. 1
Le Verrier, U.J.J. 1855, Ann. Obs. Paris
Oesterwinter, C., Cohen, C.J. 1972, Celes. Mech. 5, 317
Simon, J.L., Bretagnon, P. 1975 a, Astron. & Astrophys. 42, 259
Simon, J.L., Bretagnon, P. 1975 b, Astron. & Astrophys. Suppl. 22, 107
Simon, J.L., Bretagnon, P. 1978 a, Astron. & Astrophys. A paraître
Simon, J.L., Bretagnon, P. 1978 b, Astron. & Astroph. Suppl. A paraître

ABSTRACT: At the Bureau of Longitudes the construction of planetary theories have been developed in three directions: A general theory of the motion of the four largest planets in the solar system is in the course of development at the Faculty of Sciences at Lille by L. Duriez (1977) following the methods of V. A. Brumberg and J. Chapront (1973). Theories of the classical type with secular variations of the motions of all of the planets from Mercury to Neptune are being completed at the Bureau of Longitudes. They are constructed by P. Bretagnon and J. L. Simon (1975, 1978). The numerical complement to all of these studies, numerical integration, a representation of the solution by Tchebychev series, are being carried out by P. Rocher as concerns the motions of minor planets, and by J. Piranx for the action of Pluto on Uranus and Neptune in the framework of theories with secular variations.

DISCUSSION

Message: Qu'est quecest la raison pour ces termes du deuxieme ordre dans les demi-axes?

Bretagnon: L'equation de Lagrange pour le demi-grand axe s'ecrit:

$$\frac{da}{dt} = \frac{2}{na} \frac{\partial R}{\partial M}$$

ou $\frac{\partial R}{\partial M}$ est proportionnel à G, constante de la gravitation. Par la troisieme loi de Kepler on a: $n^2a^3 = G(1+m)$.

d'ou: $\frac{da}{dt} = \frac{2na^2}{1+m} \times 0(m)$.

On voit don que' à la deuxieme approximation, cette equation aura la forme:

$$\frac{da}{dt} = \frac{0(m^2)}{1+m} = 0\ (m^2) + 0\ (m^3) + \cdots$$

Par consequent, les termes seculaires du demi-grande axe de la deuxieme approximation sont en realite d'ordre 3 par rapport aux masses. Ils proviennent, bien sur, de la partie indirecte de la fonction perturbatrice.

Garfinkel: What is the perturbation method that you use in your theory? Is it anything like the method of von Zeipel or the method of Lie-series?

Bretagnon: We are using the old method of Leverrier.

Garfinkel: How does your theory compare with the work of Meffroy?

Bretagnon: The only comparisons we have made were with the results of numerical integration.

NEW APPROACH TO THE PLANETARY THEORY

Manabu Yuasa and Gen'ichiro Hori
Department of Astronomy, University of Tokyo, Bunkyo-ku,
Tokyo 113, Japan

1. INTRODUCTION

A new approach to the planetary theory is examined under the following procedure: 1) we use a canonical perturbation method based on the averaging principle; 2) we adopt Charlier's canonical relative coordinates fixed to the Sun, and the equations of motion of planets can be written in the canonical form; 3) we adopt some devices concerning the development of the disturbing function. Our development can be applied formally in the case of nearly intersecting orbits as the Neptune-Pluto system. Procedure 1) has been adopted by Message (1976).

2. CANONICAL RELATIVE COORDINATES FIXED TO THE SUN

We consider n+1 celestial bodies. Let their masses be $m_i(i=0,\ldots,n)$ and their coordinates referred to the center of mass be $\vec{\rho}_i(i=0,\ldots,n)$. Then the Hamiltonian F of this system can be written as

$$F = -T + V = -\frac{1}{2}\sum_{i=0}^{n} m_i\dot{\vec{\rho}}_i^{\,2} + \sum_{i>j\geq 0} \frac{k^2 m_i m_j}{\rho_{ij}} , \tag{1}$$

where T, V, k^2, and ρ_{ij} represent the kinetic energy, the potential energy, the gravitational constant of Gauss, and $|\vec{\rho}_i-\vec{\rho}_j|$ respectively. We regard m_0 as the Sun and $m_i(i=1,\ldots,n)$ as the planets. The relative coordinates $\vec{r}_i(i=0,\ldots,n)$ fixed to the Sun are introduced by putting $\vec{r}_i=\vec{\rho}_i-\vec{\rho}_0(i=0,\ldots,n)$. Next, we introduce the momenta $\vec{p}_i(i=1,\ldots,n)$ which are conjugate to the coordinates $\vec{r}_i(i=1,\ldots,n)$ as follows (Charlier 1902):

$$\vec{p}_i = \frac{\partial T}{\partial\dot{\vec{r}}_i} = \frac{\partial}{\partial\dot{\vec{r}}_i}\frac{1}{2}\sum_{i=0}^{n} m_i\dot{\vec{\rho}}_i^{\,2} = m_i\dot{\vec{\rho}}_i . \tag{2}$$

Then the Hamiltonian of the system is given by

R. L. Duncombe (ed.), Dynamics of the Solar System, 69–72.

$$F = \sum_{i=1}^{n} \{-\frac{1}{2}\frac{\vec{p}_i^{\,2}}{m_i'} + \frac{\mu_i m_i'}{r_{i0}}\} + \sum_{i>j\geq 1} (-\frac{\vec{p}_i\cdot\vec{p}_j}{m_0} + \frac{k^2 m_i m_j}{r_{ij}}) \quad , \tag{3}$$

where $m_i' = m_0 m_i/(m_0+m_i)$, $\mu_i = k^2(m_0+m_i)$, and $r_{ij} = |\vec{r}_i - \vec{r}_j|$.

Let the quantities a_i, e_i, I_i, ℓ_i, ω_i, and Ω_i be the semi-major axis, the eccentricity, the inclination, the mean anomaly, the argument of perihelion, and the longitude of the node of the motion of the i-th planet around the Sun. Then the canonical variables L_i, G_i, H_i, ℓ_i, g_i, and h_i can be defined as

$$\begin{aligned} L_i &= m_i'\sqrt{\mu_i a_i} \;, \quad G_i = L_i\sqrt{1-e_i^2} \;, \quad H_i = G_i \cos I_i \\ \ell_i &= \text{mean anomaly}, \quad g_i = \omega_i \;, \quad h_i = \Omega_i \quad . \end{aligned} \tag{4}$$

The equations of motion are

$$\frac{d(L_i,G_i,H_i)}{dt} = \frac{\partial F}{\partial(\ell_i,g_i,h_i)} \;, \quad \frac{d(\ell_i,g_i,h_i)}{dt} = -\frac{\partial F}{\partial(L_i,G_i,H_i)} \;, \tag{5}$$

with

$$F = F_0 + F_1 \;, \quad F_0 = \sum_{i=1}^{n} \frac{\mu_i^2 m_i'^3}{2L_i^2} \;, \quad F_1 = \sum_{i>j\geq 1} (-\frac{\vec{p}_i\cdot\vec{p}_j}{m_0} + \frac{k^2 m_i m_j}{r_{ij}}) \;. \tag{6}$$

The function F_1 is the disturbing function and to be represented by L_i, G_i, H_i, ℓ_i, g_i, and h_i.

3. DEVELOPMENT OF THE DISTURBING FUNCTION IN TERMS OF THE INCLINATIONS

We consider only two planets m_1 and m_2. If v_1 and v_2 are the true longitudes of the two planets, the mutual distance r_{12} is given by

$$\begin{aligned} r_{12}^2 = r_1^2 + r_2^2 - 2r_1 r_2 [& c_1^2 c_2^2 \cos(v_1-v_2) + c_1^2 s_2^2 \cos(v_1+v_2-2\Omega_2) + s_1^2 c_2^2 \cos(v_1+v_2-2\Omega_1) \\ & + s_1^2 s_2^2 \cos(v_1-v_2-2\Omega_1+2\Omega_2) + 2c_1 s_1 c_2 s_2 \{\cos(v_1-v_2-\Omega_1+\Omega_2) \\ & - \cos(v_1+v_2-\Omega_1-\Omega_2)\}] \quad , \end{aligned} \tag{7}$$

where $c_i = \cos(I_i/2)$, $s_i = \sin(I_i/2)$, $(i=1,2)$. At this stage we define

$$q \equiv (r_1^2 + r_2^2)/2r_1 r_2 (c_1 c_2 - s_1 s_2)^2 \quad , \tag{8}$$

and the inverse of the mutual distance is expressed as

$$\frac{1}{r_{12}} = \frac{1}{\sqrt{2r_1 r_2}(c_1 c_2 - s_1 s_2)} [q - \cos(v_1-v_2) - \frac{\delta}{(c_1 c_2 - s_1 s_2)^2}]^{-1/2} \quad , \tag{9}$$

with

$$\delta = s_1^2c_2^2\cos(v_1+v_2-2\Omega_1)+c_1^2s_2^2\cos(v_1+v_2-2\Omega_2)+s_1^2s_2^2\cos(v_1-v_2-2\Omega_1+2\Omega_2)$$
$$+ 2c_1s_1c_2s_2\{\cos(v_1-v_2-\Omega_1+\Omega_2)-\cos(v_1+v_2-\Omega_1-\Omega_2)\} \tag{10}$$
$$+ (2c_1s_1c_2s_2-s_1^2s_2^2)\cos(v_1-v_2) \quad .$$

By the binomial expansion of the equation (9), $1/r_{12}$ is written in the form

$$\frac{1}{r_{12}} = \frac{1}{\sqrt{2r_1r_2}(c_1c_2-s_1s_2)} \sum_{n=0}^{\infty}(-1)^n\binom{-1/2}{n}\left[\frac{\delta}{(c_1c_2-s_1s_2)^2}\right]^n \times$$
$$\times\ [q-\cos(v_1-v_2)]^{-(n+1/2)} \quad . \tag{11}$$

Furthermore, we expand $[q-\cos(v_1-v_2)]^{-(n+1/2)}$ by the 2-nd kind associated Legendre function Q_μ^ν. And we get

$$\frac{1}{r_{12}} = \sum_{n=0}^{\infty}\sum_{j=-\infty}^{\infty}\frac{2^n}{n!}(c_1c_2-s_1s_2)^{-2n}\delta^n\beta_{n+1/2}^{(j)}(q)\cos j(v_1-v_2) \quad , \tag{12}$$

where

$$\beta_{n+1/2}^{(j)} = \frac{(-1)^n}{2^n\pi}\ \frac{(q^2-1)^{-n/2}}{\sqrt{r_1r_2}(c_1c_2-s_1s_2)}\ Q_{j-1/2}^n(q) \quad . \tag{13}$$

These expansions converge regardless of the values of r_1 and r_2 except for the following two cases: 1) the case when two planets collide; 2) the case when $\Omega_1-\Omega_2=\pi$, $v_1=v_2$, $v_1+v_2-\Omega_1-\Omega_2=0$, and $r_1=r_2$. Consequently, above development can be applied formally even in the case of nearly intersecting orbits as the Neptune-Pluto system.

4. DEVELOPMENT OF THE DISTURBING FUNCTION IN TERMS OF THE ECCENTRICITIES

We use Newcomb's operator and r_1, r_2, v_1, v_2 can be expressed in terms of a_1, a_2, e_1, e_2, λ_1, λ_2, ℓ_1, ℓ_2, where λ_1, λ_2 are the mean longitudes. For the simplicity of notations, we put

$$\frac{2n}{n!}(c_1c_2-s_1s_2)^{-2n}\delta^n\cos j(v_1-v_2) = \sum_y C_{n,y}(I_1,I_2)\cos[j(v_1-v_2)+y_1v_1$$
$$+y_2v_2+y_3\Omega_1+y_4\Omega_2] \quad , \tag{14}$$

where the summation is taken in all the combinations of $y_1,..,y_4$ appeared. Then the inverse of the mutual distance can be expanded as follows:

$$\frac{1}{r_{12}} = \sum_{n=0}^{\infty}\sum_{j=-\infty}^{\infty}\sum_y\sum_{k_1=-\infty}^{\infty}\sum_{k_2=-\infty}^{\infty}\sum_{s_1=|k_1|+0,2,..}\sum_{s_2=|k_2|+0,2,..}C_{n,y}(I_1,I_2)\times$$
$$\times\ \Pi_{k_1}^{s_1}(D_1|j+y_1)\Pi_{k_2}^{s_2}(D_2|-j+y_2)e_1^{s_1}e_2^{s_2}\beta_{n+1/2}^{(j)}(q_0)\times \tag{15}$$

$$\times \cos[j(\lambda_1-\lambda_2)+y_1\lambda_1+y_2\lambda_2+y_3\Omega_1+y_4\Omega_2+k_1\ell_1+k_2\ell_2] \quad ,$$

where $D_1=a_1\cdot\partial/\partial a_1$, $D_2=a_2\cdot\partial/\partial a_2$, $q_0=(a_1^2+a_2^2)/2a_1a_2(c_1c_2-s_1s_2)^2$, and $\Pi_{k_1}^{s_1}(D_1|j+y_1)$, $\Pi_{k_2}^{s_2}(D_2|-j+y_2)$ are Newcomb's operators.

5. EVALUATIONS OF $\beta_{n+1/2}^{(j)}(q_0)$

From the equations (11) and (12) we get

$$\begin{aligned} &\frac{(2n-1)!!}{2^{2n}\sqrt{2a_1a_2}(c_1c_2-s_1s_2)}\,[q_0-\cos(v_1-v_2)]^{-(n+1/2)} \\ &= \beta_{n+1/2}^{(0)} + 2\sum_{j=1}^{\infty}\beta_{n+1/2}^{(j)}\cos j(v_1-v_2) \quad , \end{aligned} \tag{16}$$

and we can determine the values of $\beta_{n+1/2}^{(j)}$ by the numerical Fourier analysis if a_1, a_2, c_1, c_2, s_1, s_2 are given. On the other hand, the equation (13) and the recurrence formulas of Q_μ^ν give rise to the recurrence formulas of β, $D_1{}^\nu\beta$, and $D_2{}^\nu\beta$. These recurrence formulas are of much help for the evaluation of β.

The practical development of the disturbing function has been performed to the fourth order of the eccentricity and the inclination. As an application, we are trying to study the Neptune-Pluto system by a canonical perturbation method.

REFERENCES

Charlier, C. L. : 1902, "Die Mechanik des Himmels", Erster Band, pp. 234-237.

Message, P. J. : 1976, in "Long-Time Predictions in Dynamics", ed. V. Szebehely and B. D. Tapley (D. Reidel Publishing Company), pp. 279-293.

SEMI-ANALYTICAL LUNAR EPHEMERIS - THE MAIN PROBLEM

Jacques Henrard
Facultés Universitaires de Namur, B-5000 Namur, Belgium

The theory of the motion of the Moon can be approached by two types of methods. In the analytical methods, the solution is usually expanded in formal power series of the small quantities m, e and γ. It is known that the expansions in m are slowly convergent and that, in order to obtain an accuracy of the order of the meter, the expansions have to be carried to very high order (beyond 20) involving several hundred thousands terms.

In the numerical methods, the solution is sought under the form of a Fourier series solution (with about a thousand terms). The coefficients are determined by an iterative procedure.

While being conceptually more elegant (although its mathematical foundations are not firmer than those of the numerical methods), the analytical methods are wasteful. Many of the terms computed are numerically very small and the overall accuracy depends on a few very poorly convergent subseries. On the other hand, the fact that analytical methods provide a tube of trajectories rather than one particular solution is very important. With a single trajectory as usually provided by the numerical methods, it is impossible to improve on the solution by including other effects or by adjusting the constants.

It is thus natural to try to blend the advantages of the two types of techniques. Mrs. Chapront at the Bureau des Longitudes (Paris, France) is including in a numerical scheme the computation of the partial derivatives of the solution with respect to the constants of motion in order to get what one could call a semi-numerical solution. On the other hand, we are developping an analytical solution expanded not around the zero of the small parameter m, e, γ but around their nominal values. In this way, at each order, the functions are no longer exact finite Fourier series but truncated Fourier series and only numerically meaningful terms are kept. We like to call this solution a semi-analytical solution.

With these two approaches, the gap between analytical and numerical

R. L. Duncombe (ed.), Dynamics of the Solar System, 73–75.

methods seems to be bridged. Nevertheless, there remains an important difference between the two philosophies. The difficult problem of resonance presents itself quite differently in order by order methods and in iterative methods. Much more work should be done before we can comment meaningfully on this difference.

SEMI-ANALYTICAL METHOD

We propose to base our solution on a completely analytical solution of Hill's Problem that we describe elsewhere (J. Henrard, Celestial Mechanics, in print). Hill's Problem is much simpler, so that a good analytical solution is feasible without too many problems. Our solution is complete up to order 25 and we estimate its accuracy at 1.10^{-10} in longitude.

This solution is given implicitely in functions of the angular variables

$$\lambda = l + g + h \qquad p = -g - h \qquad q = -h$$

and their conjugate momenta L, P and Q which, together with the initial values of the angular variables, are the constants of the motion.

If we introduce the canonical transformation

$$L = L^{\star}\,(1 + \Delta L) \qquad P = P^{\star} + L^{\star}\,\Delta P \qquad Q = Q^{\star} + L^{\star}\,\Delta Q$$

from (λ, p, q, L, P, Q) to $(\lambda, p, q, \Delta L, \Delta P, \Delta Q)$, we can express the solution of Hill's Problem and, by substitution, the perturbation function of the Main Problem in powers of ΔL, ΔP and ΔQ. The radius of convergence of these series appears to be around $1/200$ which is much larger than what we need. We choose to truncate them at degree 4 .

The coefficients in these expansions are Fourier series in the angular variables λ, p and q. Of course, in order to manipulate these, we have to truncate them. We can allow the relative truncation error to be larger when the degree in ΔL, ΔP, ΔQ is higher, so that only a few terms are kept at degree 4 against several thousands at degree 1 . A relative simple scheme determines the truncation level of each function.

With these new canonical variables, we can apply the Lie transform method and eliminate the periodic terms from the perturbation functions. We choose to eliminate separately the monthly terms, the annual terms and the long period terms. This is not imposed by the problem but, in this way, the analysis of the propagation of the truncation errors is much simplified. In the semi-analytical method we propose, this analysis is of prime importance and replaces in some way the analysis of the convergence of the subseries in the purely analytical method.

RESULTS FOR THE MAIN PROBLEM

The elimination of the periodic terms from the hamiltonian function of the Main Problem and the expansion of the corresponding solution have been carried out on a computer.

According to our estimates, the solution is obtained with a relative accuracy of about $2.5\ 10^{-10}$ (or $5''10^{-5}$ in longitude) within a tube of solution corresponding to a relative change in eccentricity of 1.10^{-2}. If we take into acount only the first partial derivatives, the same accuracy can be kept for a relative change in eccentricity of 1.10^{-5}.

A comparison with the Analytical Lunar Ephemeris of Deprit, Henrard and Rom has been made and does not infirm our estimate of accuracy.

Our solution consists in the expansions of the three traditional functions : longitude, latitude and sine parallax, but also in the expansions of r (the distance Earth-Moon) and x / r, y / r and z / r (the unit vector in the direction of the Moon). Together with the coefficients of the Fourier series, we give the first partial derivatives with respect to the internal constants (ν - mean motion of the Moon -, E - coefficient of $\sin l$ in the longitude -, Γ - coefficient of $\sin F$ in the latitude -) and the external constants (e_2 - eccentricity of the Earth -, μ_2 - mass ratio of the system Earth-Moon -, μ_1 - mass ratio of the Moon -).

The solution is under a form directly suitable for the computation of further perturbations. Preliminary results have been obtained for the perturbation by the oblateness of the Earth by the author and for the direct perturbation of Venus by Daniel Standaert.

THE SATELLITE CASE OF THE THREE-BODY PROBLEM

Gen'ichiro Hori
Department of Astronomy, University of Tokyo, Bunkyo-ku,
Tokyo 113, Japan

1. EQUATIONS OF MOTION

In the lunar or satellite theory, the disturbing function is developed in the form in which each term consists of two factors. The first factor depends on the position of the sun, and the second one that of the satellite. The main term of the disturbing function is

$$F_2 = \nu^2 a^2 \left(\frac{a_1}{r_1}\right)^3 [C_A (A + 1 + \frac{3}{2}e^2) + C_B (B + \frac{15}{2}e^2) + C_C \cdot C] \quad , \tag{1}$$

where

$$C_A = -\frac{1}{8} + \frac{3}{8}c^2 + \frac{3}{8}s^2\cos(2f_1-2h) , \quad C_B = \frac{1}{8}s^2\cos 2g + \frac{1}{16}(1+c)^2\cos(2f_1-2g-2h)$$

$$+ \frac{1}{16}(1-c)^2 \cos(2f_1+2g-2h) , \quad C_C = \frac{1}{8}s^2\sin 2g - \frac{1}{16}(1+c)^2\sin(2f_1-2g-2h) \tag{2}$$

$$+ \frac{1}{16}(1-c)^2\sin(2f_1+2g-2h) \quad .$$

These, together with the factor $(a_1/r_1)^3$, give the solar position. The subscript 1 is referred to the sun. While, $A+1+(3/2)e^2$, $B+(15/2)e^2$, and C, where

$$A = -e^2 - 2e\cos u + \frac{1}{2}e^2\cos 2u, \quad B = -3e^2 - 6e\cos u + (3 - \frac{3}{2}e^2)\cos 2u,$$

$$C = \eta(6e\sin u - 3\sin 2u) \quad , \tag{3}$$

give the position of the satellite. In equations (2) and (3),

$$c = \cos i \;, \quad s = \sin i \;, \quad g = \omega, \quad h = \Omega - \omega_1, \quad \eta = \sqrt{1-e^2} \quad , \tag{4}$$

R. L. Duncombe (ed.), Dynamics of the Solar System, 77–83.

and f_1 is the true anomaly of the sun, u the eccentric anomaly of the satellite, ν in equation (1) being the mean motion of the sun.

In terms of Delaunay variables, the equations of motion are

$$\frac{d}{dt}(L,G,H,K) = \frac{\partial F}{\partial(\ell,g,h,k)}\ , \quad \frac{d}{dt}(\ell,g,h,k) = -\frac{\partial F}{\partial(L,G,H,K)}\ , \tag{5}$$

with the Hamiltonian

$$F = \frac{\mu^2}{2L^2} - \nu K + F_2 \quad , \tag{6}$$

where k is the mean anomaly of the sun.

2. ELIMINATION OF ℓ

Applying canonical perturbation theory (Hori 1966), the first canonical transformation results in the new Hamiltonian

$$F^* = \frac{\mu^2}{2L'^2} - \nu K' + \nu^2 a^2 (\frac{a_1}{r_1})^3 [C_A(1+\frac{3}{2}e^2) + C_B \frac{15}{2}e^2] \quad , \tag{7}$$

and the equation for the determining function, of the form

$$n\frac{\partial S_2}{\partial \ell'} + \nu\frac{\partial S_2}{\partial k} = \nu^2 a^2 (\frac{a_1}{r_1})^3 \sum_{X=A,B,C} C_X \cdot X \quad . \tag{8}$$

This equation can be integrated by the successive application of the integration by parts to find S_2 in power series in ν/n, and in the closed form both in e and e_1.

$$S_2 = \frac{\nu^2 a^2}{n}(\frac{a_1}{r_1})^3 \Sigma C_X \cdot {}^{(1)}X - \frac{\nu^3 a^2}{n^2}\Sigma \{C_X(\frac{a_1}{r_1})^3\}^{(1)} \cdot {}^{(2)}X + O(\frac{\nu^4 a^2}{n^3}) \quad , \tag{9}$$

where

$$\begin{aligned}
&{}^{(1)}A = (-2e+\frac{3}{4}e^3)\sin u + \frac{3}{4}e^2\sin 2u - \frac{1}{12}e^3\sin 3u, \quad {}^{(1)}B = (-\frac{15}{2}e+\frac{15}{4}e^3)\sin u \\
&+(\frac{3}{2}+\frac{3}{4}e^2)\sin 2u + (-\frac{1}{2}e+\frac{1}{4}e^3)\sin 3u, \quad {}^{(1)}C = \eta[-\frac{15}{4}e^2 - \frac{15}{2}e\cos u + (\frac{3}{2}+\frac{3}{2}e^2)\times \\
&\times\cos 2u - \frac{1}{2}e\cos 3u], \quad {}^{(2)}A = e^2 - \frac{3}{16}e^4 + (2e-\frac{3}{8}e^3)\cos u + (-\frac{7}{8}e^2+\frac{1}{6}e^4)\cos 2u \\
&+\frac{11}{72}e^3\cos 3u - \frac{1}{96}e^4\cos 4u, \quad {}^{(2)}B = \frac{33}{8}e^2 - \frac{27}{16}e^4 + (\frac{33}{4}e - \frac{27}{8}e^3)\cos u
\end{aligned} \tag{10}$$

$$+(-\frac{3}{4}-\frac{19}{8}e^2+e^4)\cos 2u+(\frac{5}{12}e+\frac{1}{24}e^3)\cos 3u+(-\frac{1}{16}e^2+\frac{1}{32}e^4)\cos 4u,$$

$$^{(2)}C=\eta[(-\frac{33}{4}e+3e^3)\sin u+(\frac{3}{4}+\frac{11}{4}e^2)\sin 2u+(-\frac{5}{12}e-\frac{1}{4}e^3)\sin 3u+\frac{1}{16}e^2\sin 4u],$$

and

$$\left\{C_A(\frac{a_1}{r_1})^3\right\}^{(1)}=(-\frac{1}{8}+\frac{3}{8}c^2)(-3\frac{e_1}{\eta_1})(\frac{a_1}{r_1})^4\sin f_1-\frac{3}{8}s^2\frac{1}{\eta_1}(\frac{a_1}{r_1})^4[-\frac{1}{2}e_1\sin(f_1-2h)$$

$$+2\sin(2f_1-2h)+\frac{5}{2}e_1\sin(3f_1-2h)],\quad \left\{C_B(\frac{a_1}{r_1})^3\right\}^{(1)}=\frac{1}{8}s^2(-\frac{3}{2}\frac{e_1}{\eta_1})(\frac{a_1}{r_1})^4\times$$

$$\times[\sin(f_1-2g)+\sin(f_1+2g)]-\frac{1}{16}(1+c)^2\frac{1}{\eta_1}(\frac{a_1}{r_1})^4[-\frac{1}{2}e_1\sin(f_1-2g-2h)$$

$$+2\sin(2f_1-2g-2h)+\frac{5}{2}e_1\sin(3f_1-2g-2h)]-\frac{1}{16}(1-c)^2\frac{1}{\eta_1}(\frac{a_1}{r_1})^4\times$$

$$\times[-\frac{1}{2}e_1\sin(f_1+2g-2h)+2\sin(2f_1+2g-2h)+\frac{5}{2}e_1\sin(3f_1+2g-2h)],\qquad (11)$$

$$\left\{C_C(\frac{a_1}{r_1})^3\right\}^{(1)}=\frac{1}{8}s^2(-\frac{3}{2}\frac{e_1}{\eta_1})(\frac{a_1}{r_1})^4[\cos(f_1-2g)-\cos(f_1+2g)]-\frac{1}{16}(1+c)^2\frac{1}{\eta_1}(\frac{a_1}{r_1})^4\times$$

$$\times[-\frac{1}{2}e_1\cos(f_1-2g-2h)+2\cos(2f_1-2g-2h)+\frac{5}{2}e_1\cos(3f_1-2g-2h)]$$

$$-\frac{1}{16}(1-c)^2\frac{1}{\eta_1}(\frac{a_1}{r_1})^4[-\frac{1}{2}e_1\cos(f_1+2g-2h)+2\cos(2f_1+2g-2h)+\frac{5}{2}e_1\times$$

$$\times\cos(3f_1+2g-2h)]\quad .$$

If the partial derivative with respect to e is denoted by the subscript e, we have

$$A_e=-e-2\cos u,\quad B_e=\frac{a}{r}[(-9+9e^2)\cos u-3e\cos 2u+3\cos 3u]\quad ,$$

$$C_e=\frac{a}{r\eta}[(9-\frac{33}{2}e^2)\sin u+(3e+3e^3)\sin 2u+(-3+\frac{3}{2}e^2)\sin 3u]\quad ,$$

$$^{(1)}A_e=(-2+e^2)\sin u+\frac{1}{2}e\sin 2u,\quad ^{(1)}B_e=(-9+9e^2)\sin u-\frac{3}{2}e\sin 2u+\sin 3u\quad ,$$

$$^{(1)}C_e=\frac{1}{\eta}[-\frac{9}{2}e+\frac{33}{4}e^3+(-9+\frac{33}{2}e^2)\cos u+(-\frac{3}{2}e-\frac{3}{2}e^2)\cos 2u+(1-\frac{1}{2}e^2)\cos 3u]\quad ,$$

$$^{(2)}A_e=e-\frac{3}{8}e^3+(2-\frac{3}{4}e^2)\cos u+(-\frac{3}{4}e+\frac{1}{4}e^3)\cos 2u+\frac{1}{12}e^2\cos 3u\quad ,\qquad (12)$$

$$^{(2)}B_e = \frac{9}{2}e - \frac{39}{8}e^3 + (9 - \frac{39}{4}e^2)\cos u + (-\frac{5}{4}e + \frac{9}{4}e^3)\cos 2u + (-\frac{1}{3} - \frac{1}{4}e^2)\cos 3u$$

$$+ \frac{1}{8}e\cos 4u, \quad ^{(2)}C_e = \frac{1}{\eta}[(-9 + \frac{87}{4}e^2 - \frac{15}{2}e^4)\sin u + (\frac{5}{4}e - \frac{19}{4}e^3)\sin 2u$$

$$+ (\frac{1}{3} + \frac{1}{12}e^2 + \frac{1}{4}e^4)\sin 3u + (-\frac{1}{8}e + \frac{1}{16}e^3)\sin 4u] \quad .$$

Equations (12) are of use in the derivation of short-period perturbation in ℓ and g.

3. ELIMINATION OF k

After the mean anomaly of the satellite is eliminated, the new equations of motion are

$$\frac{d}{dt}(L',G',H',K') = \frac{\partial F^*}{\partial(g',h',k)}, \quad \frac{d}{dt}(\ell',g',h',k) = -\frac{\partial F^*}{\partial(L',G',H',K')}, \tag{13}$$

with the Hamiltonian (7) or

$$F^* = \frac{\mu^2}{2L'^2} - \nu K' + \nu^2 a^2 [C^*_A(A_1 + \frac{1}{\eta_1{}^3}) + C^*_B B_1 + C^*_C C_1] \quad , \tag{14}$$

where

$$A_1 = (\frac{a_1}{r_1})^3 - \frac{1}{\eta_1{}^3}, \quad B_1 = (\frac{a_1}{r_1})^3\cos 2f_1, \quad C_1 = (\frac{a_1}{r_1})^3\sin 2f_1,$$

$$C^*_A = \sum_{j=0}^{1} c_{0,2j}\cos 2jg', \quad C^*_B = \sum_{j=-1}^{1} c_{2,2j}\cos(2jg'-2h'), \tag{15}$$

$$C^*_C = -\sum_{j=-1}^{1} c_{2,2j}\sin(2jg'-2h'),$$

and

$$c_{00} = (-\frac{1}{8} + \frac{3}{8}c^2)(1 + \frac{3}{2}e^2), \quad c_{02} = \frac{15}{16}s^2e^2, \quad c_{20} = \frac{3}{8}s^2(1 + \frac{3}{2}e^2),$$

$$c_{2\mp 2} = \frac{15}{32}(1 \pm c)^2e^2 \quad . \tag{16}$$

The second canonical transformation, $L', G', H', K', \ell', g', h', k \to L', G'', H'', K'', \ell'', g'', h'', k$, which eliminates k from the new Hamiltonian F^{**}, leads to the result:

$$F^{**} = \frac{\mu^2}{2L'^2} - \nu K'' + \frac{\nu^2 a^2}{\eta_1{}^3}C^*_A + \nu^3 a^4 [\{C^*_A, C^*_C\} \langle {}^{(1)}C_1 A_1 \rangle + \{C^*_B, C^*_C\} \langle {}^{(1)}C_1 B_1 \rangle] \quad , \tag{17}$$

and

$$S_1^* + S_2^* = \nu a^2 \sum_{X=A,B,C} C_X^* \cdot {}^{(1)}X_1 + \nu^2 a^4 [\{C_A^*, C_C^*\} \cdot {}^{(1)}D_1 + \{C_A^*, C_C^*\} \cdot {}^{(1)}E_1 + \{C_B^*, C^*_C\} \cdot {}^{(1)}F_1] \quad , \tag{18}$$

where

$$^{(1)}A_1 = \frac{1}{\eta_1{}^3}(f_1 - k + e_1 \sin f_1), \quad ^{(1)}B_1 = \frac{1}{\eta_1{}^3}(\frac{e_1}{2}\sin f_1 + \frac{1}{2}\sin 2f_1 + \frac{e_1}{6}\sin 3f_1),$$
$$^{(1)}C_1 = \frac{1}{\eta_1{}^3}[-\frac{1+2\eta_1}{6(1+\eta_1)^2}e_1{}^2 - \frac{e_1}{2}\cos f_1 - \frac{1}{2}\cos 2f_1 - \frac{e_1}{6}\cos 3f_1] \quad , \tag{19}$$

and

$$\{C_A^*, C_B^*\} = \sum_{j=-1}^{1} c_{2,2j}^* \sin(2jg''-2h''), \quad \{C_A^*, C_C^*\} = \sum_{j=-1}^{1} c_{2,2j}^* \times$$
$$\times \cos(2jg''-2h''), \quad \{C_B^*, C_C^*\} = -\sum_{j=0}^{1} c_{0,2j}^* \cos 2jg'' \quad , \tag{20}$$

with

$$c_{00}^* = \frac{9}{32}\frac{\eta c}{na^2}[2s^2 + (33+17c^2)e^2], \quad c_{02}^* = \frac{135}{32}\frac{\eta cs^2}{na^2}e^2 \quad ,$$
$$c_{20}^* = \frac{9}{32}\frac{\eta cs^2}{na^2}(2-17e^2), \quad c_{2\mp 2}^* = \pm\frac{45}{64}\frac{\eta}{na^2}(1\pm c)^2(2\mp 3c)e^2 \quad , \tag{21}$$

and finally,

$$\langle {}^{(1)}C_1A_1\rangle = -[\frac{1}{4} + \frac{1+2\eta_1}{6(1+\eta_1)^2}]\frac{e_1{}^2}{\eta_1{}^6}, \quad \langle {}^{(1)}C_1B_1\rangle = -(\frac{1}{4} + \frac{1}{6}e_1{}^2)\frac{1}{\eta_1{}^6},$$
$$\langle {}^{(1)}A_1 \cdot {}^{(1)}B_1\rangle = -[\frac{1-3\eta_1}{6(1+\eta_1)} + \frac{2}{3e_1{}^2}\ln\frac{1+\eta_1}{2\eta_1}]\frac{1}{\eta_1{}^3}, \quad ^{(1)}D_1 = -\frac{1}{2}{}^{(1)}A_1 \cdot {}^{(1)}B_1$$
$$+ \frac{1}{2}\langle {}^{(1)}A_1 \cdot {}^{(1)}B_1\rangle + \frac{1}{\eta_1{}^6}[-\frac{e_1{}^2}{48(1+\eta_1)^2}(\frac{65}{3} + \frac{130}{3}\eta_1 + 31\eta_1{}^2 + 12\eta_1{}^3) - \frac{3}{4}e_1\cos f_1$$
$$-(\frac{1}{4} + \frac{1}{6}e_1{}^2)\cos 2f_1 - \frac{5}{36}e_1\cos 3f_1 - \frac{1}{48}e_1{}^2\cos 4f_1], \quad ^{(1)}E_1 = -\frac{1}{2}{}^{(1)}A_1 \cdot {}^{(1)}C_1$$
$$+ \frac{1}{\eta_1{}^6}[-\{\frac{1+2\eta_1}{6(1+\eta_1)^2}e_1{}^2 + \frac{3}{4}\}e_1\sin f_1 - (\frac{1}{4} + \frac{1}{6}e_1{}^2)\sin 2f_1 - \frac{5}{36}e_1\sin 3f_1 \tag{22}$$
$$-\frac{1}{48}e_1{}^2\sin 4f_1], \quad ^{(1)}F_1 = -\frac{1}{2}{}^{(1)}B_1 \cdot {}^{(1)}C_1 + \frac{1}{\eta_1{}^6}[-\{\frac{1+2\eta_1}{12(1+\eta_1)^2} + \frac{7}{12}\}e_1\sin f_1$$

$$-\{\frac{1+2\eta_1}{12(1+\eta_1)^2}+\frac{7}{48}\}e_1^{\,2}\sin 2f_1 - \{\frac{1+2\eta_1}{36(1+\eta_1)^2}+\frac{1}{8}\}e_1\sin 3f_1 - (\frac{1}{16}+\frac{1}{24}e_1^{\,2})\sin 4f_1$$

$$-\frac{1}{24}e_1\sin 5f_1 - \frac{1}{144}e_1^{\,2}\sin 6f_1] \quad .$$

In my previous work on the same subject (Hori 1963), i and e_1 were assumed to be vanishingly small. In that case equation (17) is reduced to

$$F^{**} = \frac{\mu^2}{2L'^2} - \nu K'' + \frac{1}{4}\nu^2 a^2(1+\frac{3}{2}e^2) + \frac{225}{64}\frac{\nu^3 a^2}{n}\eta e^2 \quad , \tag{23}$$

and the new equations are immediately integrated. In the present general case, F** still depends on g" and h", and another canonical transformation is required for the complete solution.

Let the transformation be L', G", H", K", ℓ", g", h", k→L', G"', H"', K", ℓ"', g"', h"', k. By a proper choice of a determining function, we may have an integral

$$\frac{\mu^2}{2L'^2} - \nu K'' + \nu^2 a^2[(-\frac{1}{8}+\frac{3}{8}c^2)(1+\frac{3}{2}e^2) + \frac{15}{16}s^2e^2\cos 2g''']=\text{const.} \quad , \tag{24}$$

where

$$c = \cos i = \frac{H'''}{G'''} \quad , \quad \eta = \frac{G'''}{L'} \quad . \tag{25}$$

Such a determining function is $O(\nu/n)$, and the use of elliptic functions is inevitable if no restriction is put on the size of i and e_1.

When $\alpha=H'''/L'$ is smaller than $(3/5)^{1/2}$, libration in g"' occurs. The center of libration is at $g'''=\pi/2$ and $\eta=(5\alpha^2/3)^{1/4}$. If η is smaller than $(5\alpha^2/3)^{1/2}$ when g"' is $\pi/2$, g"' has circulation even when $\alpha<(3/5)^{1/2}$.

REFERENCES

Hori, G. : 1963, Astron. J., 68, 125.
Hori, G. : 1966, Publ. Astron. Soc. Japan, 18, 287.

DISCUSSION

Marchal: Do you obtain some cases in which the eccentricity goes until one or at least reaches large values?

Hori: I indeed can obtain large variations of the eccentricity; however, I cannot say that I reach one since my theory is not valid for collision orbits.

Message: Does the equilibrium point shown in your last figure represent a periodic solution of the third sort?

Hori: Yes, in triply primed variables.

Garfinkel: In your final result, what has become of the periodic terms involving the argument h?

Hori: Being small terms of the third order in ν, the calculation of their effect appears in the next order of the perturbation.

Garfinkel: With the use of your integral, the Hamiltonian reduces to that of the Ideal Resonance Problem, exhibiting both the libration and the circulation regimes in the argument g of the perihelion.

Hori: Yes. Indeed, it is in the form of your Ideal Resonance Problem in the variables G''' and g'''.

FORMULE D'INVERSION DE LAGRANGE ET SON APPLICATION A LA THEORIE DES PERTURBATIONS

Takeshi Inoue
Université-Kyoto-Sangyo, Kamigamo, Kyoto, 603, JAPON.

1. FORMULE DE LAGRANGE

Pour résoudre les équations littérales par le moyen des séries, Lagrange (1770) établit une méthode fort ingénieuse. Rappelons ici tout brièvement ce qu'il énonce dans ses oeuvres. Considérons une équation d'une seule inconnue y de la forme suivante :

$$\eta-y+\varepsilon\chi(y)=0, \tag{1}$$

où η et ε sont respectivement un paramètre quelconque et une petite quantité et que $\chi(y)$ est une fonction analytique dans un intervalle. Alors on peut développer une des racines de l'équation (1) qui devient égale au paramètre η lorsqu'on met la valeur de la quantité ε nulle :

$$y=\eta+\sum_{1}^{\infty}\frac{\varepsilon^n}{n!}\left[\frac{d^{n-1}}{dy^{n-1}}\{\chi(y)\}^n\right]_{\varepsilon=0}. \tag{2}$$

On peut considérer ladite équation comme une fonction aux deux variables η et y ainsi que l'expression (2) comme une fonction uniforme définie par la fonction implicite (1) (Dieudonné, 1968).

On peut aussi développer une fonction analytique K(y) dont l'argument y est donné par la relation (2) :

$$K(y)=K(\eta)+\sum_{1}^{\infty}\frac{\varepsilon^n}{n!}\left[\frac{d^{n-1}}{dy^{n-1}}\left(\frac{dK(y)}{dy}\{\chi(y)\}^n\right)\right]_{\varepsilon=0}. \tag{3}$$

Quoique la formule de Lagrange soit bien efficace quand on veut résoudre l'équation de Képler par exemple, le nombre des inconnues ou des variables est, dans la plupart des calculs perturbateurs, supérieur à un et les formules sus-dénommées ne sont plus convenables sans y rien changer. C'est donc qu'il nous faudrait les étendre afin de les faire devenir applicables au cas où il existerait plusieurs inconnues ou plusieurs variables.

R. L. Duncombe (ed.), Dynamics of the Solar System, 85–89.

2. EXTENSION AU CAS DE PLUSIEURS VARIABLES

Soient f un nombre naturel, ε une petite quantité, $\eta_1,..,\eta_f$ des paramètres et $Y_1,..,Y_f$; $\chi_1,..,\chi_f$ des fonctions analytiques dans un ensemble par rapport aux variables $y_1,..,y_f$ et aux paramètres $\xi_1,..,\xi_f$. Considérons un système de fonctions implicites sous la forme :

$$\eta_j - Y_j(y;\xi) + \varepsilon\chi_j(y;\xi) = 0, \qquad (j=1,2,..,f), \tag{4}$$

où nous avons écrit simplement y et ξ au lieu de $y_1,..,y_f$ et de $\xi_1,..,\xi_f$. Supposons ici que nous puissions uniformément résoudre le système (4) dans un ensemble convenable par repport aux variables $y_1,..,y_f$ à condition que la quantité ε soit égale à zéro comme il suit :

$$y_k = \psi_k(\xi;\eta), \qquad (k=1,2,..,f). \tag{5}$$

Dans le cas où la quantité ε n'est plus nulle, nous supposons que le système (4) soit résoluble en séries infinies développées suivant les puissances croissantes de la petite quantité ε :

$$y_k = \tilde{\psi}_k(\xi;\eta;\varepsilon) = \psi_k(\xi;\eta) + \sum_1^\infty \frac{\varepsilon^n}{n!}\psi_k^{(n)}(\xi;\eta), \qquad (k=1,2,..,f). \tag{6}$$

Nous connaissons nombre de tentatives pour obtenir une formule utile qui nous permette de calculer les fonctions $\psi_k^{(n)}(\xi;\eta)$. Parmi elles, celle de Brown et Shook (1933) serait la première, mais elle n'aboutit à rien. Bien que Mr. Percus (1964) construise une telle formule, elle serait trop compliquée pour être appliquée aux problèmes pratiques. La formule présentée par MM. Feagin et Gottlieb (1970) est bien utilisable. La base sur laquelle ils s'appuient est le théorème de Leibniz et alors on doit obligatoirement commencer le calcul des fonctions $\psi_k^{(n)}(\xi;\eta)$ de l'ordre le plus bas vers l'ordre plus élevé. Nous présenterons ainsi une formule qui nous permet de calculer n'importe quel ordre des fonctions sans savoir d'autres expressions déjà obtenues :

$$\psi_k^{(n)}(\xi;\eta) = \left[\frac{\partial^n\tilde{\psi}_k}{\partial\varepsilon^n}\right]_{\varepsilon=0} = \tilde{D}_{k_2}^{n-1}\left[\frac{\partial\psi_k}{\partial\eta_{k_1}}\tilde{\chi}^{*n}_{k_1}\right], \qquad (k=1,2,..,f;\ n=1,2,..), \tag{7}$$

où l'on effectue les calculs comme suit :

$$\tilde{D}_{k_{m+1}}^{n-m}\left[\frac{\partial\psi_k^{(n;m-1)}}{\partial\eta_{k_m}}\tilde{\chi}^{*n-m-1}_{k_m}\right] = \tilde{D}_{k_{m+2}}^{n-m-1}\left[\frac{\partial\psi_k^{(n;m)}}{\partial\eta_{k_{m+1}}}\tilde{\chi}^{*n-m}_{k_{m+1}}\right] =$$

$$= \tilde{D}_{k_{m+2}}^{n-m-1}\left[\left\{\frac{\partial^2\psi_k^{(n;m-1)}}{\partial\eta_{k_{m+1}}\partial\eta_{k_m}}\chi^*_{k_m} + (n-m+1)\frac{\partial\psi_k^{(n;m-1)}}{\partial\eta_{k_m}}\frac{\partial\chi^*_{k_m}}{\partial\eta_{k_{m+1}}}\right\}\tilde{\chi}^{*n-m}_{k_{m+1}}\right]; \tag{8}$$

$$\tilde{D}^1_{k_n} \equiv D_{k_n},\ \tilde{D}^0_{k_{n+1}} \equiv 1;\ \tilde{\chi}^{*1}_{k_n} \equiv \chi^*_{k_n};\ \frac{\partial\psi_k^{(n;0)}}{\partial\eta_{k_1}} \equiv \frac{\partial\psi_k}{\partial\eta_{k_1}}; \qquad (m=1,2,..,n-1).$$

Nous avons introduit des indices muets dans ces expressions.

3. INTRODUCTION DES OPERATEURS

Lagrange a écrit l'expression (3) sous une forme fermée à l'aide d'un opérateur différentiel. Mais, il n'y avait toujours qu'une seule inconnue dans son cas. Nous remarquons ici que même le cas où figurent plusieurs variables dans les fonctions dont il s'agit, il y a des cas qui nous laissent la possibilité de réduire le problème de l'inversion à plusieurs variables au cas d'une seule en introduisant des opérateurs convenables. Nous pouvons y profiter de la formule classique de Lagrange sans qu'on l'étende aux cas spéciaux.

Dans le but de rendre plus clairement ce que nous énonçons ici, nous appliquerons notre résultat à la théorie du mouvement d'un satellite artificiel due à la méthode de von Zeipel-Brouwer (1959). Considérons une transformation canonique d'un système hamiltonien $F(x_1,x_2;y_1,y_2;\varepsilon)$ à un autre système hamiltonien $\Phi(\xi_1,\xi_2;-,\eta_2;\varepsilon)$ engendrée par une fonction génératrice :

$$\eta_j = \frac{\partial S}{\partial \xi_j} = y_j + \frac{\partial \varepsilon S_1}{\partial \xi_j} + \dots, \qquad (j=1,2), \tag{9}$$

$$x_j = \frac{\partial S}{\partial y_j} = \xi_j + \frac{\partial \varepsilon S_1}{\partial y_j} + \dots; \qquad (j=1,2), \tag{10}$$

$$S = S(y;\xi;\varepsilon) = y_1\xi_1 + y_2\xi_2 + \varepsilon S_1(y;\xi) + \varepsilon^2 S_2(y;\xi) + \dots. \tag{11}$$

Si l'on met dans les relations (9) :

$$Y_j(y;\xi) = y_j, \quad \varepsilon\chi_j(y;\xi) = \frac{\partial \varepsilon S_1}{\partial \xi_j}; \qquad (j=1,2), \tag{12}$$

$$\varepsilon^m S_m(y;\xi) \equiv 0, \qquad (2 \leqq m), \tag{13}$$

ce n'est pas autre chose qu'un système de fonctions implicites à plusieurs variables, identique avec le système (4). On peut donc résoudre immédiatement le système (9) par rapport aux variables y à l'aide de notre formule (7). En substituant ce résultat au deuxième membre du système (10), on obtiendra la transformation canonique sous une forme explicite. Si l'on modifie légèrement la formule (7), on peut également résoudre le système (9) dans le cas où les fonctions $\varepsilon^m S_m(y;\xi)$ ne sont plus nulles.

Puisque la fonction génératrice S ne contient pas la variable indépendante, nous aurons une équation qui exprime la conservation de la valeur des hamiltoniens :

$$F(x_1,x_2;y_1,y_2;\varepsilon)-\Phi(\xi_1,\xi_2;-,\eta_2;\varepsilon)=0. \quad (14)$$

Supposons maintenant que les relations suivantes soient remplies presque partout dans l'ensemble des variables :

$$\left|\frac{\partial\Phi}{\partial\eta_2}\right| \simeq \left|\varepsilon\frac{\partial F}{\partial x_2}\right| \simeq \left|\varepsilon^2\frac{\partial F}{\partial x_1}\right|, \quad (15)$$

et que les hamiltoniens y soient analytiques. Sous ces conditions, on peut développer le premier membre de l'équation (14) suivant les puissances de la quantité ε à proximité du point $(y_1,y_2;\xi_1,\xi_2)$:

$$\{F(\xi;y)-\Phi(\xi;y)\}+(F_{x_1}+F_{x_2}p-\Phi_{\eta_2}P^*)z+\frac{1}{2}F_{x_1x_1}z^2+F_{x_1x_2}zpz+\frac{1}{2}F_{x_2x_2}(pz)^2+ \quad (16)$$
$$+\ldots-\frac{1}{2}\Phi_{\eta_2\eta_2}(P^*z)^2-\ldots=0,$$

où nous avons introduit une abréviation et des opérateurs donnés comme il suit :

$$z=\frac{\partial\varepsilon S_1}{\partial y_1} \ ; \quad p=\frac{\partial}{\partial y_2}\int dy_1, \quad P^*=\frac{\partial}{\partial\xi_2}\int dy_1. \quad (17)$$

Cela montre que nous avons pu réduire le problème au cas d'une seule variable z et nous pourrons ainsi résoudre l'équation (16) par la formule (2) en y substituant les relations suivantes :

$$\eta : (1+{}^F x_2/{}^F x_1\ p-{}^\Phi \eta_2/{}^F x_1 P^*)^{-1}\{\Phi(\xi;y)-F(\xi;y)\}/{}^F x_1, \quad (18)$$

$$\varepsilon\chi(y) : (\text{comme ci-dessus})^{-1}\{-\frac{1}{2}{}^F x_1x_1 z^2-{}^F x_1x_2 zpz-\ldots\}/{}^F x_1. \quad (19)$$

On trouvera ailleurs les démonstrations générales pour les formules que nous venons de traiter.

ABSTRACT: The article studies the application of the Lagrange series inversion formula to perturbation theories. One of the main problems is to generalize the Lagrange formula to several variables. The article also shows the need for introducing some operators to simplify the notation. To terminate, an application to the Von Zeipel-Brouwer satellite theory is given.

REFERENCES

1. Lagrange,J.L.: 1869, Oeuvres III (Gauthier-Villars, Paris), p.5.
2. Dieudonné,J.: 1968, Calcul infinitésimal (Hermann, Paris), p.250.
3. Brown,E.W. et Shook,C.A.: 1964, Planetary Theory (Dover, New York), p.40.
4. Percus,J.K.: 1964, Communications on Pure and Applied Mathematics, 17, 137.
5. Feagin,T. et Gottlieb,R.G.: 1970, Celestial Mechanics, 3, 227.
6. Brouwer,D.: 1959, Astronomical Journal, 64, 378.

A PRECISE DETERMINATION OF SOME CRITICAL TERMS IN THE SOLAR SYSTEM

Jean Chapront and Rudolf Dvorak
Bureau des Longitudes, Paris, France and
Astronomisches Institut der Universität Graz, Austria

ABSTRACT

A new form to determine the contribution of some special small divisors in perturbation theory is presented in this paper. We can avoid to calculate all the $k\lambda+k'\lambda$ as it has to be done normally (λ and λ' designate the mean longitudes of the two regarded planets). For a chosen k and k' we calculate with a very high precision the contribution to the perturbation of the elements with the aid of the Hansen's coefficients.

INTRODUCTION

In determining the long periodic perturbations in the planetary motions we adapted a specific formula for the inverse distance and have chosen complex elements as has been described by R. Dvorak (1978). To calculate the long periods in the planetary motions we have taken only the mean values of the right members in the equations of Lagrange. We can also determine as another problem the single influence of critical terms on the motions. Although we can give solutions only to the first order of the masses it is a contribution of great interest, e.g. for the direct and indirect planetary pertubations on the Moon's motion. Another paper by J.Chapront and R.Dvorak (1978) is in preparation to explain the above mentioned subject in detail. Here we shall present only the method.

THE INVERSE DISTANCE AND THE HANSEN COEFFICIENTS

We used a set of complex elements

$$\begin{aligned} &a \text{ the semi major axes} \\ &\lambda \text{ the mean longitude} \\ &z=e.\exp j\tilde{\omega} \\ &\zeta=\sin(i/2).\exp j\Omega \end{aligned} \qquad (1)$$

R. L. Duncombe (ed.), Dynamics of the Solar System, 91–93.

(j stands for $(-1)^{1/2}$) and the development of $1/\Delta$ described by R. Dvorak (1977), where the short-periodic variables $(r/a)^n.\tau^m$ and $(a'/r')^{n'}\tau'^{m'}$ ($\tau = \exp j(v+\tilde{\omega})$) have been separated:

$$\Delta^{-s} = (\frac{a'}{r'})^{-s} (2-\delta) \text{ Re} \sum_{n_i}^{N_1} A_{n_i} (\frac{r}{a})^n \tau^m (\frac{a'}{r'})^{n'} \tau'^{m'} \quad (2)$$

The A_{n_i} (functions of the semi major axes a and a' and the inclinations i and i') are numbers which are given in tables calculated in advance for all couples of planets. The Hansen coeff icients are introduced to compute the values of $(r/a)^n \tau^m$ and $(a'/r')^{n'}\tau'^{m'}$ as it has been done first by V. Brumberg (1967):

$$\begin{aligned} (\frac{r}{a})^n \tau^m &= (\frac{r}{a})^n \exp jmv \, . \exp jm\tilde{\omega} = \\ &= \sum_k X_k^{n,m} \exp jk\lambda \, . \exp j(m-k)\tilde{\omega} \end{aligned} \quad (3)$$

As we are interested only in one specific critical term of the two regarded planets, we employ the Hansen coefficient $X_k^{n,m}$ for the k fixed in advance. The same mechanism of calculation has to be applied for the $(a'/r')^{n'}\tau'^{m'}$ to be able to seperate above all a small divisor of the form $k\lambda + k'\lambda'$ in the inverse distance $1/\Delta$.

THE LAGRANGE EQUATIONS

The separation of the inverse distance $1/\Delta$ in the Lagrange equations for the variables a, λ, z and ζ gives the following final form for an element σ:

$$\frac{d\sigma}{dt} = \sum_{n=1}^{N_\sigma} (B_n^\sigma \{a_n, b_n, c_n, d_n\} + C_n^\sigma \frac{1}{r'^3}) \quad (4)$$

where the parentheses stand for the multiplication

$$\{a_n, b_n, c_n, d_n\} = \frac{1}{\Delta^s} (\frac{r}{a})^{a_n} \tau^{b_n} (\frac{a'}{r'})^{c_n} \tau'^{d_n} \quad (5)$$

This multiplication is basically one of two power series. But in our formulae this is a simple modification of the powers of (r/a) and τ as well as for (a'/r') and τ' in (2). It should be mentioned that we have to build a product of a complex number x and the real part of another complex number y (the development (2) for the inverse distance); we have to respect:

$$x \, \mathrm{Re}(y) = \frac{1}{2}(xy + x\bar{y}) \tag{6}$$

($\bar{y}$ stands for the conjugate complex number).As a consequence we are led to a simple shifting of the powers in the calculation of the two quantities $(r/a)^n \tau^m$ and $(a'/r')^{n'} \tau'^{m'}$:

$$\{a_n, b_n, c_n, d_n\} = \left(\frac{r}{a}\right)^{n+a_n} \tau^{m+b_n} \left(\frac{a'}{r'}\right)^{n'+c_n} \tau'^{m'+d_n} + \left(\frac{r}{a}\right)^{n+a_n} \tau^{m-b_n} \left(\frac{a'}{r'}\right)^{n'+c_n} \tau'^{m'-d_n} \tag{7}$$

RESULTS

We have already results of the couple Venus-Earth for some multiples of the longitudes and these values show a very good agreement with a first order theory (J.Chapront et al, 1975; N.Abu-El-Ata and J.Chapront, 1974). Because of the complexity of the programs we need computation times of some twenty minutes on a IBM 360/44, when we want a precision of 10^{-10} for a single element. We hope to be able to speed up the process in stocking many of the computed quantities (even the Hansen coefficients) which seems to be absolutely necessary for the determination of the long periods.

REFERENCES

Abu-El-Ata,N. ,Chapront,J. 1974,Astron.Astrophys. 38,57

Brumberg,V. 1967,Bull.Inst.Theor.Astron.TII, 125

Chapront,J. ,Bretagnon,P. ,Mehl,M. 1975, Celestial Mechanics 11,379

Dvorak, R. 1978, Dynamics of Planets and Satellites and Theories of their Motion, ed.V.G.Szebehely (Dordrecht: Reidel) p.57

ASTRONOMICAL MEASUREMENTS AND COORDINATE CONDITIONS IN RELATIVISTIC CELESTIAL MECHANICS

V. A. Brumberg
Institute of Theoretical Astronomy
192187 Leningrad, U.S.S.R.

ABSTRACT

The differences between the Newtonian theory and the general relativity appear both in the "relativistic effects," such as the advance of perihelion, and in the deflection of light beams that modifies the astronomic observations and the radar reflection measurements.

Determination of dynamical effects from the equations of motion and calculation of ephemerides in terms of measurable quantities on the basis of the equations of light should be performed in one and the same coordinate system. The choice of coordinate system is arbitrary. For illustration we consider complanar circular motions of the Earth and one of the inner planets in the solar gravitational field described by the generalized three-parametric Schwarzschild metric. Specific values of the metric parameters characterize just as adopted gravitational theory, so also a definite coordinate system (for example, isotropic or "standard" coordinates). Coordinates of the planets and radii of the orbits are coordinate-dependent quantities and cannot be directly reconciled with measurable quantities such as the round-trip transit times of radar signals or the angular distance between the planet and the distant fixed source (quasar). These ephemeris data may be calculated in terms of the initial measured values independently of the employed coordinate system. Relativistic ephemeris corrections should be taken into account both in radar reflection measurements and astrometric observations.

R. L. Duncombe (ed.), Dynamics of the Solar System, 95.

PART III

EPHEMERIDES, EQUINOX AND OCCULTATIONS

THE EPHEMERIDES: PAST, PRESENT AND FUTURE

P. K. Seidelmann
U. S. Naval Observatory, Washington, D. C. 20390

Over the years there has been a continuing trend toward unification of both the annually printed volumes of ephemerides and the fundamental bases for the ephemerides. Thus, from many completely independent national and private publications, which were based on a multitude of theories, there has developed a continually improving agreement on accurate astronomical constants, planetary, lunar and satellite theories and cooperative methods of printing the annual ephemerides in different languages. This trend is continuing, currently, with the adoption of a new system of astronomical constants in 1976, consideration of revisions of nutation, the planned revision of the A.E. (American Ephemeris and Astronomical Ephemeris) for 1981, and the expected introduction in 1984 of new fundamental planetary and lunar ephemerides based on the new constants and on the FK5. Currently the differences between the printed ephemerides and observations sytematically exceed 2" for Mars at some times and 6" for Neptune at all times. It is anticipated that a new set of fundamental theories will be introduced which will be based on a consistent set of astronomical constants and in agreement with the available observational data to the printed accuracy.

INTRODUCTION

From earliest times man has been interested in the Sun, Moon, planets and stars and in determining their positions. Particularly, there has been an interest in being able to predict the positions of the solar system bodies and interesting phenomena, such as eclipses, planetary groupings and directions, and the times of rising and setting.

The results of these interests were numerous mathematical, or analog, approaches to the calculations, observations

R. L. Duncombe (ed.), Dynamics of the Solar System, 99–114.

of various types, some recorded but most of limited accuracy and not documented, and some attempts to predict the positions of the "wandering" objects.

Eventually national publications of the ephemerides appeared in the various countries in different years. The following are examples of such publications and their initial years of publication: from France the Connaissance des Temps in 1679, from England The Nautical Almanac and Astronomical Ephemeris in 1767, from Germany the Berliner Astronomisches Jahrbuch in 1776, from Spain the Efemerides Astronomicas in 1791, from the United States The American Ephemeris and Nautical Almanac in 1855, from the Union of Soviet Socialist Republics the Annuaire Astronomique in 1923, from Japan the Japanese Ephemeris in 1943, and from India the Indian Ephemeris and Nautical Almanac in 1958. Initially, these publications were based on various theories of the motions of the bodies, different astronomical constants, different geographical coordinate systems and different time scales.

INTERNATIONAL COOPERATION

Formal cooperation dated from the International Meridian Conference held in Washington, D. C. in October 1884, when the Greenwich meridian and the universal day were adopted. At the Conference Internationale des Etoiles Fondamentales in Paris in May 1896 the fundamental constants for nutation, aberration, solar parallax and lunisolar and planetary precession were adopted. Active cooperation between the preparers of the national ephemerides began in October 1911 with the Congres International des Ephemerides Astronomiques in Paris. The distribution of calculations between the principal ephemeris offices (France, Germany, Great Britain, Spain and the United States) was recommended and the constants for the flattening of the Earth and the semidiameter of the Sun at unit distance were adopted. In the United States the recommendation required official approval by an Act of Congress (37 Stat. L., 328, 342). In 1919 the International Astronomical Union (IAU) was founded, including Commission 4 for Ephemerides.

In the Draft Report for the Twelfth General Assembly, held in Hamburg in 1964, D. H. Sadler indicated that the functions of Commission 4 on Ephemerides are twofold: "firstly, to ensure that the published ephemerides fully meet the requirements of astronomers and other users; and secondly, to coordinate the work of the offices of the national ephemerides to ensure consistency, economy of effort, and efficiency." The history of cooperation in the IAU on ephemerides has been directed toward these functions, with the primary emphasis o the first function being on the constants used and the bases for the ephemerides. A brief review of the efforts in these two areas should be helpful.

The initial agreement concerning constants predated the IAU and has been mentioned previously. At a meeting on fundamental constants for astronomy that was held in Paris during March 1950 the definition of ephemeris time was recommended and the lunar ephemeris was brought into accordance with the solar ephemeris with respect to ephemeris time. These recommendations were adopted in 1952 by Commission 4 of the IAU at the Eighth General Assembly in Rome.

At the Twelfth General Assembly (Hamburg, 1964) it was reported that IAU Symposium No. 21 (Paris, May 1963) concluded that a change in the conventional IAU system of constants could no longer be avoided. The inconsistencies and inadequacies of the system of that time, the better values of some constants from recent determinations, and deficiencies revealed by discussions of high accuracy observations indicated the need for new constants. A list of constants proposed by the Working Group on the System of Astronomical Constants was adopted and recommended for use at the earliest practicable date in the national and international astronomical ephemerides. These constants were introduced in the ephemerides for 1968. It was also noted that the constants of precession and planetary masses had not been changed and that consideration should be given to their future improvement.

In August, 1970, IAU Colloquium No. 9 on the IAU System of Astronomical Constants was held in Heidelberg, and recommended the establishment of three Working Groups on Planetary Ephemerides, Precession, and Units and Time Scales. The recommendations were adopted and the Working Groups were established at the 1970 IAU General Assembly in Brighton, England. A Working Meeting on Constants and Ephemerides was held in October, 1974, in Washington, D. C. to draft a proposed report of the Working Groups. The chairmen of the Working Groups met in September, 1975, and June, 1976, in Herstmonceaux and Washington, respectively. The Report and Recommendations, known as the _Joint Report of the Working Groups of IAU Commission 4 on Precession, Planetary Ephemerides, Units and Time Scales_, were adopted by the IAU in August, 1976, at the meeting in Grenoble.

In 1976 a Working Group on Cartographic Coordinates and Rotational Elements of Planets and Satellites was established by IAU Commissions 4 and 16. In 1977 a Working Group on Nutation was established by Commission 4. These groups are expected to provide recommendations for consideration at the 1979 General Assembly in Montreal.

Concerning the second primary function of Commission 4 on ephemerides, specifically the coordination of the efforts of the offices, there has been a progression from the distribution of calculations among the various offices, which was out-dated by the advent of high speed computers, to the unification of printing and the exchange of reproduction proofs for printing by the various countries.

As early as 1932 at the Fourth General Assembly of the IAU in Cambridge, Mass., Dr. L. J. Comrie, Director of the British Nautical Almanac Office, suggested that duplicate printing in the national volumes of ephemerides be discussed, particularly with respect to the apparent places of stars. At that time Professor Herrero suggested that the ideal would be an international almanac. At the Fifth General Assembly (Paris, 1935) an agreement was reached for a single publication of the Apparent Places of Stars, initially to be printed in Great Britain and now printed in Germany.

At the Ninth General Assembly (Dublin, 1955) an International Fundamental Astronomical Ephemeris (IFAE) was discussed. The IFAE was to be a single publication, under the auspices of the IAU, containing the fundamental astronomical ephemerides to the fullest accuracy. The national ephemerides could then be much smaller and cater more directly to the practical astronomer. While there were many in favor of this proposal, there was the practical difficulty that the sales were not likely to cover the cost of printing, due to the anticipated, required, free distribution. Also, there was the difficulty for astronomers purchasing books published in other countries, the required cooperation of almost all national ephemerides and the loss of flexibility. Thus, full agreement and adoption was not possible. However, the use of reproduction proofs for the printing by various countries of material prepared by a single source was an attractive alternative. At the same time, it was announced that agreement had been reached for the unification of the *American Ephemeris and Nautical Almanac* and *The Nautical Almanac and Astronomical Ephemeris* beginning with the year 1960. It was hoped that other ephemeris offices would make use of the considerable saving of composition and proof reading by reproduction by photoligraphy.

At the Tenth General Assembly (Moscow, 1958) it was reported that the *Astronomisch - Geodatisches Jahrbuch*, introduced for the year 1949, had ceased publication after the 1957 edition and that the *Berliner Astronomisches Jahrbuch*, introduced for the year 1776, would cease publication after the edition for 1959. As a result of these savings of composition and printing costs the Astronomisches Rechen-Institut was able to take over composition and publication of the

Apparent Places of Fundamental Stars. Thus, while it was sad to see a publication of long standing cease, it indicated an increased international cooperation and permitted a beneficial transfer of functions.

G. M. Clemence and D. H. Sadler reported at the Eleventh General Assembly (Berkeley, 1961) that an important step towards unification of the National Ephemerides had taken place in the unification of the American and British volumes. They stated further that:

> "Many dearly-held, but essentially unimportant, standards and prejudices have had to be sacrificed on both sides; it is surprising how quickly these lose their former importance in the satisfaction of a comprehensive agreement. That same co-operation, goodwill and confidence exists between all the national ephemeris offices, and, although differences of language will introduce some further difficulties, there is no obstacle to complete unification that will not be overcome in course of time."

Between 1962 and 1964 the Japanese and Russian ephemerides began using the advanced proofs from the Astronomical Ephemeris.

In October 1974 plans were begun to revise the organization, content and basis for the American Ephemeris and Nautical Almanac/Astronomical Ephemeris. It was decided that a single unified printing in English will be made in the United States and that it will be available from both Her Majesty's Stationery Office in England and from the Superintendent of Documents in the United States. Since the United States legal code requires the title The American Ephemeris and Nautical Almanac, a bill was introduced into Congress to modify the legal code to permit a change in the name of the publication. Also, it was decided that the organization and content would be changed for the 1981 edition. A list of the principal modifications follows:

1. Replace the hourly apparent lunar ephemeris by daily short power series which permit the direct determination of the lunar position for any time.
2. Eliminate all 1st differences.
3. Eliminate Independent Day Numbers.
4. Eliminate fixed tables for unit conversions.
5. Give longitudes and latitudes of the Moon and planets to $0\overset{\circ}{.}01$ accuracy only.
6. Organize the volume into sections which have their own pagination.

7. Provide times of sunrise, sunset, moonrise, and moonset for southern latitudes.

8. Provide times of civil twilight in addition to astronomical twilight.

9. Expand the list of occultations to include lunar occultations of radio sources and planetary occultations of stars.

10. Include transformation matrices for reduction of apparent places.

11. Include an ephemeris of the barycenter of the solar system.

12. Include a BIH polar motion table.

13. Include physical ephemerides for all planets, commensurate with current knowledge.

14. Give satellite ephemerides to observing accuracy for all satellites, generally for the entire year.

15. Provide selected minor planet ephemerides to 1" accuracy.

16. Expand the star list to about 1600 stars.

17. Include standard lists of variable stars, radio sources, pulsars and x-ray sources.

18. Give locations of observatories to reduced precision in the annual Observatories List, and periodically publish for each observatory a list of accurate astronomical or geodetical locations of each instrument.

19. Introduce a rewritten explanation with a glossary of terms used in the volume.

20. Introduce tables of new values of astronomical constants as appropriate.

21. Introduce chapters for a new version of the Explanatory Supplement as supplements to the A.E. prior to the publication of the new volume.

Additionally, it is planned that the new constants, the reference system of the FK5, and new fundamental ephemerides based on the new constants will be introduced in the 1984 edition.

FUTURE FORM OF EPHEMERIDES

With the widespread use of computers, which began in the 1950's, ephemerides in machine readable form were required. Initially a cooperative arrangement was established so that duplicate copies of machine readable data were retained at and available from the Royal Greenwich Observatory, the Astronomisches Rechen-Institut and the U. S. Naval Observatory (USNO). As more data became available from numerous other sources, IAU Commission 4 established the International Information Bureau on Astronomical Ephemerides at the Bureau des Longitudes in Paris, France, to provide information on the availability of machine readable data.

The availability and common use of electronic calculators and mini-computers, particularly those that are programmable, has introduced a desire for ephemerides and astronomical data in the form of equations or polynomials instead of the tabular form. This interest stimulated the experimental publication of the *Almanac for Computers* (Kaplan et al 1976, Doggett et al 1977). This almanac contains Chebyshev polynomials, which can be truncated to any desired accuracy, for the apparent right ascension, declination, true distance, semi-diameter and ephemeris transit of the Sun, Moon and planets. These data are given to the accuracy of the A.E., but another section contains navigational data to the accuracy of 0!1. The *Almanac for Computers* also contains the mean positions of navigational stars and expressions for determining their apparent positions, Fourier series expressions for the planets, which provide limited accuracy for a period of several centuries, and expressions for various astronomical and navigational computations. Also, a useful collection of formulae for computing astronomical data with hand-held calculators has been prepared by B. D. Yallop (1978). This Technical Note includes methods for calculating the position of the Sun and Moon, and solving the navigational triangle.

To satisfy the requirements for ephemerides to limited accuracy for planning purposes prior to the annual publication of the national ephemerides, a new volume, *Planetary and Lunar Coordinates*, is being introduced. The initial volume for 1980-1984 continues the series of publications *Planetary Coordinates* but is broader in scope and includes heliocentric and geocentric planetary coordinates and lunar coordinates.

The annual publication, *Astronomical Phenomena*, is being revised, expanded and produced in cooperation with Her Majesty's Nautical Almanac Office. Among the material to be added are a diary of geocentric solar system phenomena, increased information on the visibility of the planets, a low-precision ephemeris of the Sun, and tables for computing times of moonrise and moonset for southern latitudes. The extensive tables of Polaris will be deleted; in their place will be a single table giving the daily position of Polaris and σ Octantis.

For archival purposes, microfiche can be stored in a minimum of space and readily accessed for astronomical data in the past. However, in most cases for current use, the inconvenience, equipment requirements and illumination conflicts make microfiche inferior to the published volumes.

For the forseeable future, there appears to be a continuing requirement for the published national ephemerides in addition to the machine readable and formulae or polynomial expression versions of astronomical data.

PRESENT FUNDAMENTAL EPHEMERIDES

Presently, the fundamental ephemerides forming the bases for the annually published astronomical data are based on the following sources: the ephemeris of the Sun is based on Newcomb's Tables of the Sun (1895); the ephemeris of the Moon is based on Brown's theory as given in the Improved Lunar Ephemeris (1954); the ephemerides of the inner planets are derived from Newcomb's tables of these planets (1895-8) with Ross's corrections applied for Mars (1917); and the outer planet ephemerides are based on a numerical integration by Eckert, Brouwer, and Clemence (1951), with inner planet perturbations by Clemence (1954). These ephemerides are, therefore, based on theories from thirty to eighty years old and on a heterogeneous combination of astronomical constants adopted over the same period of time. In addition, the theory of relativity and the variability of the rotation of the Earth have been discovered during this period.

New observational techniques have provided much more accurate observational data and a means of measuring the distance to some of the solar system objects. Thus, radar and laser ranging techniques have produced a requirement for more accurate ephemerides in order to make observations.

The optical observational data available also indicate discrepancies between the theories and the observed positions of the objects and these discrepancies are summarized in Table 1. Table 2 is a summary of some known characteristics of the differences between the published ephemerides and the observations. These variations are the symptoms of some known, and probably some unknown, theoretical problems. It is recognized that the basic constants used for the ephemerides do not represent the best current knowledge, particularly for planetary masses and for the precession constant. The Ephemeris Time Scale prior to the 1976 redefinition does not differentiate between coordinate and proper time and, while defined based on the Sun, it is determined from observations of the Moon. The location and motion of the equinox utilized for the different bodies varies and it is determined basically only from observations. The theory of nutation is recognized as having inadequacies which can cause periodic errors in the observations.

TABLE 1.

Current Average Observed Minus Ephemeris Differences

	Right Ascension s	Declination "
Mercury	+0.14	-0.3
Venus	+0.10	-0.1
Mars	-0.12	0.0
Jupiter	-0.03	+0.3
Saturn	0	-0.3
Uranus	+0.03	-0.2
Neptune	-0.48	+0.8
Sun	+0.10	+0.1

Note: These values are based on observations obtained with the Six-inch Transit Circle of the U. S. Naval Observatory during the 1975-6 period.

TABLE 2

Known Differences between Published Ephemerides and Observations

Planet	Discrepancy and References
Earth	The longitude may be in error by 0".5
Mars	Right ascension may be in error by as much as $0^s.2$ at opposition
Jupiter	Periodic errors reaching about 0".3 in longitude and 0".5 in latitude (Klepczynski et al, 1970)
Saturn	Periodic errors reaching about 0".5 in longitude and 0".7 in latitude (Klepczynski et al, 1970)
Uranus	-0".3/century secular error in latitude (Duncombe & Seidelmann, 1978)
Neptune	Secular runoff in longitude, -6".0 in 1975; periodic error in latitude (Duncombe & Seidelmann, 1978)

THE NEW FUNDAMENTAL EPHEMERIDES

Recognizing the current deficiencies in the published ephemerides, an effort was initiated to correct the fundamental ephemerides for the annual national publications. The first step in this effort was the determination and adoption of a new set of astronomical constants, representing the best values currently available; a revision of the ephemeris time scale, which acknowledges the availability of atomic clocks and the existence of relativistic theories; and a new definition of the standard epoch, equinox and fundamental reference frame, which is necessary for the preparation of the FK5. This step was accomplished at the Sixteenth General Assembly of the IAU (Grenoble, 1976; Duncombe et al, 1977).

It was recognized at that time that several other matters require correction and should be changed, if possible, with the introduction of the new system of constants, new ephemerides and the FK5. These matters include the theory of nutation, which is currently based on a rigid Earth model and has errors with respect to observations which can accumulate to $0^{s}.007$ in right ascension. For this purpose a Working Group on Nutation was established by Commission 4 and it is expected to submit a report and recommendations at the 1979 General Assembly of the IAU. Additionally, Commissions 4 and 16 have established a Working Group on Cartographic Coordinates and Rotational Elements of Planets and Satellites to recommend the definitions and values of the physical ephemerides for the Sun, planets and satellites. Also, it will be necessary for the IAU in 1979 to adopt the location of the equinox and equator for the FK5.

Given the new constants and definitions, new fundamental ephemerides can be prepared. Bearing in mind the widespread interest and participation in this work by such institutions as the Astronomisches Rechen-Institut, Bureau des Longitudes, Her Majesty's Nautical Almanac Office, Institute of Theoretical Astronomy, Japanese Hydrographic Office and the University of Tokyo, Jet Propulsion Laboratory, U. S. Naval Observatory, and others, several guiding principles have been established. First, we do not wish to have published ephemerides based on an isolated or single computation. Rather, we are seeking to have available several sources of the ephemerides, calculated by different methods and techniques so that intercomparisons can be performed to ensure the accuracy, rather than the internal precision, of the published ephemerides. While the different ephemerides may not agree completely, we should understand the cause of the difference and specify the assumptions and theory underlying the published values. Second, all available observational data with appropriate weighting should be utilized, directly or indirectly, in the preparation of the ephemerides. Third, the bases, constants, theories and reference frame for all the ephemerides should be consistent and specified. Fourth, in addition to the ephemerides that are published to the required accuracy, machine readable ephemerides of greater precision, covering extended periods of time, both in the past and in the future, should be available.

In pursuit of these new ephemerides and of the specified objectives, the following activities are in progress at the USNO:

(1) The occultation observations from 1830 to 1955 have been collected and placed in machine readable form. A corrected star catalog of the Zodiacal Zone has been prepared for the analysis of the occultation data. The analysis will determine the values of delta T, i.e. the difference between Universal Time and Ephemeris Time over the period 1830 to 1955. Her Majesty's Nautical Almanac Office is cooperating in this project and has analyzed the same occultation data, employing the N30 star catalog, and determined delta T values over this period. The two partially independent analyses will be compared to provide a best set of values for delta T over the period.

(2) All planetary observations are being collected and systematically reduced to the FK4 coordinate system. The assistance of the observatories still making planetary observations has been sought in order to distribute the effort required and to facilitate the use of knowledge that is only available at the observing site. Emphasis is being placed on collecting the observations of the inner planets, at least back to 1900, and the outer planets back to 1830. Currently, all observations of Mars, Neptune, and Pluto are available based on investigations by Laubscher (1970), Jackson (1974), and Cohen, Hubbard and Oesterwinter (1967), respectively. All observations of Uranus, except from Greenwich have been collected and reduced to the FK4 by Jackson (USNO). Many of the observations of Jupiter and Saturn were collected at the same time as the Uranus observations. The observations of the Sun, Mercury, and Venus are being collected.

(3) A program for integrating in terms of elliptical elements has been developed by George Kaplan (USNO) and, by carrying increased precision for the mean anomaly and argument of perihelion, the precision can be increased on the computer such that it is adequate for all solar system objects for the time period of interest.

(4) A program of numerical integration of the Moon and planets in heliocentric coordinates has been pursued in collaboration with Dr. Claus Oesterwinter at the Naval Surface Weapons Center in Dahlgren, Virginia.

(5) The most recent Jet Propulsion Laboratory Development Ephemeris covering an extended period of time is available for comparison and will be utilized as one of the possible sources of the ephemerides.

(6) A program to determine highly precise ephemerides over a fixed period of time by means of Chebyshev polynomials has been developed by LeRoy Doggett (USNO). This independent approach to calculating ephemerides has already proven to be a very effective means of detecting signatures of imprecision in the other methods of calculating ephemerides. The differences in the methods are such that truncation, round-off, interpolation, insufficient step size and other discrepancies can be detected at the precision of several parts in 10^{13}, although the causes are not always obvious.

(7) An analytical manipulation system has been developed by K. F. Pulkkinen and T. C. Van Flandern (USNO) which is capable of performing algebraic processes on terms involving numerous algebraic factors and trigonometric terms with many angular arguments. This manipulation system has been used to calculate expressions for positions of the planets to limited precision (1' of arc), for a period of several centuries, and it is being used to calculate analytical partial derivatives for determining the corrections to the initial conditions.

(8) Previous investigations concerning the ephemerides of the outer planets have indicated the presence of unexplained systematic trends in the observational residuals (Duncombe and Seidelmann, 1978). Since this could be due to catalog errors in the previous reduction of the observational data, or to an unresolved problem in the fitting of the ephemerides to the observations, an early comparison of the observations of Uranus and Neptune was initiated. Also, T. Corbin (1977) investigated a daily correction approach for the Uranus and Neptune observations from the Paris Observatory in order to see if a significant improvement in the observational residuals could be achieved by such a painstaking and detailed reduction procedure.

(9) Numerical general theories are being attempted by Seidelmann (1977) using the methods of Musen and Carpenter (1963) and Carpenter (1963, 1965, 1966, 1966). By a combination of iterative and ordered calculations, the general theories for the inner planets are currently being developed, but it is likely that these theories will not be available in time for inclusion in the new ephemerides.

(10) The algebraic manipulator mentioned previously will eventually be utilized in an attempt to develop im-

proved planetary theories by means of the Airy method, whereby, based on the specified equations of motion, corrections are determined to a given, or approximate, theory.

(11) It is also planned that the algebraic manipulator will be used to develop analytical expressions for the planetary part of the lunar theory.

CONCLUSION

The progress in international cooperation, as reviewed here, continues to proceed toward minimizing the duplication of effort in computations, composition and printing of astronomical data. Publication of a single international astronomical ephemeris might be the eventual goal of this cooperation.

Currently, preparations are being made for a revised version of the Astronomical Ephemeris/American Ephemeris and Nautical Almanac, which involves cooperation among Germany, France, Great Britain and the United States and which will be printed only in the United States.

The various steps necessary for the computation and production of new fundamental ephemerides are in progress. New astronomical constants have been adopted and other changes will be proposed to the IAU in 1979. It is our aim and hope to introduce the FK5 and the new fundamental ephemerides, based on new and improved theories, into the national ephemerides of 1984.

REFERENCES

Carpenter, L., 1963, NASA Technical Note D-1898.
Carpenter, L., 1965, NASA Technical Note D-2852.
Carpenter, L., 1966, NASA Technical Note D-3078.
Carpenter, L., 1966, NASA Technical Note D-3168.
Clemence, G. M., 1954, Astron. Pap. Amer. Ephemeris XIII, part V.
Cohen, C. J., Hubbard, E. C. and Oesterwinter, C., 1967 Astron. J. 72, 973.
Corbin, T., 1977, Dissertation, University of Virginia.
Doggett, L. E., Kaplan, G. H., Seidelmann, P. K., 1977, Almanac for Computers, 1978. Washington, D. C.
Duncombe, R. L., Fricke, W., Seidelmann, P. K. and Wilkins, G. A., 1977, Proceedings of the Sixteenth General Assembly, Grenoble 1976, p. 56.
Duncombe, R. L. and Seidelmann, P. K., 1978, IAU Colloquium No. 41, Reidel, in press.

Eckert, W. J., Brouwer, D. and Clemence, G. M., 1951, Astron. Pap. Amer. Ephemeris XII.
Jackson, E. S., 1974, Astron. Pap. Amer. Ephemeris XXII, part II.
Kaplan, G. H., Doggett, L. E. and Seidelmann, P. K., 1976, U. S. Naval Observatory Circular 155, Almanac for Computers, 1977, Washington, D. C.
Klepczynski, W. J., Seidelmann, P. K. and Duncombe, R. L., 1970, Astron. J. 75, 739.
Laubscher, R. E., 1970, Dissertation, Yale University.
Musen, P. and Carpenter, L., 1963, J. Geophys. Res. 68, 2727.
Newcomb, S., 1895, Astron. Pap. Amer. Ephemeris VI, part I.
Newcomb, S., 1895-8, Astron. Pap. Amer. Ephemeris VI, part II, III, IV.
Ross, F. E., 1917, Astron. Pap. Amer. Ephemeris IX, part II.
Seidelmann, P. K., 1977, Celestial Mechanics, in press.
Yallop, B. D., 1978, NAO Technical Note No. 46, Royal Greenwich Observatory.
____________ 1954, Improved Lunar Ephemeris 1952-1959, U. S. Government Printing Office, Washington, D. C.
____________ 1961, Explanatory Supplement, Her Majesty's Stationery Office, London.
____________ 1978, Planetary and Lunar Coordinates 1980-1984, Draft Edition, Royal Greenwich Observatory.

Protocols of the Proceedings of the International Conference held at Washington for the purpose of fixing a Prime Meridian and a Universal Day. October 1884. Washington, D. C., 1884.

Proces-Verbaux of the Conference Internationale des Etoiles Fondamentales de 1896. Paris, Bureau des Longitudes, 1896.

Congres Internationale des Ephemerides Astronomiques tenu a l'Observatoire de Paris du 23 au 26 Octobre 1911. Paris, Bureau des Longitudes, 1912. A full account, with English translation of the resolutions, is given in M.N.R.A.S., 72, 342-345, 1912.

Colloque International sur les Constantes Fondamentales de l'Astronomie. Observatoire de Paris, 27 Mars--ier Avril 1950. Colloques Internationaux du Centre National de la Recherche Scientifique, 25, 1-131, Paris, 1950. The proceedings and recommendations are also available in Bull. Astr., 15, parts 3-4, 163-292, 1950.

Proceedings of IAU Symposium No. 21 (Paris, 1963) on "The System of Astronomical Constants", Bull. Astron. 25, 1-324, 1965.

Proceedings of IAU Colloquium No. 9 (Heidelburg, 1970) on "The IAU System of Astronomical Constants", Celestial Mechanics, 4, no. 2, 128-280, 1971.

The reports and recommendations of Commission 4 of the International Astronomical Union have been published as follows:

Trans. I.A.U.,	Assembly	
I, 159, 207; 1923.	Rome	1922
II, 18-19, 178, 229; 1926.	Cambridge, England	1925
III, 18, 224, 300; 1929.	Leiden	1928
IV, 20, 222, 282; 1933.	Cambridge, Mass.	1932
V, 29-33, 281-288, 369-371;1936.	Paris	1935
VI, 20-25, 336, 355-363; 1939.	Stockholm	1938
VII, 61, 75-83; 1950.	Zurich	1948
VIII, 66-68, 80-102; 1954.	Rome	1952
IX, 80-91; 1957.	Dublin	1955
X, 72, 85-99; 1960.	Moscow	1958
XI, A, 1-8; 1962. B, 164-167, 441-462; 1962.	Berkeley	1961
XII, A, 1-10; 1965. B, 101-105, 593-625; 1966.	Hamburg	1964
XIII, A, 1-9; 1967. B, 47-53, 178-182; 1968.	Prague	1967
XIV, A, 1-9; 1970. B, 79-85, 198-199; 1971.	Brighton	1970
XV, A, 1-10; 1973. B, 69-72; 1974.	Sydney	1973
XVI, A1, 1-7; 1976. B, 31, 49-67, 1977.	Grenoble	1976

DISCUSSION

Aoki: What kind of relativistic theory are you using to make up the ephemeris?
Duncombe: Einstein.

Aoki: Which method do you want to use for ephemeris calculations, the analytical method or the numerical integral?
Van Flandern: We hope that the results by both methods will agree with each other.

Aoki: If not, which do you prefer?
Van Flandern: We will keep working until they agree.

Everhart: Has some thought been given to publishing velocities as well as positions in the ephemeris? With both positions and velocities available, then an investigator may use his own numerical integrator for values at desired times. The tabular interval for such position-velocity data could be large, say 200 days (1950.0 frame).

Van Flandern: The printed ephemerides are in polar coordinates, and not intended to serve that purpose. The machine-readable ephemerides provide such position and velocity data. Consideration will be given to including such a table in the printed volumes as well. It would be based on the equinox and epoch J2000.

PRESENT STATUS OF THE ASTRONOMICAL EPHEMERIS

Sh. Aoki
Tokyo Astronomical Observatory
Mitaka, Tokyo, 181 Japan

A. M. Sinzi
Hydrographic Department of Japan
Tsukiji-5, Chuo-ku, Tokyo, 104 Japan

The IAU (1976) System of Astronomical Constants and a new set of fundamental theories will expectedly be introduced into the international and national ephemerides for the volumes of 1984 onwards. In order to avoid any confusion in the future, it is necessary to manifest the character of the data published in the current volumes of the Astronomical Ephemeris = American Ephemeris, both abbreviated as A.E. With this end in view, computer programs for the calculations of the ephemerides of the Sun and inner planets based on the Newcomb's (1895, 1898) Tables have been prepared at Tokyo Astronomical Observatory (TAO) and Hydrographic Department of Japan (JHD) independently of each other using different computers and hence different types of FORTRAN. JHD has further prepared the programs for the Moon's ephemerides based on the Brown-Eckert theory and has reproduced the Eckert, Brouwer and Clemence's numerical integrations of the outer planets. Fundamental ephemerides thus calculated are compared with those data tabulated in the A.E. for the year of 1975, as an example, in the present paper.

In the Introduction (pp.9 - 20) of the Newcomb's (1895) Tables of the Sun (hereafter called Tables), he gives the basic data and formulas (hereafter called basis) which were used to tabulate the individual values in the main body (called tabulation) of the Tables. However, the actual data in the tabulation are not necessarily consistent with the basis. In addition to the effects due to the rounding-off and the omission of small terms in the tabulation, two causes have been found to explain the discrepancy between the basis and the tabulation: (i) Clemence (1943) pointed out that, in the calculation of the perturbations in longitude by Venus, Newcomb had not used some terms listed in the basis but had purposely adopted other terms which are given in the column "Tables VIII and XII" in Table I of the Clemence's paper, although Clemence could not find the ground of these terms. (ii) Kinoshita et al. (1974) found that the tabulation of the perturbations in longitude by Jupiter might have been calculated as if each numerical coefficient for $j = 5$ were situated by one rank upper in Table D of the basis. The error in the tabulation thus amounts to $\pm 0''.07$ by (i) and $\pm 0''.03$ by (ii). The discrepancy in the perturbations in longitude by

R. L. Duncombe (ed.), Dynamics of the Solar System, 115–120.

Mars between the basis and the tabulation amounts to $\pm 0''.04$ and seems somewhat systematic but its cause has not been found.

In the basis (p. 10) series expansions of the equation of the center and the logarithm of the radius vector are presented and they are almost consistent with the respective tabulations. In this connection, it is remarked that, when we simply apply the Kepler's equation, the coefficients of the series of Log R are practically consistent with the value of the eccentricity e in the basis (p. 9), while in the series of the equation of the center, there appears a discrepancy by $(0''.004 + 0''.007\,T)$ $\sin 3g$ which may yield a discrepancy by $\pm 0''.010$ in λ after 1980.

In principle, both programs prepared at the JHD and the TAO are strictly based on the basis with following modifications for longitude: (a) "Tables VIII and XII" of the Clemence's (1943) paper are adopted for the perturbations by Venus, and (b) the equation of the center and Log R are calculated by solving the Kepler's equation with the value of e in the basis. Although there are several differences between the programs of the JHD and those of the TAO, the discrepancies between their calculated data are less than $\pm 0''.001$ in λ and β, and $\pm 0.000\,000\,01$ A.U. in the radius vector, namely one-tenth in the units of the respective printed last figures in the A.E.

The programs for Mercury, Venus and Mars have been prepared in the same manner as those for the Sun, being based on the respective bases of the Newcomb's (1895, 1898) Tables with Ross' (1917) correction for Mars. JHD-data and TAO-data agree well with each other within $\pm 0''.001$ in λ and β, and $\pm 0.000\,000\,01$ A.U. in the radius vectors.

We may thus esteem that both JHD- and TAO-programs for the Sun and the inner planets are strictly consistent with Newcomb's theories. Thereupon the JHD-data are compared with the A.E.-data. Discrepancies AE - JHD in the ecliptic coordinates for the Sun and Mercury are illustrated in Figure 1 as an example. Each pair of horizontal broken-lines in the figure indicates the one-half of the printed last decimal in the A.E. Since the JHD-data are expressed down to the lower decimals than the A.E.-data, individual dots inside the broken-line pair indicate that those A.E.-data agree with the JHD-data exactly at the printed last decimals. For λ and β of the Sun, only the dots on the respective abscissae correspond to the exact agreement at the printed last decimals. Frequency distributions of the discrepancies AE - JHD are presented in Table 1. Its horizontal argument is the discrepancy at the printed last decimals which are listed in the second column.

Most of the A.E.-data of the inner planets seem passable, excepting for the systematic bias in the radius vector of Mercury. We should not underestimate the discrepancies in the Sun's coordinates, any error of which may affect significantly the calculations of the geocentric coordinates of the planets. An example of the transfer of errors is exhibited in Figure 2. Its abscissa denotes the actual discrepancy Δr_M in the geocentric distance r_M of Mercury, and the ordinate denotes the

Table 1. Frequency distribution of AE - JHD

		last decimal in AE	(Units : last decimals in AE)										
			≤-5	-4	-3	-2	1	0	+1	+2	+3	+4	+5
Sun	λ	$0''.01$	24*	40	55	49	56	24	40	42	21	10	4
	β	$0''.01$			18	34	76	73	89	39	26	10	
	R	E - 7**				13	89	114	119	23	7		
	X***	E - 7				2	49	110	131	58	14		
	Y	E - 7			7	75	89	108	66	18	1		
	Z	E - 7				8	89	191	70	6			
Mercury	λ	$0''.1$					19	312	35				
	β	$0''.1$					8	348	10				
	R	E - 7		1	16	107	161	71	10				
Venus	λ	$0''.1$					10	156	17				
	β	$0''.1$					12	162	9				
	R	E - 7					19	139	25				
Mars	λ	$0''.1$					2	79	11				
	β	$0''.1$					7	81	4				
	R	E - 7					9	52	27	4			
Sun	α	$0^{s}.01$					32	315	18				
	δ	$0''.1$					29	299	37				
Mercury	α	$0^{s}.01$					60	278	27				
	δ	$0''.1$					25	298	41	1			
	r	E - 7		1	22	62	108	96	53	20	3		
Venus	α	$0^{s}.01$					32	284	47	2			
	δ	$0''.1$					36	275	53	1			
	r	E - 7		3	8	15	42	89	93	53	41	19	2
Mars	α	$0^{s}.01$					29	310	26				
	δ	$0''.1$					26	314	25				
	r	E - 7			5	30	64	77	85	63	31	10	
Jupiter	α	$0^{s}.001$					47	240	78	1			
	δ	$0''.01$					55	230	81				
	r	E - 7		3	18	61	87	62	79	46	11	1	
Saturn	α	$0^{s}.001$					42	255	69				
	δ	$0''.01$					44	275	47				
	r	E - 7		3	30	87	116	96	28	5	1		
Uranus	α	$0^{s}.001$					31	287	48				
	δ	$0''.01$					110	239	17				
	r	E - 6					206	160					
Neptune	α	$0^{s}.001$					29	277	59				
	δ	$0''.01$					104	234	17				
	r	E - 6					198	167					
Pluto	α	$0^{s}.001$						222	144				
	δ	$0''.01$					131	224	11				
	r	E - 6					48	316	2				

* -7:5, -6:5, -5:14. ** E - 7 stands for 10^{-7} A.U.
*** Values for the nearest beginning of year. Those for 1950.0 exhibit similar frequency distributions.

effect δr_M due to the discrepancies in the Sun's coordinates and the Mercury's heliocentric coordinates to the calculation of its geocentric distance, i.e.

$$\Delta r_M = r_M(AE) - r_M(JHD),$$

$$\delta r_M = a\delta\lambda_S + b\delta\beta_S + c\delta R_S + d\delta\lambda_M + e\delta\beta_M + f\delta R_M,$$

where $a = \partial r_M / \partial \lambda_S, \quad \delta\lambda_S = \lambda_S(AE) - \lambda_S(JHD),$ etc.

We can find a clear correlation between Δr_M and δr_M and a systematic bias.

Numerical integrations of the heliocentric rectangular coordinates of the outer planets have yielded satisfactorily identical data with those by Eckert, Brouwer and Clemence (1951). Only less than 0.5% of data differ from the Eckert et al's data by ± 1 at the printed last

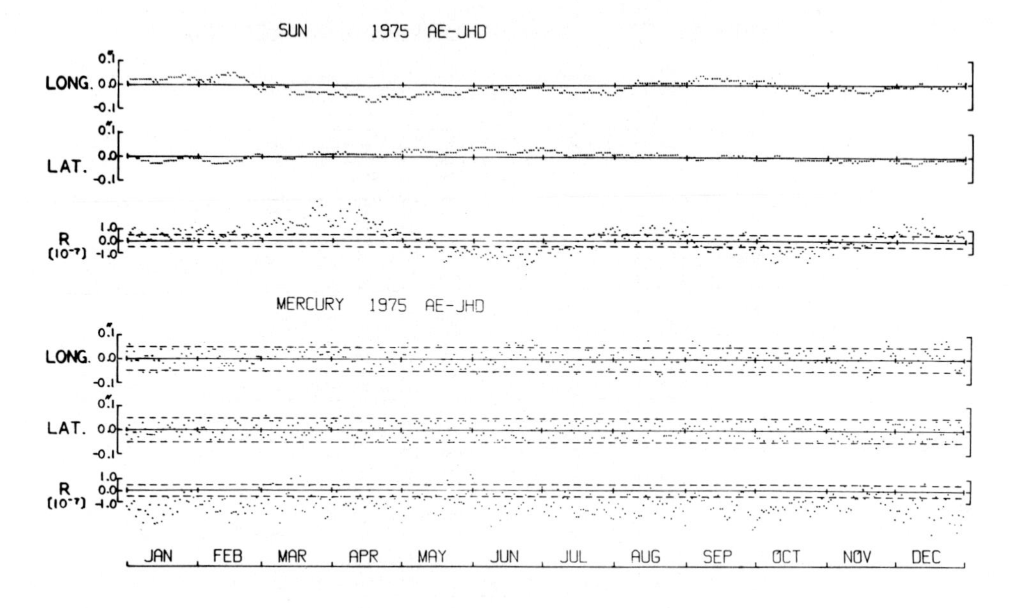

Figure 1. (above)

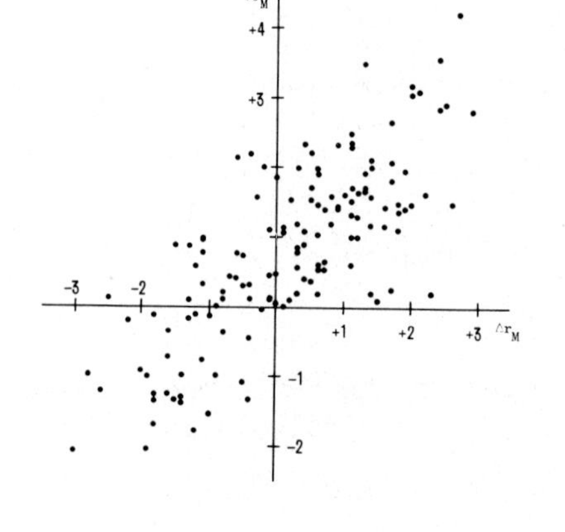

Figure 2. (left)

Figure 3. (below)

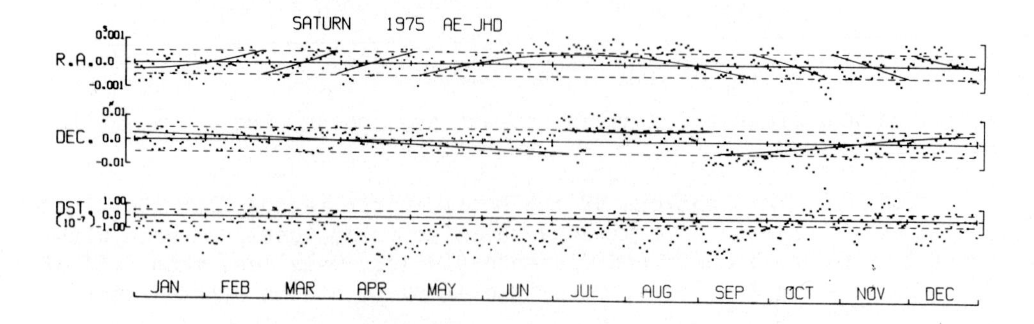

decimal, 0.000 000 001 A.U. Since the heliocentric spherical coordinates in the A.E. do not include the Clemence's (1954) correction due to the perturbations by the inner planets, the comparison with the JHD-data has not been made for these quantities. Discrepancies in the geocentric ephemerides are rather significant. Examples are shown in Figure 3. The discrepancies in α and δ may be mostly caused by (i) discrepancies in the Sun's ephemerides, and (ii) inappropriate procedure of correction in the A.E. due to the change in the value of aberration constant. The latter effect is shown by discontinuous lines in Figure 3. Values of AE - JHD in the geocentric distances of Uranus and Neptune both concentrate strongly in the range between 0 and -1 in the unit of the printed last decimal, suggesting that the A.E. tabulates these quantities by omitting the figures lower than the printed last decimal.

(Note) Numerical integrations for the principal minor planets by Duncombe (1969) have not been traced. Their reduction to the geocentric ephemerides yields higher values in declination than the A.E. by 0.″1 constantly when the declination is negative. Erratum for the data from 1972 to 1980 is thus being noticed in the every volume of the A.E. Until the volume for 1971, the A.E. tabulated the geocentric ephemerides deduced from the Herget's (1962) integrations and agrees well with the JHD-calculations.

Programs of j = 2 ephemerides of the Moon have been prepared by Inoue (1977) following the theory by Eckert, Walker and Eckert (1966). The agreement between the USNAO-data (Van Flandern, 1976) and the JHD-data is so satisfactory that their discrepancies are $\pm$0.″0002 in λ and β, and $\pm$0.″000 002 in π to be compared with their respective last decimals in the A.E. (0.″01 in λ and β, and 0.″0001 in π). Among all half-day ephemerides between 1973 and 1980, the discrepancies AE - JHD by $\pm$1 at the printed last decimals of the A.E. occur at 8% of data for λ, 21% for β and 5% for π. These discrepancies may be mostly caused by the defects in the A.E.-programs, for example, the incomplete treatment of the corrections for the change in the value of the Earth's flattening.

Programs for nutation are based on the Woolard's (1953) formulas. The JHD-data have been compared with the USNAO-data provided by Van Flandern (1976). Discrepancies between them are less than $\pm$0.″000 003 in longitude and $\pm$0.″000 001 in obliquity. We may thus esteem that the JHD- and the USNAO-data are both completely strict to the Woolard's formulas. On the other hand, about 30% of the A.E.-data are differ from the JHD-data by $\pm$1 at the printed last decimals (0.″001) both in longitude and obliquity. These discrepancies may be attributed to the simplification adopted in the A.E.-calculation schema as explained by Wilkins (1954).

The discrepancies AE - JHD surveyed above might be regarded as passable in the former days. However, with the general availability of computers, we can now calculate the ephemerides as precisely as we wish, whenever the basic theory is given. Agreement between the JHD- and the USNAO-data is its good example. On the other hand, it may be practically difficult to calculate the data which are exactly identical with those printed in the current volumes of the A.E. Hence, we consider that the international ephemerides in the future should be calculated at least at two organizations independently using different types of computer.

The present investigation has been carried out as a collaborative work by many members of the Tokyo Astronomical Observatory and the Hydrographic Department of Japan. The authors are greatly indebted to them. A fuller version of this work will be published in the "Report of Hydrographic Researches", No. 14 (1979) under the names of A.M. Sinzi, K. Inoue, Y. Kubo, Sh. Aoki, H. Kinoshita and H. Nakai.

REFERENCES

Clemence, G.M.: 1943, "Astron. J." 50, pp. 127 - 128.
Clemence, G.M.: 1954, "Astron. Pap. Amer. Eph." Vol. XIII, Pt. V.
Duncombe, R.L.: 1969, "Astron. Pap. Amer. Eph." Vol. XX, Pt. II.
Eckert, W.J., Brouwer, D. and Clemence, G.M.: 1951, "Astron. Pap. Amer. Eph." Vol. XII.
Eckert, W.J., Walker, M.J. and Eckert, D.: 1966, "Astron. J." 71, pp. 314 - 332.
Herget, P.: 1962, "Astron. Pap. Amer. Eph." Vol. XVI, Pt. III.
Inoue, K.: 1977, "Rep. Hydrogr. Res." No. 12, pp. 59 - 94.
Kinoshita, H., Nakai, H. and Aoki, Sh.: 1974, "Ann. Tokyo Astron. Obs." 2nd Ser. Vol. XIV, No. 1.
Newcomb, S.: 1895, "Astron. Pap. Amer. Eph." Vol. VI, Pts. I, II, III.
Newcomb, S.: 1898, "Astron. Pap. Amer. Eph." Vol. VI, Pt. IV.
Ross, F.E.: 1917, "Astron. Pap. Amer. Eph." Vol. IX, Pt. II.
Van Flandern, T.C.: 1976, private communication.
Wilkins, G.A.: 1954, "Improved Lunar Ephemeris" by W.J. Eckert, R. Jones and H.K. Clark, U.S. Gov. Printing Off., Washington,D.C., pp. 420 - 422.
Woolard, E.W.: 1953, "Astron. Pap. Amer. Eph." Vol. XV, Pt. I.

DISCUSSION

Henrard: What is the JHD-ephemeris? Is it a new theory?
Sinzi: It is strictly based on the Brown-Eckert theory.

Duncombe: I should like to congratulate Dr. Aoki and Dr. Sinzi on their very thorough investigation. It very clearly illustrates why we need a new basis for the ephemerides in the A.E.

OBSERVATIONS OF MARS 1950-1976 COMPARED TO EPHEMERIDES

R.L. Duncombe, The University of Texas at Austin
Y. Kubo, Hydrographic Department of Japan, Tokyo, Japan
P.K. Seidelmann, U.S. Naval Observatory, Washington, D.C.

The present basis for the ephemeris of Mars in the National Ephemerides is the theory of S. Newcomb (1898) as amended by the corrections of Ross (1917). These amendments by Ross, however, are empirical in nature and therefore the present ephemeris of Mars does not have a strictly gravitational basis. In order to provide a gravitationally consistent basis for the ephemeris of Mars, Clemence (1949,1961) constructed a new general perturbation theory based on the final elements of Mars as derived by Newcomb for the epoch 1850. To test the adequacy and accuracy of this new theory, Clemence compared it against 87 observations from 1802-1839 and 1931-1950. This provided provisional values of the constants (without secular variation) for his new theory. These provisional elements and Clemence's theory were used to produce a heliocentric ephemeris of Mars for the period 1800-2000 (Duncombe and Clemence 1960, Duncombe 1964).

Laubscher (1971) discussed all of the observational data of Mars from 1751 to 1969, and derived the definitive constants of Clemence's theory of Mars. These definitive constants were used in conjunction with Clemence's theory by Kaplan, Pulkkinen and Emerson (1975) to produce a new geocentric ephemeris of Mars. Soon after this new geocentric ephemeris of Mars was published, it was noticed that the meridian circle observations made by the six-inch Transit Circle, U.S. Naval Observatory, while agreeing in right ascension, showed marked discordances in declination. To locate the source of this discrepancy it was decided to compare the observations in the period 1950-1976, with three different geocentric ephemerides: One based on the theory of Newcomb, as amended by Ross which appears in the American Ephemeris, henceforth referred to as the A.E.; a second, based on the evaluation of Clemence's new theory with the provisional elements by Duncombe, and thirdly, a geocentric ephemeris based on the evaluation of Clemence's theory using the definitive values of the elements derived by Laubscher. It should be noted that the heliocentric ephemeris by Duncombe was computed with elements for the mean epoch 1850 without inclusion of secular terms.

R. L. Duncombe (ed.), Dynamics of the Solar System, 121–127.

These geocentric ephemerides were compared against 424 observations of Mars made with the six-inch Transit Circle of the U.S. Naval Observatory (Adams, et al 1964, 1967, 1968; Klock, et al 1970, 1973) and the Danjon Astrolabes at Paris, San Fernando and Grasse (CERGA) (Debarbat 1977, Pham-Van, et al 1978), covering the twelve oppositions from 1950-1976. The sum of the square of the residuals in right ascension and declination compared to these three ephemerides are shown for each opposition and for the entire period in Table I. It is evident that the residuals in right ascension are generally smaller for both the Duncombe ephemeris and the Laubscher ephemeris as compared to the Newcomb ephemeris in the A.E. In declination, however, the results are markedly different. Here the Newcomb ephemeris in the A.E. shows better agreement with the observations than either the Duncombe ephemeris or the Laubscher ephemeris. Figures I, II and III show in detail the comparison of the observations in declination with the three ephemerides for the opposition of 1971. It is at this opposition that we have the largest values of the declination residuals, although the same tendency can be seen at every opposition. These figures show that in declination, the Laubscher ephemeris is noticeably inferior to the Newcomb ephemeris. The Duncombe ephemeris computed for the epoch 1850 without secular terms falls between them.

Since the Newcomb ephemeris in the A.E. represents the declination observations fairly well, it was decided to form the difference in declination between the Laubscher ephemeris and the A.E. These differences are illustrated in Figure IV, for period 1966-1976. The largest differences in this figure correspond to the points in the orbit of Mars where the latitude is greatest, i.e., the points 90° from the nodes. To further check this result, a comparison was made of the differences in the heliocentric latitude of the Laubscher ephemeris minus the A.E. and the Duncombe ephemeris minus the A.E. These differences are shown in Figure V. It seems apparent from this figure that the discrepancy shown by the observations arises from the value of the inclination adopted in the Laubscher ephemeris. The Duncombe ephemeris based on elements without secular variations is seen to fall part way between Laubscher's ephemeris and the ephemeris in the A.E. This seemed to indicate that the error might be in the secular change of the inclination adopted by Laubscher.

To determine the empirical correction to Laubscher's inclination, at the mean epoch of the observations, residuals were formed against several ephemerides containing arbitrary corrections to the inclination. Minimizing the sum of the square of the residuals in declination formed from each ephemeris provided the optimum correction to the inclination at each opposition. The final value of the correction at epoch 1965.03 is determined to be $-0\overset{''}{.}51 \pm 0\overset{''}{.}05$. Following this lead, Seidelmann examined the original computations of Laubscher's analysis, and traced the error to an incorrect algebraic sign of the secular term in the expression for the correction to the inclination.

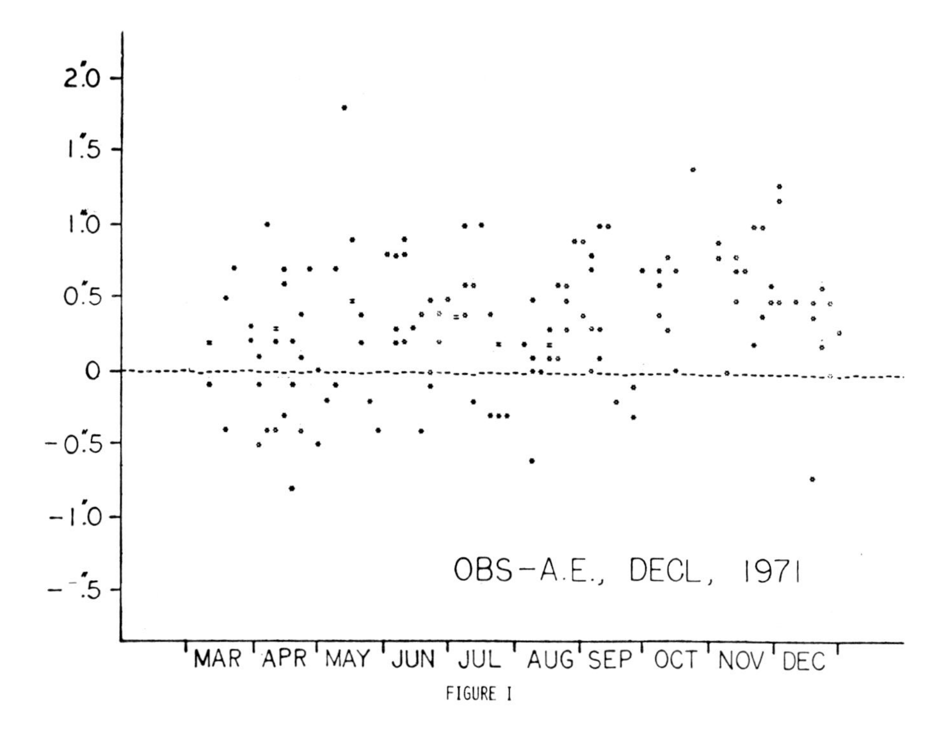

FIGURE I

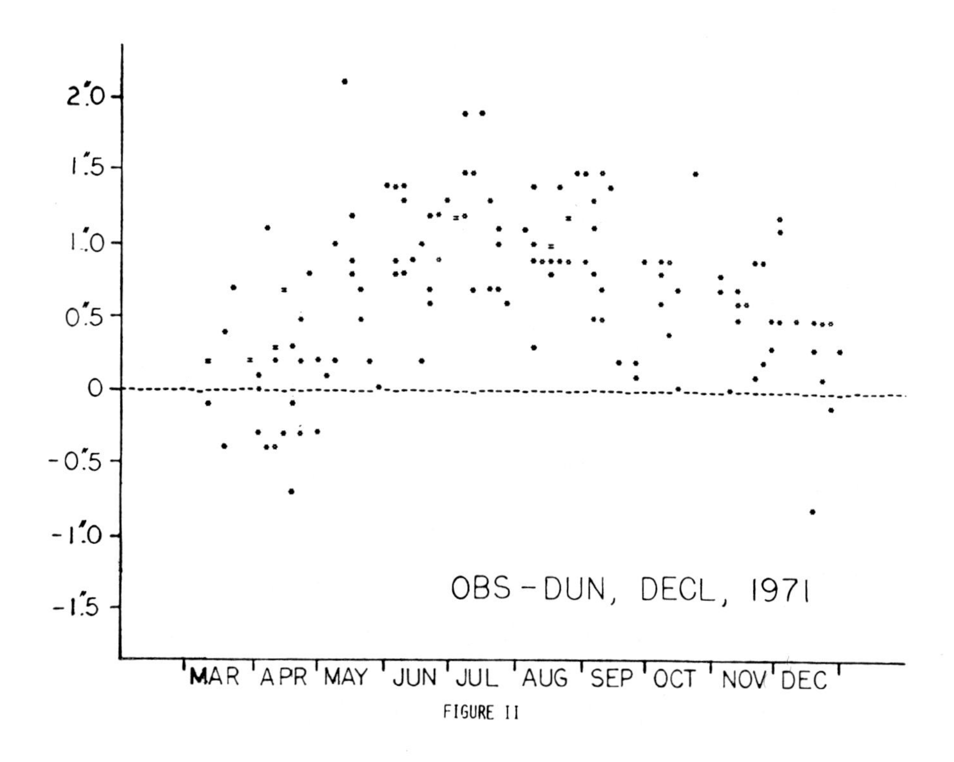

FIGURE II

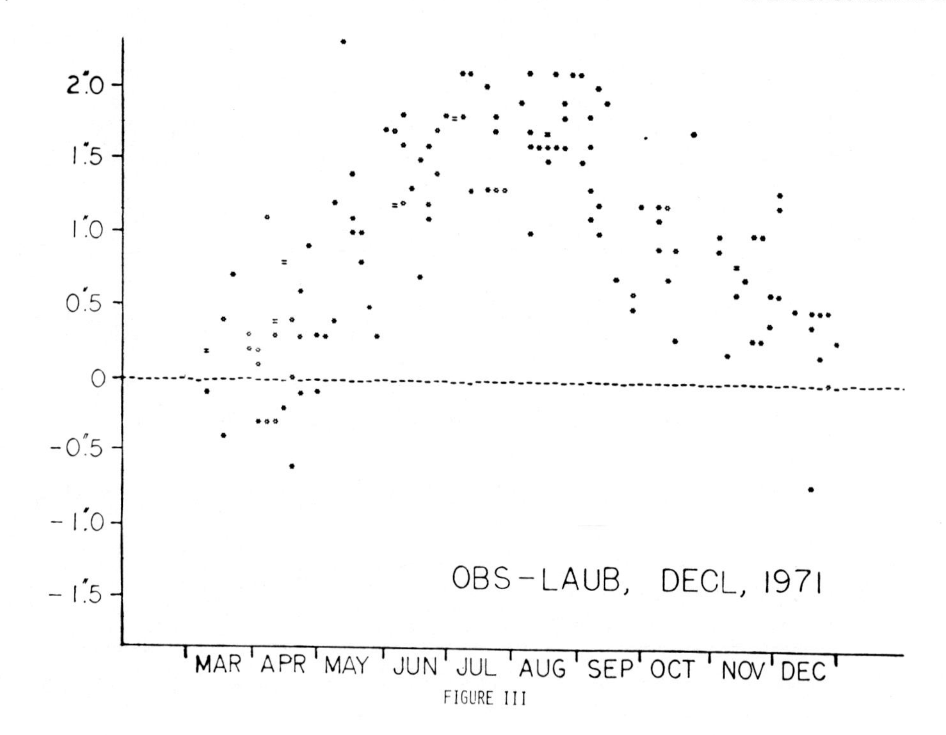

FIGURE III

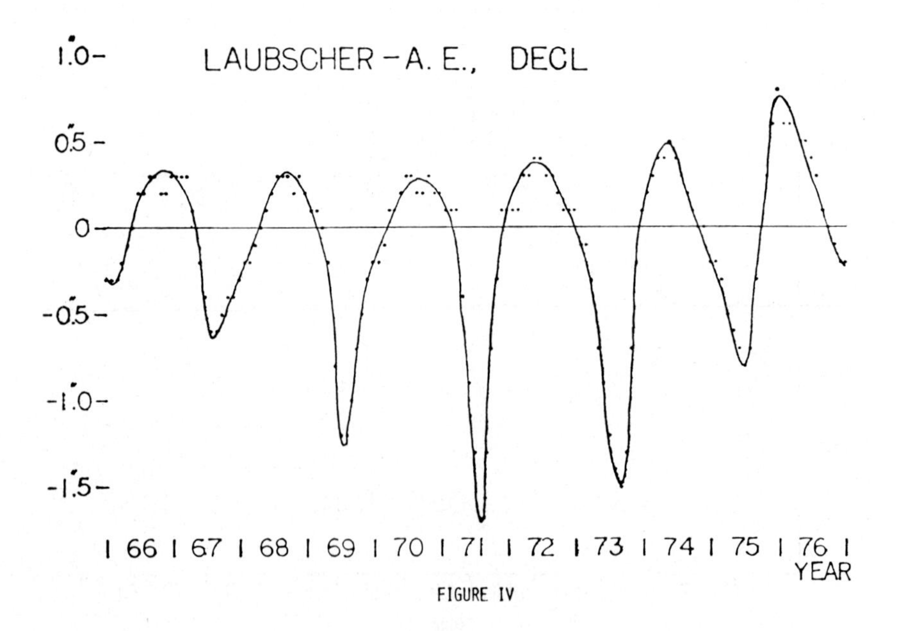

FIGURE IV

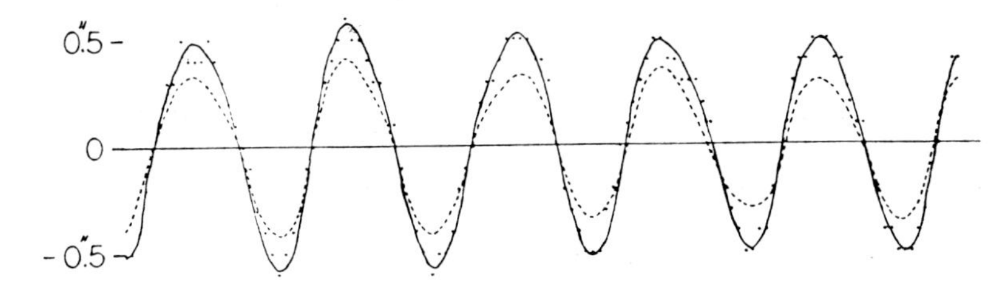

FIGURE V

TABLE I.

Opposition	N	Right Ascension AE	DUN	LAUB, O	LAUB, C	LAUB, N	Declination AE	DUN	LAUB, O	LAUB, C	LAUB, N
1950	20	$0^{s}.035$	$0^{s}.017$	$0^{s}.043$	$0^{s}.030$		$6''.39$	$6''.85$	$8''.42$	$5''.11$	
1952	22	0.025	0.015	0.022	0.023		3.19	3.32	3.68	3.02	
1954	25	0.123	0.024	0.019	0.017		6.83	23.47	37.34	21.80	
1956	25	0.457	0.252	0.129	0.051	$0^{s}.049$	10.76	15.52	38.97	8.59	$8''.62$
1958	29	0.556	0.259	0.133	0.143	0.112	10.33	10.60	9.64	6.03	5.28
1960	24	0.240	0.136	0.118	0.117	0.087	3.12	9.25	16.43	2.69	3.10
1963	34	0.299	0.142	0.193	0.122	0.098	16.52	5.74	12.71	7.73	7.83
1965	29	0.119	0.071	0.122	0.064	0.058	4.90	11.51	18.08	1.80	1.99
1967	33	0.037	0.029	0.056	0.040	0.048	3.30	4.90	6.34	2.26	2.38
1969	17	0.037	0.033	0.034	0.040	0.050	1.62	3.90	6.81	1.50	1.59
1971	133	0.482	0.577	0.449	0.404	0.395	28.69	86.24	196.06	23.06	22.13
1975-76	33	0.780	0.121	0.070	0.092	0.063	2.74	10.66	19.57	2.64	2.84
SUM 1950-76	424	$3^{s}.190$	$1^{s}.676$	$1^{s}.388$	$1^{s}.143$		$98''.39$	$191''.96$	$374''.05$	$86''.23$	
SUM 1956-76	357	$3^{s}.007$	$1^{s}.620$	$1^{s}.304$	$1^{s}.073$	$0^{s}.960$	$81''.98$	$158''.32$	$324''.61$	$56''.30$	$55''.76$

LAUB, O : Original Laubscher,
LAUB, C : Laubscher Differentially Corrected,
LAUB, N : New Laubscher

The correct expression is $\Delta i = (0''.077 \pm 0''.007) + (+0''.197 \pm 0''.010)T$, where T is reckoned from 1850.0. Comparison of observations against Laubscher's ephemeris with this correction to the inclination is shown in Table I in both right ascension and declination under the heading Laub,C. This correction to Laubscher's value of inclination produced improved agreement with the observations.

A new geocentric ephemeris of Mars based on Laubscher's elements with the correct expression for Δi was prepared at the U.S. Naval Observatory. Comparison of the observations against this new ephemeris is shown in Table I in both right ascension and declination under the heading of Laub,N. The slight difference between this column and the preceding one in both right ascension and declination is due to the use of a different ephemeris of the sun in the two cases. Since the new ephemeris did not extend to the oppositions of 1950, 1952 and 1954, the sums of the square of the residuals of the observations compared to all of these ephemerides have been formed again for the period 1956-1976 and are shown at the bottom of Table I in order to compare with the results from the new ephemeris. It is evident that the new ephemeris provides improved representation of the observations in both right ascension and declination. This confirms that Clemence's theory of Mars, evaluated with the correct constants, provides a superior standard for the comparison of observations. This analysis illustrates the value of, and the necessity for, consistent series of observations of the principal planets. Without the observations made at Paris, San Fernando, Grasse, and the U.S. Naval Observatory, it would have been extremely difficult, if not impossible at the present epoch, to pin down the source of the error in the ephemeris of Mars.

REFERENCES

Adams, A.N., Bestul, S.M. and Scott, D.K.:1964, U.S. Nav. Obs., Sec. Ser., Vol XIX, pt I.

Adams, A.N. and Scott, D.K.:1967, U.S. Nav. Obs. Circ. No. 115.

Adams, A.N. and Scott, D.K.:1968, U.S. Nav. Obs. Sec. Ser., Vol XIX, pt II.

Adams, A.N. and Scott, D.K.:1968, U.S. Nav. Obs. Circ. No. 118.

Clemence, G.M.:1949, Astron. Pap., Amer. Eph., Vol. XI, pt II.

Clemence, G.M.:1961, Astron. Pap., Amer. Eph., Vol. XVI, pt II.

Debarbat, S.:1977, Vistas in Astron., Vol 21, pp 93-106.

Debarbat, S., Pham-Van, J. and Sanchez, M.: "Ephemerides et position de Mars en 1975-1976," unpublished.

Duncombe, R.L. and Clemence, G.M.:1960, U.S. Nav. Obs. Circ. No. 90.

Duncombe, R.L.:1964, U.S. Nav. Obs. Circ. No. 95.

Kaplan, G.H., Pulkkinen, K.F. and Emerson, B.:1975, U.S. Nav. Obs. Circ. No. 151.

Klock, B.L. and Scott, D.K.:1970, U.S. Nav. Obs. Circ. No. 127.

Klock, B.L. and Scott, D.K.:1973, U.S. Nav. Obs. Circ. No. 143.

Laubscher, R.E.:1971, Astron. and Astrophys., Vol 13, pp 426-436.

Newcomb, S.:1898, Astron. Pap., Amer. Eph., Vol VI, pt IV.

Pham-Van, J., Choplin, H., Delmas, C., Grudler, P., Guallino, G., Mignard, F. and Vigouroux, G.:1978, Astron. and Astrophys. Suppl., Vol 31, pp 171-173.

Ross, F.E.:1917, Astron. Pap., Amer. Eph., Vol IX, pt II.

DISCUSSION

Fricke: Was there any reason for excluding from your discussion the observations of Mars made with the 9"TC around 1940? Are these observations of minor quality than those made with the 6"TC?

Duncombe: We confined our analysis to observations made from 1950 to 1975. The 9"TC observations do not meet the standard set by the 6"TC.

EPHEMERIS OF MARS

Hiroshi KINOSHITA and Hiroshi NAKAI
Tokyo Astronomical Observatory,2-21-1 Osawa,Mitaka,Tokyo,Japan

Newcomb's (1898) Tables of Mars with Ross's corrections have been used to compute its ephemeris in national ephemerides. Clemence (1949 and 1961) constructed a more precise theory of Mars and determined preliminary mean elements using a limited number of observations. Laubscher (1971) compared Clemence's theory with about 11000 observations by meridian circles from 1851 through 1969 and about 800 observations by radars and derived an improved set of mean elements. Duncombe and Kubo (1977) compared the ephemeris of Mars, in which Clemence's theory and Laubscher's elements were adopted, with the observations made by the meridian circle at Washington in 1972 and found large systematic residuals, which were larger than residuals by Newcomb's theory. Therefore, Duncombe and Kubo suspected that Laubscher's treatment of data was erroneous. On the other hand we had undertaken a reconstruction of Clemence's theory. In the meantime Kubo and Seidelman independently have found Laubscher's error in the determination of the inclination.

Clemence used the method of Hansen, that is the method modified by Hill in his theories of Jupiter and Saturn. We have followed Clemence's procedure as closely as possible to repeat his calculations and to obtain the same results as Clemence's calculations. One of the distinguished features of Hansen's method is that his method is applicable to orbits of higher eccentricity and inclination; numerical values for the mean elements other than mean anomalies are substituted in any stage of calculations. This feature inevitably introduces mixed secular terms in the expressions of perturbations, and, therefore, a solution by this method is only applicable to a limited time interval. In fact Clemence's goal was to attain a precision of $0''.01$ in the geocentric position of Mars for five centuries or more before and after 1850. This feature also introduces another problem in the calculation of the second-order perturbations; the term of the longitude having any argument $jl'+kl$ (l' and l are mean anomalies) are multiplied by $n^3/(jn'+kn)^3$, n and n' being the mean motions of l and l'. If the term with $jl'+kl$ is a long periodic one, this factor causes the loss of too many significant figures. In Hansen's method both long- and short-periodic perturbations are treated simultaneously, and the necessary places of decimals are set by the term of the

R. L. Duncombe (ed.), Dynamics of the Solar System, 129–132.

longest period among the perturbations. Hence in order to determine both long- and short-periodic terms with the same accuracy, all the calculations have to be carried to the same number of decimals and require keeping a lot of terms in the intermediary calculations and costing a lot of time and memory of a computer.

All the calculations in our work are carried out by Poisson-Series-Processor developed by Nakai (1978). PSP manipulates operations of addition, multiplication, partial differentiation, integration, and so on for the following type of Poisson series:

$$PS=\sum_{ij} C_{ij} P_1^{j_1} P_2^{j_2} P_3^{j_3} P_4^{j_4} P_5^{j_5} P_6^{j_6} \times {}^{\cos}_{\sin}(i_1 l_1 + i_2 l_2 + i_3 l_3 + i_4 l_4 + i_5 l_5 + i_6 l_6) \quad ,$$

where C_{ij} are numerical coefficients (rational or floating), j's are powers of P, and l's are angular variables. PSP adopts Hash coding method, which drastically reduces the computing time of multiplication of the two Poisson series, because this method does not require a large amount of sorting and moving the data in core and does not take a fair amount of storage space.

Let Δ be the mutual distance of two planets and α be a/a', where the primed quantity relates always to the outer planet. If the eccentric anomalies are taken as the independent variables, the series expansion of $(\Delta/a')^2$ can be written in a closed form. Let $(\Delta/a')^2=1+E$, where E becomes zero for $\alpha=0$, and denote a constant part of E by E_c and a periodic part by E_p:

$$(\Delta/a')^2=(1+E_c)D^2 \quad , \quad D^2=1+E_0 \quad , \quad \text{and} \quad E_0=E_p/(1+E_c) \quad .$$

In order to obtain a series expansion of 1/D with respect to the eccentric anomalies, we use the following two kinds of iteration formula:

$$\text{a)} \ \frac{1}{\delta_1} = 1-\frac{1}{2}E_0+\frac{3}{8}E_0^2 \quad , \qquad \text{b)} \ \frac{1}{\delta_1} = 1-\frac{1}{2}E_0+\frac{3}{8}E_0^2 \quad ,$$

$$E_n = \frac{5}{8}E_{n-1}^3 - \frac{15}{64}E_{n-1}^4 + \frac{9}{64}E_{n-1}^5 \quad , \qquad \varepsilon_n = \left(\frac{1}{\delta_n}\right)^2 D^2 - 1 \quad ,$$

$$\frac{1}{\delta_{n+1}} = \frac{1}{\delta_n}\left(1-\frac{1}{2}E_n+\frac{3}{8}E_n^2\right) \quad , \qquad \frac{1}{\delta_{n+1}} = \frac{1}{\delta_n}\left(1-\frac{1}{2}\varepsilon_n+\frac{3}{8}\varepsilon_n^2\right) \quad .$$

The order of magnitude of both E_n and ε_n is $E_0{}^{3n}$. In case (a), as number n increases, the amount of calculations usually decreases. However, rounding-off errors accumulate. On the other hand, in case (b), as number increases, the amount of calculations does not decrease, but rounding-off errors do not accumulate. In the actual calculations, we obtain an approximate series expansion by (a), then improve it by (b). Taking into account higher powers of E_n or ε_n, we can make more rapidly convergent formula than (a) or (b). However, our experience shows that such iteration formulae are usually time consuming because of increase of multiplications of Poisson series.

In developing a'/Δ in terms of the eccentric anomalies by (a) and (b) and transforming it into a series in terms of the mean anomalies by means of Besselian functions, we use ten places of decimals, which are larger by two places than Clemence's value. a'/Δ for the Earth includes 2173 terms. If this series is truncated at eight places of decimals, the number of terms reduces to 1377, which is less by twenty than in Clemence's theory. The largest difference in the numerical coefficients of the trigonometric series reaches up to 75×10^{-8} for the term with $35l_M-29l_E$, where l_M and l_E are mean anomalies of Mars and the Earth, respectively. Our results are in good agreement to the ninth digit with the results obtained by double Fourier analysis. Also we check the accuracy of the series expansion by substituting numerical values of the mean anomalies into the series expansion and a'/Δ for the Earth. The maximum difference is about 4×10^{-7} for the series truncated at 10^{-8}.

Using series expansions for a'/Δ and adopting the same formulae as Clemence did, we obtain the first-order perturbations of $n\delta z$, ν, and $u/\cos i$, which are perturbations in the mean anomaly, the radius vector, and the latitude, respectively. In any calculation, we keep at least one more digit than Clemence did. The following table shows the terms, which have the largest difference between Clemence's results and ours (K. and N.).

	$\times10^{-4}$	Clemence	K. and N.	Diff.	$n_M/(in+jn')$	
	$n\delta z$	missing	- 19	19	-38.3	$\cos(17l_E-32l_M)$
Earth	$u/\cos i$	-295	-296	1	1.3	$\cos(2l_E-3l_M)$
	ν	73	75	2	- 1.1	$\cos(8l_E-16l_M)$
	$n\delta z$	-471	-453	18	20.6	$\cos(l_M-6l_J)$
Jupiter	$u/\cos i$	0	- 35	35	0.5	$\sin 2l_M$
	ν	- 75	- 72	3	- 1.0	$\sin(-2l_M+6l_J)$

The large difference in the periodic perturbation of $n\delta z$ might have been caused by small divisors due to its long periodic. The discrepancy in $u/\cos i$ by Jupiter seems to be a typographical error, because this term is listed in the table in which Clemence compared his results with Newcomb's theory. The difference in mixed secular terms factored by nt does not exceed 7×10^{-6} of the second of arc, which appears in $n\delta z$ by Saturn.

Clemence compared his theory, which includes second- and third-order perturbations, with the results obtained by the numerical integration by Herget. The largest difference in the orbital longitude was $0''.042$, which cannot be explained by the above discrepancies in the first-order perturbations between Clemence's results and ours. The calculations of second-order perturbations are being undertaken.

REFERENCES

Clemence,G.M.:1949,"Astron.Papers Amer.Ephemeris"11,pt.2.
Clemence,G.M.:1961,"Astron.Papers Amer.Ephemeris"16,pt.2.
Duncombe,R. and Kubo,Y.:private communication.
Laubscher,R.E.:1971,"Astron. and Astrophys."13,pp.426-436.
Nakai,H.:1978,"Tokyo Astron. Obs. Report"18,pp.242-259.
Newcomb,S.:1898,"Astron.Papers Amer.Ephemeris"6.

DISCUSSION

Henrard: Would you comment on the advantage of Fourier analysis over iteration procedure for the computation of the distance between two planets?

Kinoshita: Yes. A double Fourier analysis is the most efficient way to obtain a trigonometric expansion of the disturbing function in terms of the mean anomalies. However, for the next stage of calculation, we definitely need a formula manipulation system.

WHERE IS THE EQUINOX ?

W. Fricke
Astronomisches Rechen-Institut, Heidelberg, Germany, F.R.

1. INTRODUCTION

Within the work being carried out at Heidelberg on the establishment of the new fundamental reference coordinate system, the FK5, the determination of the location of celestial equator and the equinox form an important part. The plane of the celestial equator defined by the axis of rotation of the Earth and the plane of the ecliptic defined by the motion of the Earth about the Sun are both in motion due to various causes. The intersection of the equator and the ecliptic, the dynamical equinox, is therefore in motion. Great efforts have been made in the past to determine the location and motion of the dynamical equinox by means of observations of Sun, Moon and planets in such a manner that the dynamical equinox can serve as the origin of the right ascension system of a fundamental catalogue. The results have not been satisfactory, and we have some important evidence that the catalogue equinox of the FK4 is not identical with the "dynamical equinox". Moreover, is has turned out that the difference $\alpha(\mathrm{DYN}) - \alpha(\mathrm{FK4}) = E(T)$ depends on the epoch of observation T. Duncombe et al. (1974) have drawn attention to the possible confusion between the catalogue equinox and dynamical equinox; they mention the difference between two Earth longitude systems, one established by the SAO using star positions on the FK4 and the other one established by the JPL using planetary positions measured from the dynamical equinox. This is undoubtedly one legitimate explanation of the difference, even if other sources of errors may also have contributed.

Obviously the location of the FK5 equinox will have to be determined together with the equator point from modern observations of members of the solar system, as far as these allow us to determine corrections to the zero points of the FK4 systems of right ascension and declination. The problems arising in this task shall be described here. This discussion shall be restricted to the location and motion of the equinox, since it has been known for some time that the equator point offers less problems. In the FK4 the equator point is determined with a mean error of $\pm 0\overset{''}{.}021$ and no significant secular change was found.

R. L. Duncombe (ed.), Dynamics of the Solar System, 133–143.

2. NEWCOMB'S EQUINOX AND ITS DEFICIENCIES

Newcomb (1872) has carried out a discussion of the observations of the Sun from 1756 to 1869 resulting in a determination of the equinox to which the right ascensions of clock stars were to be referred. This equinox, commonly called N_1, was adopted by Newcomb (1882) as the zero point in right ascension in his "Catalogue of 1098 Standard Clock and Zodiacal Stars". This equinox has often been used as a reference up to the present day. In his determination Newcomb followed on the lines of the pioneer work of Bessel (1830) who had made one equinox determination from Bradley's observations 1750-1762 and another one from his own observations from 1822 to 1835 at Königsberg. On this basis Bessel presented his famous "Tabulae Regiomontanae" which give the mean and apparent places of 36 "Maskelyne Stars" from 1750-1850. In fact, Bessel's "Tabulae" was used during a large part of the 18th century as a fundamental reference coordinate system. It is therefore quite natural that Newcomb presented his result of equinox determinations as a correction in the sense $E = \alpha_{Obs} - \alpha_{Bessel}$. Figure 1 shows Newcomb's corrections E obtained from 26 catalogues which had included observations of the Sun.

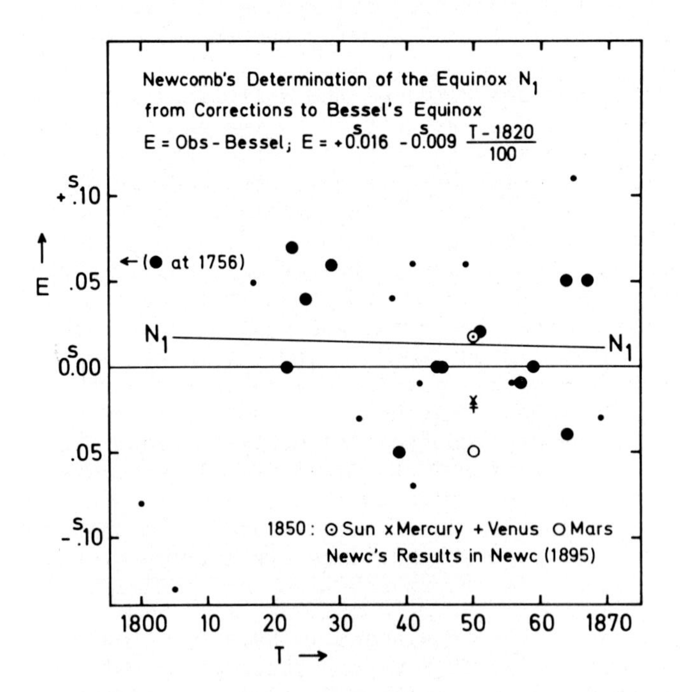

Figure 1. Newcomb's corrections E to Bessel's equinox resulting in N_1.

The straight line represents Newcomb's equinox N_1 as the result of his least squares solution (small and large filled circles denote values of E of weight 1-2, and 3-5, respectively). The secular change is, however, not significant. The values E given for 1850 for the Sun, Mercury, Venus, and Mars are results obtained by Newcomb (1895, p. 96) from all available observations up to 1890. The most surprising feature of the figure is the smallness of the deviation of Newcomb's N_1 from Bessel's equinox.

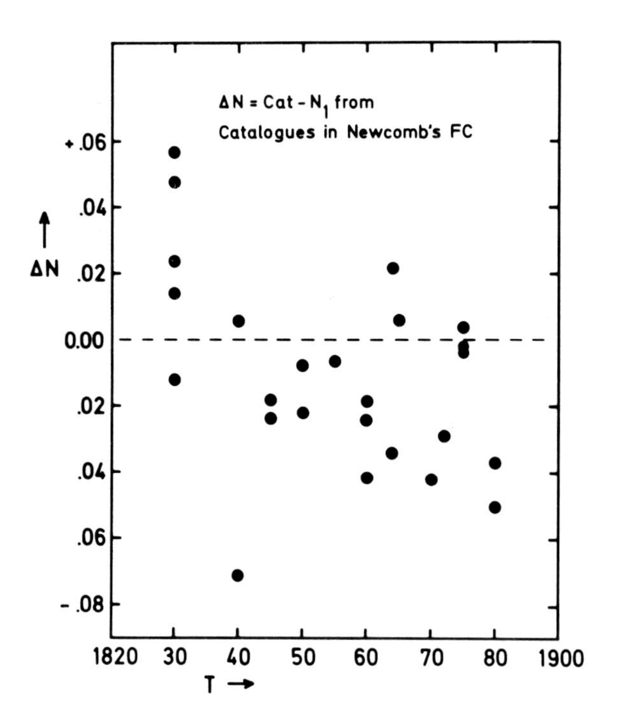

Figure 2.
Evidence of a secular variation of ΔN from catalogue comparisons with Newcomb's FC

In the formation of a fundamental catalogue Newcomb (1898) adopted the equinox N_1 as the zero point of right ascensions and the value of the general precession (Newcomb 1898) he had derived before. N_1 is therefore the catalogue equinox of Newcomb's FC. The comparison of N_1 with 28 catalogues has yielded values of ΔN from 1830 to 1880 which are shown in Figure 2. The catalogue comparisons had often to be made with a small number of bright clock stars, and in some cases the values $\Delta N = \alpha(\text{Cat}) - \alpha(N_1)$ refer to catalogues which had adopted $\alpha_{(\text{Dyn})}$ from observers who had established the zero point with observations of the Sun. The essential feature of Figure 2 is the secular variation of ΔN which has led Newcomb to the conclusion that there is an equinox motion which he has not taken into account in his FC.

3. EVIDENCE OF AN EQUINOX MOTION ?

Besides the evidence of a secular variation of ΔN provided by equinox determinations of the 19th century, one may obtain information on the size of the effect from an investigation of proper motions of stars given in a fundamental catalogue. For this purpose we have chosen Newcomb's FC (1899) whose basic material has exhibited the corrections ΔN given in Figure 2. The catalogue equinox is N_1, and it is the first fundamental catalogue, in which the consequences of a variation of ΔN should have become apparent. This catalogue has never before been investigated for this purpose, probably because Boss (1910a) soon afterwards had completed the PGC and determined corrections to Newcomb's precession on the basis of PGC proper motions. For the investigation of Newcomb's FC, which has very recently been completed by H. Scholl and myself, we have used the proper motions of 248 stars which occur in Fricke's (1967, 1977) basic material for the determination of corrections to Newcomb's precession. This material offers the advantage of a direct comparison with results obtained on the basis of the FK4. Our results are given in Table 1 under the headings "Newc FC" and "FK4", while the results obtained from the whole PGC by Boss (1910b) are under the heading "PGC". From the values under the heading "Fricke (1967, 1977)" the quantity Δp_1 = 1"10 per century has been adopted in the IAU (1976) System of Astronomical Constants; and the value Δe = 1"23 is under consideration as a correction to the FK4 proper motions in the formation of the FK5.

Table 1. Centennial values of precessional corrections from proper motions in Newc FC and from other sources

Source	Newc FC	PGC	FK4	Fricke (1967)
n	248		248	512
Δp_1	0"53 ± "40	0"88	1"08 ± .25	1"10 ± "15
Δe	1.09 .41	1.21	1.21 .24	1.23 .16

The essential result of our investigation of Newc FC is the revelation of an equinox motion Δe = 1"09 ± 0"41 (m.e.) in fair agreement with the result by Boss. This means that the whole amount of Δe found in the FK4 is already apparent in the proper motions based on observations made in the 19th century, as we had expected from the data of Figure 2. The result is in accordance with one obtained by Laubscher (1976) which is independent of hypotheses inherent in proper motion solutions. The question whether there is an equinox motion may thus be answered in the sense that zero point errors in right ascension have produced, indeed, a zero point error in the proper motions of the stars. We may therefore exclude the possibility of any unknown physical effect being responsible for the equinox motion.

It may be noted that Newcomb (1898) derived Δe = 0"30 in his famous determination of precession, and that this result has often been misinterpreted, even by Boss (1910). It has no other meaning than that it indicates errors in two equinox determinations, namely, from Bradley's observations at mean epoch 1755 and from Airy's at 1860. Newcomb's work was based on the proper motions of Auwers' catalogue (1888) of Bradley Stars, and these were derived from systems of positions at 1755 and 1860. For more detailed information, reference is made to a rediscussion of Newcomb's determination of precession by Fricke (1971). One may be amazed that such a small value of Δe resulted, and it is not surprising that Newcomb has commented on this result in stating that (with respect to his system N_1) the value Δe = + 1"00 is "legitimately worthy of discussion".

4. MAGNITUDE EQUATIONS IN POSITIONS AND PROPER MOTIONS

Systems of positons and proper motions can be strongly distorted by magnitude equations, i.e. by systematic errors depending on the magnitude. Küstner (1902) was the first to prove that if a transit circle is equipped with a micrometer with fixed threads and observations are made with full aperture, the transits of bright stars are observed too early and of faint stars too late. He found that magnitude equations can be avoided by the use of an impersonal micrometer with moving threads (hand- or motor driven) and by the use of a set of screens in front of the objective for reducing the light of bright stars to the magnitude of well observable faint stars. While every observer knows this now, there has been little concern so far about the consequences of the effect on equinox determinations of the 19th century.

From measurements of the magnitude equation in observations made at Lund from 1943 to 1945 with fixed threads and full aperture, Reiz (1951) has derived

$$\Delta\alpha_m = \alpha_{FK3} - \alpha_{Obs} = 0\overset{s}{.}004 - 0\overset{s}{.}0152\ (m - 4.5)$$

for his observations from $m = 2.5$ to about 8.5. This is a typical "personal equation"; other observers may find smaller or even greater effects. In Figure 3 this equation reduced to the FK4 is given as a dashed line. The full line in Figure 3 represents the magnitude equation of the

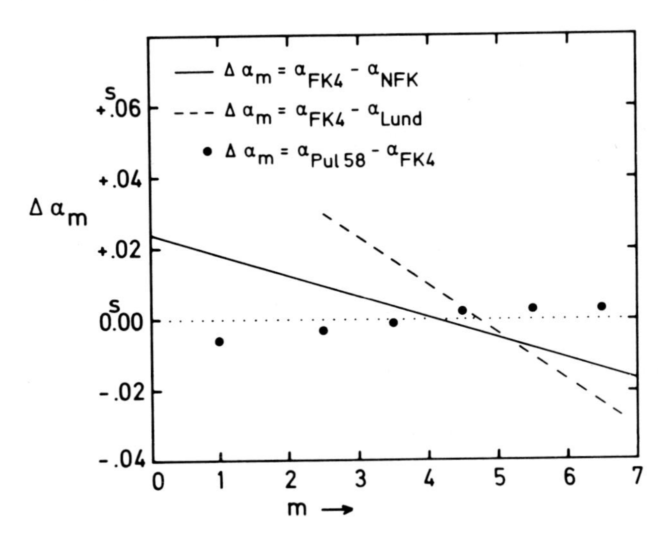

Figure 3. Magnitude equations in right ascension

NFK with respect to the FK4. The NFK (Neuer Fundamentalkatalog) is the second catalogue in the FK series. It was formed on the basis of 19th century observations by Auwers and completed by Peters (1907); it is the first fundamental catalogue in which the authors have taken magnitude equations into account whenever possible. The full line resulted from a comparison between the NFK and Küstner's observations made with impersonal micrometer and screens and a reduction to the FK4 (the reduction was made in using $\Delta\alpha_m = \alpha$(FK3 - NFK) by Nowacki (1935) and the table $\Delta\alpha_m$ (FK4 - FK3) given in the FK4). In fact, the FK4 is very near to Küstner's observations as far as a magnitude equation is concerned. The full circles in Figure 3 represent the results of recent observations by Nemiro et al. (1977) presented in the catalogue "Pulkovo 58" which is claimed to be free of magnitude effects. Since the latest result obtained at the US Naval Observatory are similar to those of "Pulkovo 58" one may conclude that the FK4 is still affected by a small magnitude equation as was to be expected. In declinations no general trend arising from magnitude equations was found.

From the existence of magnitude equations in all older observations of right ascension we conclude that the equinox determinations before 1900 were affected correspondingly, such that the right ascensions of the Maskelyne Stars ($\overline{m} = 2.0$) were measured too small by $0\overset{s}{.}02$ to $0\overset{s}{.}03$ and of Airy's clock stars ($\overline{m} = 3.9$) by at least $0\overset{s}{.}005$ to $0\overset{s}{.}010$. From 1830 to 1900 there was no significant change in the techniques of observation with transit circles, but there was a considerable change in the choice of the clock stars from very few stars brighter than m = 1, then to about 30 stars of mean magnitude 2.0, and finally to about 200 stars of mean magnitude 4.0. Figure 4 shows the distribution in magnitude of the clock

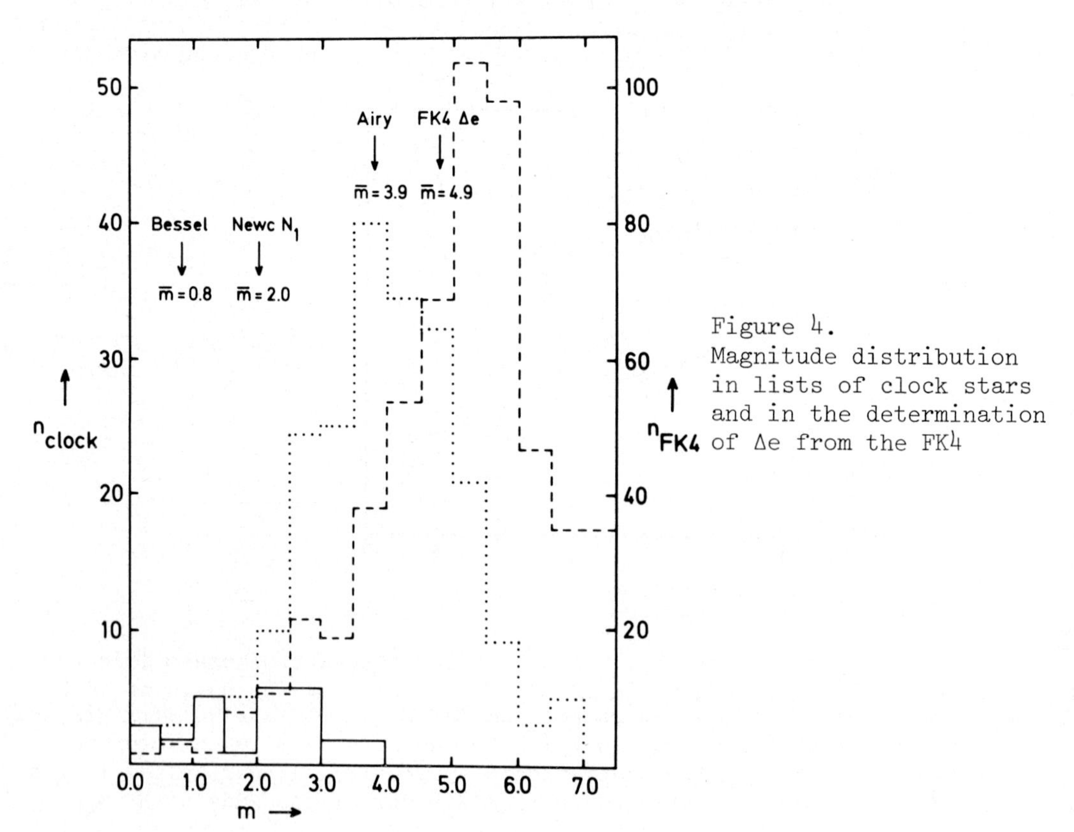

Figure 4.
Magnitude distribution in lists of clock stars and in the determination of Δe from the FK4

stars used in the determination of Bessel's equinox and of Newcomb's N_1; the Figure contains the distribution of Airy's clock stars and of the FK4 stars, which Fricke (1977) used in the determination of precession and of Δe.

One may wonder whether proper motions of stars in the FK4 system are affected by a magnitude equation such that the equinox motion Δe determined by Δk according to $\Delta k = \Delta n \cot\varepsilon - (\Delta\lambda + \Delta e)$ differs for bright and faint fundamental stars (Δn, $\Delta\lambda$ are the corrections to the general precession in declination and to planetary precession, respectively, and ε the obliquity at mean epoch). H. Scholl and the author have investigated this question in using the proper motions of the 518 FK4/FK4 Sup stars applied by Fricke (1967, 1977) in determining precession. The same procedure as in his determination has been used with some simplifications (equal weight for μ_α, μ_δ; no distances taken into account). Table 2 shows our results for Δn and Δk from solutions for equal numbers N of stars with mean magnitudes $\overline{m} = 3.95$ and 5.86, respectively. For comparison the result of our simplified procedure is given for the whole set of 518 stars.

Table 2. Solutions for Δn, Δk from proper motions of bright and fainter FK4/FK4 Sup stars. Units: arcsec per century. Errors: standard deviations.

N	256	256	512
m	0.3 to 5.2	5.4 to 7.7	0.3 to 7.7
<m>	3.95	5.86	4.90
Δn	+ .43 ± .09	+ .46 ± .10	+ .45 ± .07
Δk	− .25 .10	− .18 .12	− .21 .08

Table 2 shows the satifactory result that the corrections do not differ for bright and faint stars, and, hence, that the equinox motion in FK4 does not depend on the magnitude within the prevailing accuracy. Unfortunately, it not practicable to carry out the same analysis for Newcomb's FC.

5. MODERN EQUINOX DETERMINATIONS FROM OBSERVATIONS OF THE SUN AND PLANETS

The catalogue equinox of the FK3 determined by Kahrstedt (1931) is $N_1 - 0^{s}.05$ for 1913; it is a constant in FK3 for every given epoch. It resulted from determinations based on observations of the Sun, Mercury, and Venus around 1900. In the FK4 no change of the equinox was made because no significant correction appeared to be possible: For the GC, Boss (1937) adopted the equinox $N_1 - 0^{s}.04$, and Morgan (1952) adopted the same in N30 as a constant for every epoch.

Most of our present knowledge on the location of the equinox is based on the determinations which are given in Figure 5; this is a plot

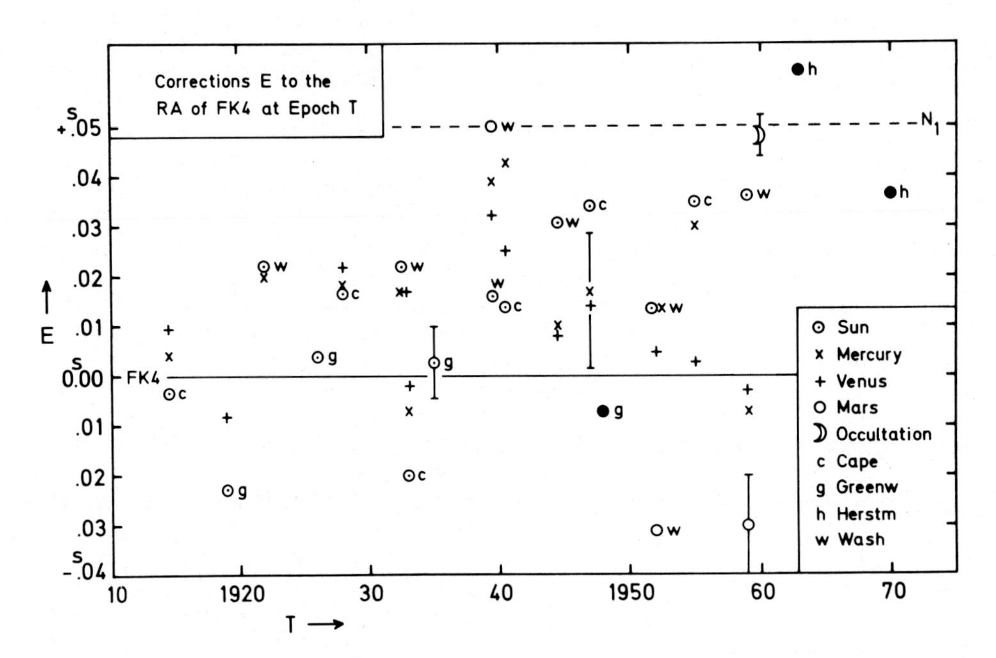

Figure 5. Corrections E from observations of different objects

of corrections E to the FK4 either as given directly by the observers or reduced to the FK4. The dashed line is Newcomb's N_1. Included in Figure 5 are the results from observations of the Sun, Mercury, Venus, Mars, and of an equinox determination by van Flandern from lunar occultations between 1950 and 1969. Filled circles are the results of observations of the Sun at Greenwich and Herstmonceux as given by Blackwell (1977) without details of the analysis. Error bars indicate typical mean errors of results from the various objects. The problems arising in the discussion of this material are the following: (a) Various results need corrections for the variation in the rate of rotation of the Earth; (b) some of the results require corrections for differences in RA observations made in day-light and at night, and (c) the relation between the observations and the FK4 has to be verified in some cases. As a rule, observers state too lightly a system to which their results are referred, and in rare cases only can this statement be confirmed. Another problem is the averaging of the results from different objects. One may, as Newcomb did, consider the Sun, Mercury, and Venus as the most suitable objects for equinox determinations and in this case average the individual results with due weights; in general, this procedure is not allowed. As can be seen in Figure 5, the observations of Mars at mean epoch 1940 (made with the 9" TC in Washington) have yielded a result which is discordant with those for the mean epochs 1952 and 1959 (obtained from observations with the 6" TC in Washington). For reasons given in the next section, it may well be advisable to use the observations of Mars only in combination with observations of minor planets. A comprehensive review of the methods for the determination of zero points and periodic errors of star catalogues including the various types of equations of condition has been given by Duma (1974). In considering the corrections determined at different epochs from the Sun alone, and in taking into account the variation of the rate of the Earth's rotation and the difference between day and night observations, Blackwell (1977) found an increase of E by about $0\overset{s}{.}05$ in the observations of the 20th century.

6. EQUINOX DETERMINATIONS FROM OBSERVATIONS OF MINOR PLANETS

Reference is made to Jackson's (1967) analysis of observations of minor planets for the purpose of determining the equinox. This work must be considered as an important manual in this field, in particular, since the criteria to be fulfilled by the observations either were not sufficiently known from the beginning or were not seriously considered. The criteria are: the observations have to be made in a well-defined system whose relation to a fundamental system must be known; a program must have observations of several minor planets because there are rather high correlations between the corrections of the individual elements of a minor planet orbit and corrections of the Earth's orbit and the coordinate system; the observations must be made at several oppositions and extending as far from opposition as possible.

From observations of Ceres, Pallas, Juno, and Vesta that fulfilled the criteria in an optimal way, Jackson obtained the correction $E = +0\overset{s}{.}002 \pm 0\overset{s}{.}015$ to the FK4 at epoch 1950.0. This result is equal to the average of Blackwell's value $E = -0\overset{s}{.}007$ at 1948 from Greenwich observations of the Sun and the value $E = +0\overset{s}{.}010$ at 1952 from Washington 6" TC observations of the Sun, Mercury and Venus by Adams et al. (1964); even if one takes $E = +0\overset{s}{.}013$ for 1952 from observations of the Sun alone, the average remains surprisingly good. In our opinion this is the only result so far known to us from minor planets fulfilling the conditions for corrections to the FK4.

Nikol'skaya (1972) has proposed a method for solving ill-conditioned equations in making use of a matrix calculus developed by Gavurin (1962). Duma and Koval (1977) have applied this method to the normal equations with 12 unknowns for each of the first four minor planets with the effect that, for example, from Vesta observations the value $E = -0\overset{s}{.}079 \pm 0\overset{s}{.}028$ obtained by the method of least squares is changed into $E = +0\overset{s}{.}005 \pm 0\overset{s}{.}002$. From a study of the method H. Scholl and I have come to the conclusion that there are no objections against Gavurin's matrix calculus but that it has been misinterpreted in the applications. From the matrix of the normal equations a matrix of Eigenvalues can be formed, and the Eigenvalues will greatly differ when strong correlations exist between the unknowns. In this case the matrix of Eigenvalues is split up into two parts, one matrix with large Eigenvalues and another one with small. Then the solution for the unknowns is the orthogonal sum of two vectors, where one is significantly determined and the other is highly insignificant. Nothing can be said against this. The authors, however, in considering the significantly determined part as a refined solution, in our opinion forget that they present that portion of the values of the unknowns which is significant, and this may be a very small part. There is apparently no way out of the difficulties in a solution of a problem, if one is dealing with unsuitable observations.

REFERENCES

Adams, A.N., Bestul, S.M., Scott, D.K.: 1964, Publ. US Naval Obs. Vol. 19, Part 1

Auwers, A.: 1888, Neue Reduktion der Bradleyschen Beobachtungen. Dritter Band. St. Petersburg

Bessel, F.W.: 1830, Tabulae Regiomontanae. Regiomonti Prussorum

Blackwell, K.C.: 1977, Monthly Notices R.A.S. 180, 65 P

Boss, L.: 1910, Astron. J. 26, 111

Boss, L.: 1937, General Catalogue. Carnegie Institution, Washington

Duma, D.P.: 1974, Determination of Zero Points and periodic Errors of Star Catalogues (In Russian). Kiev

Duma, D.P., Koval, R.N.: 1977, Pis'ma Astron. Zh. 3, 238 (English transl. Sov. Astron. Letters 3 (3)

Duncombe, R.L., Seidelmann, P.K., Van Flandern, T.C.: 1974, In "On Reference Coordinate Systems for Earth Dynamics". IAU Colloqu. No.26. p. 223. Eds. B. Kolaczek, G. Weiffenbach. Warsaw, Poland

Fricke, W.: 1967, Astron. J. 72, 1368

Fricke, W.: 1971, Astron. Astrophys. 13, 298

Fricke, W.: 1977, Veröff. Astron. Rechen-Institut, Heidelberg No. 28

Gavurin, M.K.: 1962, Zh. Mat. Mat. Fiz. 2, 287

Jackson, E.S.: 1968, Astron. Pap. Washington Vol. 20, Part 1

Kahrstedt, A.: 1931, Astron. Nachr. 244, 33

Küstner, F.: 1902, Astron. Nachr. 158, 129

Laubscher, R.E.: 1976, Astron. Astrophys. 51, 13

Morgan, H.R.: 1952, Astron. Pap. Washington Vol. 13, Part 3

Nemiro, A.A., Plyugina, A.I., Tavastsherna, K.N., Shishkina, V.I.: 1977, Trudy Glav. Astron. Obs. Pulkovo, Ser. 2, Vol. 82, 4

Newcomb, S.: 1872, Washington Observations for 1870. App.III. Washington

Newcomb, S.: 1882, Astron. Pap. Washington Vol. 1, Part 4

Newcomb, S.: 1895, The Elements of the Four Inner Planets and the Fundamental Constants of Astronomy, p. 96. Washington

Newcomb, S.: 1898, Astron. Pap. Washington Vol. 8, Part 1

Newcomb, S.: 1899, Astron. Pap. Washington Vol. 8, Part 2

Nikol'skaya, T.K.: 1972, Bull. Inst. Theor. Astron. 13, 148

Nowacki, H.: 1935, Astron. Nachr. 255, 301

Reiz, A.: 1951, Annals Obs. Lund No. 11

Van Flandern, T.C.: 1971, Astron. J. 76, 81

DISCUSSION

Aoki: Have you already corrected the reference system for the change of precession ($\Delta P_1 = 1''.1$/cent), when you have the Δe from the planetary observations?

Fricke: No.

Aoki: I think that we must change the system by introducing the corrections to precessions and equinox motion, simultaneously, for the introduction of a new fundamental system.

Fricke: You are quite right.

Aoki: Then, in order to have the equinox motion from planetary observations, it is necessary to have a system already corrected for the correction to the precession. Is it not true?

Fricke: This is not true, because a change in precession alone (without taking Δe into account) does not change the result of precession reductions; in other words, in a precession reduction an error in precession is in practice completely compensated by the corresponding error in the proper motions.

THE ROTATION OF THE MARS PLANET

N. Borderies and G. Balmino
Groupe de Recherches de Géodésie Spatiale - Toulouse - France

1. INTRODUCTION

In a non-inertial reference frame, of which the motion is given, the differential equations describing the rotation of a planet around its center of mass can be derived either under the form of Euler type equations, or from certain relations between angular velocity vectors or even angles. Unfortunately these equations are not well adapted to an analytical integration. Nevertheless, with some simplifications, Ward (1973), and Christensen (1977), by considering the motion of the spin axis in a non-inertial reference frame attached to the moving orbital plane, have found very large periodic variations in the obliquity of the Mars planet over a period of the order 1.2 10^{+5} years, which would be of great importance to the climatic history of the planet.
These oscillations cannot be attributed to the relative motion with respect to the orbit, but actually they follow geometrically from a resonance-type phenomenon which occurs basically in an inertial space.

On these basis and for the purpose of making a comprehensive investigation of the precession and nutations of the Mars planet, we have choosen to conduct it in an inertial system of coordinates axis (R).

2. METHOD

We have used the Euler dynamical equations with usual notations. Assuming that $\mathbf{r}$ remains constant, the method of variation of parameters leads to solve the differential systems :

$$\begin{cases} \dot{f} = J \, (L \cos st + M \sin st) \\ \dot{g} = J \, (L \sin st - M \cos st) \end{cases} \quad (1)$$

$$\begin{cases} \dot{\psi} = [f \sin(\varphi + st) - g \cos(\varphi + st)] / \sin\theta \\ \dot{\theta} = f \cos(\varphi + st) + g \sin(\varphi + st) \end{cases} \quad (2)$$

R. L. Duncombe (ed.), Dynamics of the Solar System, 145–150.

with $J = \frac{A + B}{2\,A\,B}$ and $s = \frac{1}{2}\left(\frac{C-A}{B} + \frac{C-B}{A}\right) r$

L and M can be derived from the expansion of the gravitational potential energy U of the present bodies (Borderies, 1977)under the form :

$$L = [\,Y\;(U) + X\;(U)\,]\;/2i, \qquad M = [\,Y\;(U) - X\;(U)\,]\;/2$$

with $i^2 = -1$, and where the operators X and Y satisfy the following properties :

$$X\;(E_{1m}^{m'}) = (1 - m'+1)\;E_{1m}^{m'-1}$$

$$Y\;(E_{1m}^{m'}) = (1 + m'+1)\;E_{1m}^{m'+1}$$

$E_{1m}^{m'}$ is a Euler function.

These properties allow us to explicit completely the systems (1) and (2) in the most general case. In particular this has been done when the acting bodies, assumed to be punctual, were the Sun, Jupiter and the Earth. In the case of solar torques, a model (6,6) of harmonic coefficients has been considered for Mars. For nutations due to Jupiter and the Earth we have limited ourselves to C_{2o}. The systems (1) and (2) are solved analytically by successive approximations. To zero order and with the secular part of U we obtain the precession. For the approximation to first order we take θ constant, ψ and φ linear, and mean orbital elements among which I and Ω have been derived from Brouwer and Van Woerkom (1951). For the next approximation, the main periodic variations are brought in the right hand side members of the equations.

3. INFLUENCE OF THE SUN

The precession rate produced by solar torques and then the nutations due to the direct effect of the Sun combined with the secular motions of Mars orbit have been computed by the above-mentionned method, and for any couple of gravity coefficients $(C_{1m'}, S_{1m'})$. With the current value of θ and a mean value for I, it is found that $\dot{\psi} = -711''$. per century.

As it was expected, potential coefficients other than C_{20} produce nutations of much smaller amplitudes and tesseral harmonics cause short-periodic oscillations.

The noteworthy result is the resonant nutation with the coefficient C_{2o} and the argument $\Omega - \psi$, which is explained by the fact that the small factor $\dot{\Omega} - \dot{\psi}$ occurs in the denominator of the expressions of the amplitudes. This sets the problem of the mean value θ_0 of θ.
As yet we have not made this determination. However we have computed the nutations of Mars for several values of θ_0, ranging from 20° to 30° and with a mean value Io of I. The obtained amplitudes for the resonant

nutations are listed in table 1.

θ_0	B_1	A_1
20°	8° 39' 49"	- 3° 38' 5"
22°	7° 19' 43"	- 3° 32' 19"
24°	6° 11' 19"	- 3° 26' 12"
26°	5° 12' 9"	- 3° 19' 46"
28°	4° 20' 29"	- 3° 13' 5"
30°	3° 34' 59"	- 3° 6' 11"

Table 1 - Amplitudes A_1 and B_1 of the resonant nutation on θ and ψ respectively for several values of θ_0

For the next approximation, the main variations of θ , ψ and I are brought in the right-hand side members of the equations which are then expanded as Taylor series to first order around mean values.

The nutations resulting from coupling with the variations $\Delta\theta = A_o \cos V_o$ and $\Delta\psi = B_o \sin V_o$, and depending on C_{20}, are shown in Table 2. The very important motion in longitude with argument $\Omega - \Psi$ is the effect of the large variation of θ on the node of Mars equator with respect to the xy - plane of (R). The largest terms correspond to the arguments $\Omega - \psi$, $2\Omega - 2\psi$ and $3\Omega - 3\psi$. As expected, it is found that the other nutations have small amplitudes.

T (x 10^4 years)		First Iteration	Second Iteration $V_0 = \Omega - \Psi$	Second Iteration $V_0 = 2\Omega - 2\Psi$
16.86	A_Ψ	5° 3′ 58″	4° 40′ 21″	2″
	A_Θ	- 3° 18′ 47″	27″	- 4″
8.43	A_Ψ	3′ 6″	18′ 13″	1′ 5″
	A_Θ	- 1′ 32″	- 5′ 49″	0
5.62	A_Ψ		5″	6″
	A_Θ		- 2″	- 2″

Table 2 - Amplitudes A_ψ and A_θ of the nutations with arguments $\Omega - \psi$, $2\Omega - 2\psi$, $3\Omega - 3\psi$. T is the period.

The long periodic behaviour of Mars orbit inclination can be known from the theory of the secular variations of the orbital elements of the principal planets by Brouwer and Van Woerkom (1951). In this theory secular variations of the orbital plane of a planet are expressed as :

$$P = \sin I \sin \Omega = \sum_{j=1}^{8} N_j \sin(-s'_j t + e_j) \tag{3}$$

$$Q = \sin I \cos \Omega = \sum_{j=1} N_j \cos(-s'_j t + e_j) \tag{4}$$

Where I and Ω are referred to the ecliptic and equinox of 1950.0 .
A Fourier analysis of the function I (t) has been achieved and has shown that the larger variations can be fairly well represented by 5 superimposed sinusoïds of which we have determined the amplitudes and phases by a least-squares process. The obtained values are given in Table 3. For mean value of I we have found Io = 3°.57

K	H_K (ARC SECOND/ YEAR)	M_K (DEGREES)	D_K (DEGREES)
1	18.744	0.67	32
2	8.100	0.39	170
3	6.990	0.24	307
4	2.221	0.17	259
5	1.110	1.66	221

Table 3. Frequencies (h_k) and corresponding amplitudes (M_k) and phase constants (d_k) for the inclination of Mars.

An oscillation $\Delta I = M \cos v_1$ of the inclination (with $v_1 = ht + c$) results in nutations depending on C_{2o} and given in table 4. The largest amplitudes are obtained with the arguments $\Omega - \psi - v_1$, $\Omega - \psi + v_1$, $2\Omega - 2\psi - v_1$, $2\Omega - 2\psi + v_1$ and v_1. They are listed in this order for each value of h.

H (ARC SECOND/ YEARS)	T (x 10^4 YEARS)	A_ψ	A_θ
	4.90	8′ 15″	- 5′ 24″
	11.72	- 19′ 43″	12′ 54″
18.744	3.80	16″	- 8″
	38.44	- 2′ 39″	1′ 19″
	6.91	- 5′ 43″	0
	8.21	8′ 3″	- 5′ 16″
	312.95	- 5° 6′ 34″	3° 20′ 29″
8.100	5.52	13″	- 7″
	17.82	43″	- 21″
	16.00	- 7′ 42″	0
	8.83	5′ 19″	- 3′ 29″
	186.17	1° 52′ 13″	- 1° 13′ 23″
6.990	5.80	9″	- 4″
	15.46	23″	- 11″
	18.54	- 5′ 30″	0
	13.08	5′ 35″	- 3′ 39″
	23.71	10′ 8″	- 6′ 37″
2.221	7.37	8″	- 4″
	9.85	10″	- 5″
	58.36	- 12′ 15″	0
	14.73	1° 1′ 26″	- 40′ 10″
	19.71	1° 22′ 10″	- 53′ 44″
1.110	7.86	1′ 21″	- 40″
	9.09	1′ 33″	- 46″
	116.73	- 3° 59′ 13″	0

Table 4 - Nutations inferred from periodic variations of I

The expressions for nutations depending on C_{2o} and arising from the torque exerted on Mars by another planet have also been developped by the method of variation of parameters. These nutations have been computed in the case of the Earth and Jupiter. The dominant effects come from the latter planet though the amplitude remains very small ($\Delta\psi = 1''.1$, $\Delta\theta = 0''7$ for the largest, with period $1.2\ 10^6$ years)

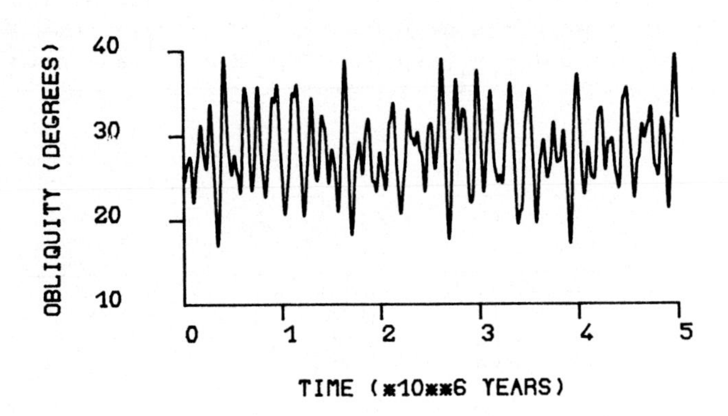

Fig. 1 Variations in the obliquity of Mars

4. CONCLUSION

Finally, we have plotted on fig.1 the behaviour of the obliquity ϵ with respect to the moving orbital plane of Mars. The periodic variations of ϵ look much more complex than those exhibited by Ward, due to the larger number of terms we took into account and to our more rigourous approach. Nevertheless, a term with an amplitude of around 10° and period of 1.2 10^6 years also exists in our results and we definitly think that such a phenomenon has now to be studied for its big impact on the planet climatic evolution.

REFERENCES

BORDERIES, N , 1977 :Mutual gravitational Potential of N Solid Bodies *Cel. Mech. in press.*

BROUWER, D.and VAN WOERKOM, A.J.J., *1951 - Astron. Pap. Amer. Ephemeris Naut. Alm., Vol XIII, pt II, 81.*

WARD, W.R. ; *1973, Science, Vol. 181, pp. 260-262*

CHRISTENSEN, E.J., 1977: Rotation of Mars, *J.P.L., Report 9 00.796*

SOLAR SYSTEM OCCULTATION PREDICTIONS USING AUTOMATED MICRODENSITOMETRY TECHNIQUES

P. J. Shelus and G. F. Benedict
University of Texas at Austin

1. Introduction

The development of automatic, computer-controlled, electronic equipment, under the direction of an intelligent user, has significantly affected the output of scientific knowledge from astronomical research. This increased output is as much a result of the easing and/or elimination of heretofore tedious tasks as it is a result of vast increases in data acquisition rates via observing systems which make more efficient use of in-coming electromagnetic radiation. One such automated system (Shelus et al, 1977) has been implemented by the present authors to predict occultations of stars by Solar System objects. In the past these predictions have been a time consuming task which was made even more onerous since only a very few observable phenomena were found. Perhaps of even more importance is that, since candidate objects to be occulted typically were obtained from star catalogs, such a search was incomplete. Note that phenomena involving stars fainter than normal catalog limits are certainly relevant in the cases of Uranus, Neptune, Pluto, minor planets and the natural satellites of the major planets.

Hardware consists of a Boller and Chivens (Photometric Data Systems) 1010A microdensitometer, controlled by a Digital Equipment Company PDP-8/e computer with 20K words of core memory; peripherals are two 830K word capacity disks and a fast floating point processor with extended precision capability. A description of an early realization of the hardware can be found in Wray and Benedict (1974). Wide-field photographic plate material is a set of glass copies of the National Geographic Society-Palomar Observatory Sky Survey (Minkowski and Abell, 1963). Reference stars are obtained from the Smithsonian Astrophysical Observatory Star Catalog (Smithsonian Institution, 1966). Planetary ephemerides are obtained from the various Developmental Ephemerides exported regularly by the Jet Propulsion Laboratory (Devine, 1967; O'Handley et al., 1969; Standish et al., 1976). Note that the system is completely independent of the source of plate material, reference star positions and/or ephemerides. The present specific

R. L. Duncombe (ed.), Dynamics of the Solar System, 151–156.

choices are a result of ready availability.

2. ESTABLISHMENT OF THE BASIC ASTROMETRIC REFERENCE SYSTEM

To outline the occultation prediction process a typical session will be described. The first step consists of determining how many and which reference stars lie on a selected wide-field plate; these objects will form the basic reference frame to be impressed upon the plate. One provides the right ascension and declination of the region of interest and the Palomar Sky Survey plate or plates on which the area of interest resides are identified and a machine-readable list of SAO stars contained in the selected region is produced. This list contains right ascensions and declinations as well as rectangular (x,y) coordinates of each object with respect to the plate center. These data establish a disk file of standard stars. A plot of the selected area is also generated. Typically, 50-150 SAO stars are provided over an approximately 25 cm square.

By inspection of the standard star list or of the plot a pair of stars widely spaced in x and at nearly the same y are identified. The plate is aligned on the microdensitometer platen using these two stars as a baseline; alignment is not critical to less than 50μ. The PDS is driven by the PDP-8/e to the position of every n^{th} (n is nominally 5 or 10) reference star on the disk file of standard stars. A single hand measurement is made at each location and an on-line step-wise regression program provides preliminary standard astrometric plate constants.

Using these preliminary plate constants, the PDS is then driven to every reference star, usually to within the image of the star on the plate. This provides easy identification in crowded fields and, for all stars, relieves the operator from any identification worries. Depending on the operator's assessment of any star image he may (a) skip measuring it altogether (image too large or excessively comated, image too close to another), (b) hand position n times and average (image too large for automatic raster scanning), or (c) allow the PDS to scan the image and obtain a "first-moment" position (this is essentially a simple center of gravity technique, where the image density is substituted for mass). When all stars have been measured the regression program provides final plate constants. The solutions and (o-c) residuals are presented for operator review. The R.M.S. error of a single SAO position for such a reduction is usually on the order of 0.3-0.6 arcsec in both right ascension and declination. Seven plate constants, including the so-called Schmidt term, are used for the Sky Survey astrometric reductions. The time required to align the plate, perform the measures and obtain final plate constants is of the order of two hours.

Once a final reference frame has been obtained and the operator has deemed it to be a satisfactory one, the system prompts the user for additional information. The two possible tasks in our current programs are the prediction of stellar occultations by Solar System objects and the establishment of secondary reference star systems for the astrometric reduction of small field plates. The former task will be dealt with here; the latter has been discussed elsewhere (Benedict et al., 1978).

3. THE STELLAR OCCULTATION PREDICTION PROBLEM

Having impressed a standard reference frame upon a wide-field photographic plate as described in the previous section the PDP-8/e drives the microdensitometer across the plate in a path which mimics the Solar System object of interest and inventories the area of that path for occultation candidates. For this task a polynomial representation of the astrometric right ascension and declination is continually evaluated. Going from (α,δ) to standard coordinates (ξ,η) by standard techniques (Smart, 1960) and then to measured (x,y) through the plate constants formerly derived provides the information needed to drive the microdensitometer. A preliminary pass generates 40-micron x-positions and 500-micron y-positions for every candidate object with a transmission less than some user input limit. This limit for stars fainter than about 10th magnitude is proportional to magnitude. A second pass, made only at the positions obtained from the first pass, produces 40-micron positions of each candidate object in two coordinates. The microdensitometer then is driven to each candidate object for final measurement. These measurements are made in a manner identical to that used for the original measurements of reference stars outlined in the previous section.

The output from this last step is a list of candidate objects which are typically within 35 arcsec (when using Sky Survey plates for scanning purposes) of the Solar System object and an approximate time of closest approach. These candidates are then scanned to produce instrumental magnitudes. When two color plates are available, the scanning of the second plate will provide, in addition to improved coordinates of each candidate object, an instrumental color. If faint $(V>10)$ standards can be found on a plate, these instrumental magnitudes can be transformed to give V and B-V with an accuracy of about ± 0.1 magnitude. The list of candidate objects is then processed to give a time history of separation for each event.

4. RESULTS

Since, in practice, faint local standards are not generally available, at the present time our final results consist of astrometric positions of each candidate star, time of closest approach, and a photographic finding chart centered on the candidate star. We feel that for fainter stars such a finding chart is much preferable to a simple position and an uncertain magnitude and color estimate. The production of finding charts involves the use of a system designed, built and implemented by our Radio Astronomy group (Hemenway et al., 1977). As an example of the results obtainable with this system we present four objects to be occulted by Saturn and several of its satellites in the 1979-1980 time frame. Geocentric positions, times of closest approach and minimum center-to-center separations are presented in Table I; the associated finding charts are given in Plate I. A complete list for the interval 1979-1985 will appear elsewhere.

5. DRIFT

All measuring engines suffer from a lack of positional stability. Extensive testing of our microdensitometer has shown the instrument to suffer from a drift which is monotonic and reasonably linear in nature, the problem being more severe on the Y-axis than on the X-axis. Two types of drift checks are used, the first being an edge position algorithm (Wray and Benedict, 1974) and the other uses the same first moment digital centering techniques as found in the system described above. Exhaustive mechanical and electronic diagnosis has reduced the rate of drift to about 2 microns per hour in Y and less than one quarter of a micron per hour in X. We have adopted the following procedure to allow results of the stated accuracy and precision to be obtained routinely. Just after the preliminary reference frame has been measured, the operator selects a magnitude 14 or 15 star near the center of the plate. The system measures the position of this star ten times and derives an average to be used as a zero point. As the occultation prediction procedure is carried out, this zero point star is monitored at intervals of about twenty minutes. After each check the position is compared with the zero point and the read out registers of the microdensitometer updated (if necessary) to bring the measured position back into coincidence with the zero point. The zero point correction is seldom more than one micron in either x or y. This dynamic correction process has proven to be quite effective even for lengthy runs involving meal breaks or other computer/operator malfunctions.

6. REMARKS

The high speed with which this task can be performed is due mainly to the following factors. Star images on the Palomar Sky Survey plates are not Gaussian; they tend to be flat-topped or even doughnut shaped. This justifies the use of faster, relatively unsophisticated first moment methods to derive positional information. Next, the operator is not required to do any bookkeeping or spend an inordinate amount of time in star identification and/or plate marking; the PDP-8/e does this for him. Set up time is kept to a minimum since one has the ability to generate preliminary plate constants to take care of mis-alignment. Finally, the fast floating point processor previously

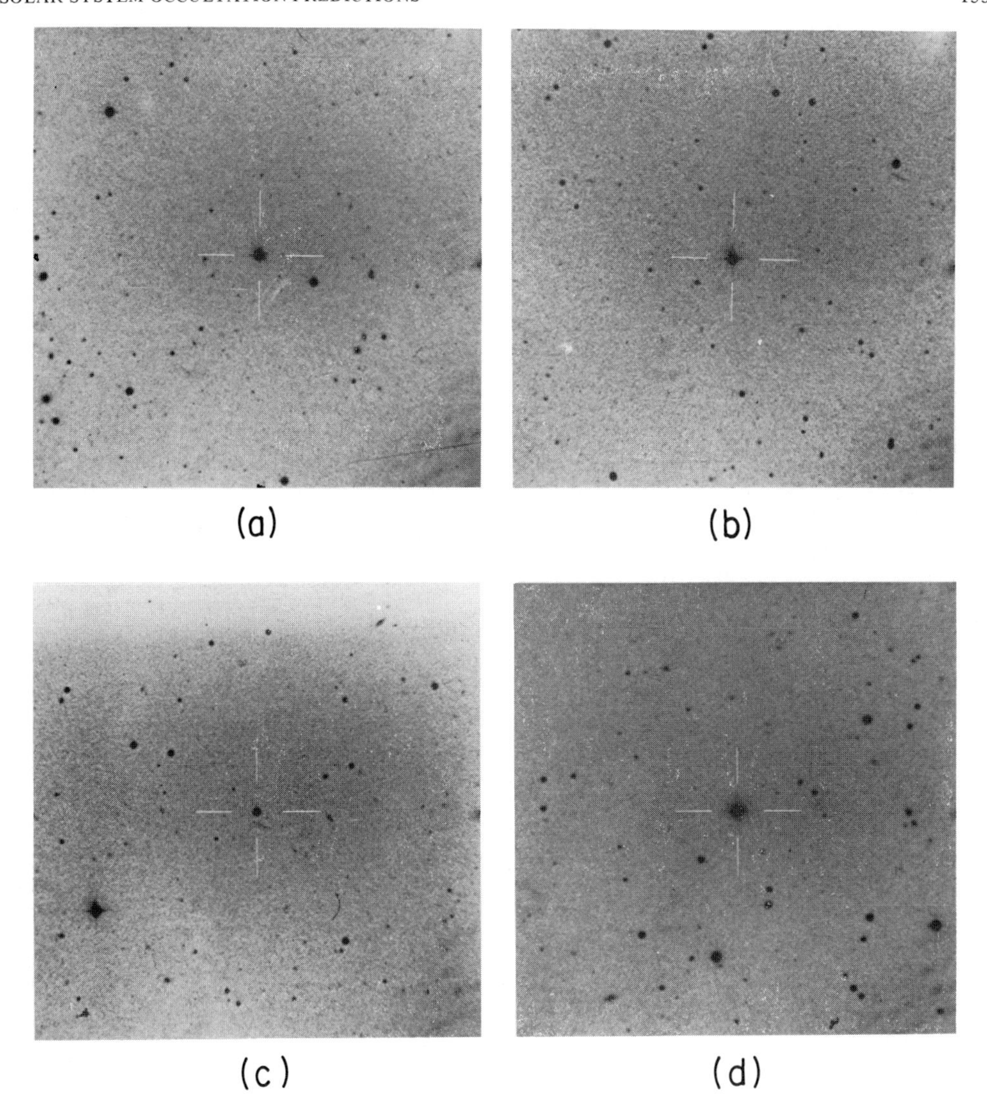

Figure 1. Selected occultation candidate finding charts (each reticle mark is 50" long).

Table 1. Geocentric phenomena of selected Saturn events.

	$\alpha_{1950.0}$	$\delta_{1950.0}$	Universal time	Δ	P.A.
a.	$11^{h}01^{m}29^{s}.61$	$+8^{\circ}17'15''.5$	1979 Jan $08^{d}09^{h}50^{m}$	$5''.0$	$326^{\circ}.9$
b.	$10^{h}59^{m}18^{s}.60$	$+8^{\circ}34'59''.0$	1979 Jan $23^{d}22^{h}28^{m}$	$9''.4$	$153^{\circ}.6$
c.	$10^{h}59^{m}03^{s}.02$	$+8^{\circ}37'08''.9$	1979 Jan $25^{d}07^{h}56^{m}$	$22''.6$	$153^{\circ}.9$
	(possible Tethys event)		1979 Jan $25^{d}12^{h}44^{m}$	$0''.9$	$321^{\circ}.7$
d.	$10^{h}47^{m}18^{s}.98$	$+9^{\circ}41'18''.6$	1979 Jul $08^{d}10^{h}50^{m}$	$6''.7$	$157^{\circ}.1$

mentioned is indispensable; its extended precision facility allows plate solutions identical to those obtained with a CDC 6400/6600 system to be generated in about 30 seconds.

7. ACKNOWLEDGEMENTS

We wish to thank E. C. Silverberg and D. S. Evans for indicating that meaningful astrophysical results can be obtained with new technology for the occultation of stars much fainter than presently found in fundamental catalogs.

8. REFERENCES

Benedict, G. F., Shelus, P. J., and Mulholland, J. D.: 1978, "Astron. J." (accepted for publication, August).
Devine, C. J.: 1967, "JPL Tech. Rept. 32-1181."
Hemenway, P. D., Torrence, G. W. and Wolf, C. S.: 1977, "Bull. American Astron. Soc." 9, 598 (abstract).
Minkowski, R. L. and Abell, G. O.: 1963, "Stars and Stellar Systems (Vol. III)", ed. by K. Aa. Strand, 481-487.
O'Handley, D. A., Holdridge, D. B., Melbourne, W. G., and Mulholland, J. D.: 1969, "JPL Tech. Rpt. 32-1465."
Shelus, P. J., Benedict, G. F., and Evans, D. S.: 1977. "Bull. American Astron. Soc." 9, 439 (abstract).
Smart, W. M.: 1960, "Text-Book on Spherical Astronomy," Cambridge University Press.
Smithsonian Institution (1966) Positions and Proper Motions of 258,997 Stars for the Epoch and Equinox of 1950.0 (Vol. I-IV), U.S. Gov't Ptg. Office, Wash., D.C.
Standish, E. M. Jr., Keesey, M. S. W., Newhall, X. X.: 1976, "JPL Tech. Rept. 32-1603."
Wray, J. D. and Benedict, G. F.: 1974, "SPIE," 44, 137.

PART IV

SATELLITES AND RINGS

REVIEW OF THE THEORIES OF MOTION OF THE NATURAL SATELLITES

P. J. Message
Department of Applied Mathematics and Theoretical Physics,
Liverpool University, Liverpool, England

Most of the natural satellites of the planets of the solar system may be put into one of three main groups, according as to which of three main influences dominate the perturbation of their motion from Keplerian motion about the primary planet. The first of these is the attraction of the Sun, which governs the perturbations of the Moon's motion about the Earth, and those of the outer satellites of Jupiter (satellites VI to XIII), and Saturn's satellite Phoebe. The second is the departure of the gravitational field of the planet from that of a spherically symmetric body (the "figure terms"), and this governs the perturbations of the two satellites of Mars, Jupiter's satellite Amalthea (V), Neptune's satellite Triton, is probably the most important influence on Uranus' satellites, and is important, though not dominant, for the inner satellites of Saturn. The third influence is the mutual attraction of the satellites themselves. An order of magnitude argument suggests that periodic perturbations from this cause could scarcely be expected to be measureable from Earth, were it not that the frequent appearance of small-integer near-commensurabilities of pairs of orbital periods, and the consequent argumentation of the associated perturbations by a variety of types of resonance effects, in the systems of Jupiter and Saturn, causes mutual perturbations to dominate the orbital theories of three of the four great satellites of Jupiter, and six of the nine satellites of Saturn, and enables the masses of most of the satellites involved to be determined with otherwise unexpected relative precision (in some favourable cases, of the order of one per-cent) from Earth based data. Let us now consider the satellite systems of each of the outer planets in a little more detail.

The two small satellites of Mars, Phobos and Deimos, were discovered by Hall at Washington in 1877. Their orbits were found to be very small (Phobos has an orbital period of about 7 2/3 hours), and very nearly circular. Interest was aroused by Sharpless (1945) whose analysis of the positional observations indicated an acceleration of the longitude of Phobos, of an amount very difficult to reconcile with either resistive drag or tidal friction. Wilkins (1968) found that the observations could be fitted adequately without the accelerative term, and Sinclair (1972), analysing observations from 1877 to 1969, found the value obtained for the acceleration to be very sensitive to the way in which the observations

R. L. Duncombe (ed.), Dynamics of the Solar System, 159–166.

were selected, so that its reality was not conclusively established. Shor (1975), including also more recent data, concluded that the observations were better fitted with the acceleration than without it. Positional data obtained by the spacecraft Mariner IX (Born and Duxbury, 1975) in general confirmed the orbital parameters obtained by Sinclair and Shor.

Turning now to Jupiter's system, the innermost satellite Amalthea (satellite V) was discovered by Barnard at the Lick Observatory in 1892. Its orbit is very nearly circular, and Van Woerkom (1950), in his discussion of observations made between 1892 and 1921, and in 1949, concluded that the eccentricity was so small that the apse motion could not be determined. P.V. Sudbury (1969), who used also observations made in 1954, 1955, and in 1967, found a lack of continuity between the expressions representing the node longitude for the two periods 1892 to 1921, and 1949 to 1967, for which observational data were available. Let us now consider the outer satellites. There is a group of four satellites, consisting of VI and VII, discovered by Perrine at the Lick Observatory in 1904 and 1905, respectively, X, discovered by Nicholson at Mount Wilson in 1938, and XIII, discovered by Kowal at the Hale Observatories in 1974. These move in very similar direct orbits, under the influence of strong solar perturbations. It was shown by Ross (1906, 1907) and Bobone (1935, 1937) that Delaunay's lunar theory may be successfully applied to these orbits. Hansen's method was applied to satellite X by Charnow, Musen, and Maury (1968). The outermost group of Jupiter's satellites comprises VIII, discovered by Melotte at Greenwich in 1908, and IX, XI, and XII, discovered by Nicholson in 1914, 1938, and 1951, respectively. These move in retrograde orbits at very similar distances from Jupiter. The solar perturbations are so great that these orbits show substantial departures from ellipses, and numerical integrations (Cowell and Crommelin, 1908, Grosch, 1947, and Herget, 1968) show that the instantaneous Keplerian elliptic parameters vary widely within a single period. Brown (1923) investigated the use of Delaunay's method, but later (1930, also Brown and Brouwer, 1937) developed a method using the true longitude as independent variable, and the reciprocal of the distance from Jupiter as one of the dependent variables, for application to satellite VIII. This method was applied by Hori (1957, 1958) to satellite IX. An approach using multiple Fourier analysis was made by Kovalevsky (1959) to satellite VIII.

Before considering the four great satellites, which offer the most intricate resonance situation in the solar system, let us turn to the system of Saturn, in which there are a number of resonance cases each involving just two satellites. The first satellite found was Titan, discovered by Huygens in 1655. The satellites Iapetus, Rhea, Tethys, and Dione were found by Cassini, in 1671, 1672, 1684, and 1684, respectively. Mimas and Enceladus were found by W. Herschel in 1789, Hyperion by the Bonds and Lassell within two days in 1848, and Phoebe was found by Pickering in 1898. A long series of positional measurements was carried out at Pulkova, and later at Babelsberg, by H. Struve and G. Struve, who analysed the data and constructed orbital theories of Mimas, Enceladus, Tethys, Dione, Rhea, Titan, and Iapetus (H. Struve; 1888,

1898, G. Struve; 1926, 1933). More recently, Jeffreys (1953, 1954) made improvements to the theories and reestimated the masses of most of the satellites, and further improvements, including the addition of later observations, by Kozai (1955, 1956, 1957, 1976; all satellites but the most recent three found), Garcia (1972: Tethys, Dione, Rhea and Titan), Rapaport (1973, 1976: Enceladus and Dione), and Sinclair (1974: Iapetus, 1977: all satellites but the most recent three). It is good to be able to report on the renewed activity in recent years in systematic positional measures of Saturn's satellites; in some cases the interval during which no observations had been made had grown to be longer than the period for which such observations were available. We have seen the effect of such a lacuna in the case of Jupiter V.

Let us now consider the particular types of orbital motion exhibited in Saturn's system. The case of Titan shows no peculiar features; the important perturbations are the precessions of the apse and of the orbit plane arising from the effect of Saturn's figure, the attraction of Iapetus and of the Sun. Iapetus, unlike the other satellites, does not remain near the plane of Saturn's equator and rings, and it could not, since the perturbations on it due to Titan and to the Sun are comparable. To first order, the secular part of the disturbing function is constant, and this implies that the normal to the orbit plane describes an approximately elliptic cone, the period of the motion being about 3000 years. The plane normal to the axis of this cone is called the Laplacian plane, and use of it as reference plane considerably simplifies the theory of the perturbations. The motion of Phoebe is governed largely by solar perturbations, and a Delaunay type theory was applied by Ross (1905) and revised by Zadunaisky (1954).

The first resonance case identified in Saturn's system was that involving Titan and Hyperion, first recognized as such by Newcomb (1891), who showed that the observed retrograde motion of the apse is reconciled with perturbation theory since the argument $\theta=4\lambda_H-3\lambda_T-\tilde{\omega}_H$, of what is in normal cases a periodic term of the disturbing function, in this case does not change monotonically, but oscillates, or "librates", about the value 180°. An alternative way of describing this is to say that the forced oscillations, due to the attraction of Titan, in the radial distance and longitude, are in this case, because of the argumentation due to the resonance, of larger amplitude than the free oscillations usually identified with the eccentricity of the orbit, so that the observed apses are the maxima and minima of the radial distance under the action of Titan, and the mean eccentricity is given by the amplitude of this forced oscillation. Motion of this type with no free oscillation would, if Titan's orbit were circular, correspond to a periodic solution of Poincaré's second sort in the restricted problem of three bodies, of such a type that, at each conjunction of Hyperion with Titan, Hyperion is at aposaturnium. In fact there is a free oscillation, giving mainly a libration in longitude of about 9° amplitude, and of period about 21 months. Also the eccentricity of Titan's orbit introduces an additional forced long-period oscillation, of period equal to that of the relative motion of the two apses (about 18 3/4 years), and of amplitude about 13°.8 in the apse longitude of Hyperion, and 0.024 in the eccentricity. The observations made between 1887 and 1922 were reduced in the treatment given by Woltjer (1928), who found that the mean apse motion and period of libration in longitude gave inconsistent estimates of the mass of Titan. Jeffreys (1953) suggested that terms in the square of Titan's eccentricity, which Woltjer did not calculate, are probably significant, and I have found that, to reconcile the observational data, it is in fact necessary to construct a second-order perturbation theory. I have done this using a Lie series transformation, of the type developed by Hori (1966), to separate the long-period and short-period parts of the motions, and I hope soon to be able to give an improved estimate of the mass of Titan.

In the case of the pair Enceladus and Dione, each produces significant perturbations on the other's motion, (so that the general, rather than the restricted, problem of three bodies must be used as a model), and there are two critical arguments, $\theta=2\lambda_{Di}-\lambda_{En}-\tilde{\omega}_{En}$ and $\theta'=2\lambda_{Di}-\lambda_{En}-\tilde{\omega}_{Di}$, of the disturbing functions. However the resonance is not so close relative to the disturbing mass-ratios as in the case of Titan's action on Hyperion, and this has the consequence that a good approximation to the long-period motion is provided by the equations linear in the rectangular-type variables $h=e_{En}\sin\theta$, $k=e_{En}\cos\theta$, $h'=e_{Di}\sin\theta'$, and $k'=e_{Di}\cos\theta'$. There are two independent free oscillations about the appropriate periodic orbit of Poincaré's second sort in the general problem of three bodies. In the case of Enceladus, the observed eccentricity corresponds to a forced oscillation, as in the case of Hyperion, while that of Dione is a free oscillation, as in the case usually encountered in orbital motion.

A further type of resonance is exhibited by Mimas and Tethys, whose motion is a free oscillation about a periodic solution of the general problem of three bodies of Poincaré's third sort, that is, near to commensurability, and not coplanar. The critical argument is $\theta=4\lambda_{Te}-2\lambda_{Mi}$ $-\Omega_{Te}-\Omega_{Mi}$, and librates about 0°. In addition to the mutual perturbations, there is appreciable precession of the orbit planes due to the oblateness of Saturn. The appropriate periodic solution is one in which the orbit planes precess at such a rate that conjunctions and oppositions of the two satellites always occur when they are at their furthest from the plane perpendicular to their total angular momentum (which is a constant) since, with this plane as reference plane, $\Omega_{Te}=\Omega_{Mi}+180°$, so that then $\theta=4(\lambda_{Te}-\lambda_{Mi})+2(\lambda_{Mi}-\Omega_{Mi})-180°$.

The case of Rhea is an example of a libration not associated with a resonance of orbital periods. Woltjer (1922) gave a treatment of the libration of the apse of Rhea about that of Titan, and this was rediscussed by Hagihara (1927), who showed that this was not dominated by the near 2:7 resonance of orbital period with Titan. Here the amplitude of the forced oscillation exceeds that of the free one in the theory of the secular variations for Rhea.

Let us now return to the four great satellites of Jupiter, which were discovered by Galileo (and perhaps independently by Marius) in 1610.

There are here two resonances of the 2:1 type ($n_1=2n_2$, and $n_2=2n_3$), but in addition we have the relation $n_1-2n_2=n_2-2n$ satisfied to at least observational accuracy. Laplace showed that the critical angle $\lambda_1-3\lambda_2+2\lambda_3$ could librate about 180°. Observations have shown that the amplitude of the libration is certainly very small. Sampson (1921) derived a theory based on a system of cylindrical polar co-ordinates, and discussed the available observational material, and produced tables which have been of long use in predicting the positions of the satellites, their eclipses and occultations by Jupiter, and transits across its disc. In recent years modification of some of the time constants have been made to maintain the agreement with observation (see Peters, 1973), and Lieske (1974, 1977) has undertaken a program of larger scope of improvement of the theory, including the 3:7 near-commensurability between satellites III and IV. A fundamentally different approach was made by de Sitter (1918, 1925) who proposed the use of an intermediary orbit which, for the satellites I, II, and III, consists of a periodic solution of the appropriate four-body problem, constructed by taking an exact solution of the perturbation equations in which the disturbing functions retain their secular and critical terms only, and in which the free oscillations in both the eccentricity and orbit plane variables are absent, and are taken account of in the calculation of the variations from the intermediary orbit. The angular arguments chosen are such that no small divisors arise in the treatment of the periodic perturbations. Sinclair (1975) showed how the broad features of the long-period motions could be described in an illuminating way by the use of the four critical arguments $\theta_1=2\lambda_2-\lambda_1-\tilde{\omega}_1$, $\theta_2=2\lambda_2-\lambda_1-\tilde{\omega}_2$, $\theta_3=2\lambda_3-\lambda_2-\tilde{\omega}_2$, and $\theta_4=2\lambda_3-\lambda_2-\tilde{\omega}_3$. The two 2:1 commensurabilities are not too close to prevent a good approximation being given by the use of the linear equations for rectangular variables of the type used for Enceladus and Dione, and the equations for the set $h_1=e_1 \sin\theta_1$, $k_1=e_1 \cos\theta_1$, $h_2=e_2 \sin\theta_2$, $k_2=e_2 \cos\theta_2$, $h_3=e_2 \sin\theta_3$, $k_3=e_2 \cos\theta_3$, $h_4=e_3 \sin\theta_4$, and $k_4=e_3 \cos\theta_4$, with the main secular variation terms included from the outset, show readily the periodic solution proposed as intermediary by de Sitter, and enable the equations for the variations to be set up easily in linear form. De Sitter (1931) analysed the data provided by Gill's heliometer measurements, as well as visual and photometric observations of eclipses, and found values for the coefficients of the main long-period terms, and frequencies and amplitudes of the larger free oscillations. The resulting equations for the masses of the satellites and the oblateness of Jupiter are far from consistent, however. Aksnes and Franklin (1974, 1975, 1976) have analysed the rich data provided by photometric observations of mutual occultations during the passage of the Earth through the plane of the orbits in 1973. Aksnes has recently begun the reformulation of the theory on the lines proposed and begun by de Sitter, giving hope that the potential advantages of that promising approach may be reaped.

Coming now to the system of Uranus, the satellites Titania and Oberon were discovered by W. Herschel in 1787, and Ariel and Umbriel by Lassell at Liverpool in 1851, and the fainter satellite Miranda by Kuyper at the McDonald Observatory in 1948. Dunham (1971) analysed the photographic observations made from 1905 to 1916, and from 1948 to 1966.

The earlier visual observations did not add to the determinations of the orbits. The orbits of the four brighter satellites are all very nearly circular, and very nearly coplanar with Uranus' equator. The apse motion of Ariel alone is properly determinable; that of Titania barely so. Consequently the masses are unknown, apart from a very rough estimate of that of Titania. The approximate relation $n_A - n_U - 2n_T + n_0 = 0^\circ.00341$ per day is not of the type leading to libration, though the relation $n_M - 3n_A + 2n_0 = -0^\circ.08$ per day may give rise to detectable perturbations of Miranda of period 12.5 years (Whitaker and Greenberg, 1973, Greenberg, 1975).

Neptune's large satellite, Triton, was discovered by Lassell at Liverpool in 1846. The perturbations of its orbit are dominated by the oblateness of Neptune, and Eichelberger and Newton (1926), improving Newcomb's theory, found that the normal to its orbit plane describes a circular cone, the semi-vertical angle being found by Gill and Gault (1968) to be 18.86 degrees, and the period of description 580.83 years. The second satellite, Nereid, was discovered by Kuyper in 1949, and its orbit has the high eccentricity of 0.76.

REFERENCES

Aksnes, K. : 1974, Icarus, 21, 100.
Aksnes, K. and Franklin, F. A. : 1975, Astron. J., 80, 56.
Aksnes, K. and Franklin, F. A. : 1975, Nature, 258, 503.
Aksnes, K. and Franklin, F. A. : 1976, Astron. J., 81, 464.
Born, G. and Duxbury, T. : 1975, Cel. Mech., 12, 77.
Bobone, J. : 1935, Astron. J., 45, 189.
Bobone, J. : 1937, Astron. Nachr., 262, 321.
Brown, E. W. : 1923, Astron. J., 25, 1.
Brown, E. W. : 1930, Yale Obs. Trans., 6, 65.
Brown, E. W. and Brouwer, D. : 1937, Yale Obs. Trans., 6, 189.
Charnow, M., Musen, P., and Maury, J. : 1968, Journ. Astronaut. Sci., 15, 303.
Cowell, P. H. and Crommelin, A. E. : 1908, Monthly Notices Roy. Astron. Soc., 68, 576.
de Sitter, W. : 1918, Ann. d. Sternw. Leiden, 12, Pt. 1.
de Sitter, W. : 1925, Ann. d. Sternw. Leiden, 12, Pt. 3.
de Sitter, W. : 1931, Monthly Notices Roy. Astron. Soc., 91, 706.
Dunham, D. : 1971, Ph.D. thesis, Yale University.
Eichelberger, W. S. and Newton, A. : 1926, Astron. Pap. Amer. Ephem., No. 9, Pt. 3.
Garcia, H. A. : 1972, Astron. J., 77, 684.
Gill, J. and Gault, B. : 1968, Astron. J., 73, 595.
Greenberg, R. : 1975, Icarus, 24, 325.
Grosch, H. J. : 1947, Astron. J., 53, 180.
Hagihara, Y. : 1927, Ann. Tokyo Obs., Appx. 17.
Herget, P. : 1968, Astron. J., 73, 737.
Hori, G. : 1957, Publ. Astron. Soc. Japan, 9, 51.
Hori, G. : 1958, Proc. Japan Acad., 34, 263.
Hori, G. : 1966, Publ. Astron. Soc. Japan, 18, 287.

Jeffreys, H. : 1953, Monthly Notices Roy. Astron. Soc., 113, 81.
Jeffreys, H. : 1954, Monthly Notices Roy. Astron. Soc., 114, 433.
Kovalevsky, J. : 1958, Astron. J., 63, 452.
Kozai, Y. : 1955, Proc. Japan Acad., 31, 6.
Kozai, Y. : 1956, Ann. Tokyo Obs., Ser. 2, 4, 191.
Kozai, Y. : 1957, Ann. Tokyo Obs., Ser. 2, 5, 73.
Kozai, Y. : 1976, Publ. Astron. Soc. Japan, 28, 675.
Lieske, J. H. : 1974, Astron. Astrophys., 31, 137.
Lieske, J. H. : 1977, Astron. Astrophys., 56, 333.
Newcomb, S. : 1891, Astron. Pap. Amer. Ephem., 3, 347.
Peters, C. F. : 1973, Astron. J., 78, 951.
Rapaport, M. : 1973, Astron. Astrophys., 22, 179.
Rapaport, M. : 1976, Astron. Astrophys., 51, 51.
Ross, F. E. : 1905, Harvard Ann., 53, No. 6.
Ross, F. E. : 1906, Lick Bull., 4, 110.
Ross, F. E. : 1917, Astron. Nachr., 174, 359.
Sampson, R. A. : 1921, Mem. Roy. Astron. Soc., 63.
Sharples, B. P. : 1945, Astron. J., 51, 185.
Shor, V. A. : 1975, Cel. Mech., 12, 61.
Sinclair, A. T. : 1972, Monthly Notices Roy. Astron. Soc., 155, 249.
Sinclair, A. T. : 1974, Monthly Notices Roy. Astron. Soc., 169, 591.
Sinclair, A. T. : 1975, Monthly Notices Roy. Astron. Soc., 171, 59.
Sinclair, A. T. : 1977, Monthly Notices Roy. Astron. Soc., 180, 447.
Struve, G. : 1926, Veroff. Berlin Ber., 6, Pt. 2.
Struve, G. : 1933, Veroff. Berlin Ber., 6, Pt. 5.
Struve, H. : 1898, Publ. Obs. Cent. Nicholas, 11.
Sudbury, P. V. : 1969, Icarus, 10, 116.
van Biesbroeck, G. : 1951, Astron. J., 56, 110.
Van Woerkom, A. J. J. : 1950, Astron. Pap. Amer. Ephem., 13, Pt. 1.
Whitaker, E. A. and Greenberg, R. J. : 1973, 165, 15.
Wilkins, G. A. : 1968, in G. Colombo (ed.), Modern Questions of Celestial Mechanics, Edizione Cremonese, Rome, p. 221.
Woltjer, J. : 1922, Bull. Astron. Inst. Netherlds., 1, 175.
Woltjer, J. : 1928, Ann. Sternw. Leiden, 16, Pt. 3.
Zadunaisky, P. E. : 1954, Astron. J., 59, 1.

DISCUSSION

Szebehely: Referring to the general versus the restricted problem, I assume you speak about the system, planet + satellite + satellite.

Message: Yes.

Szebehely: It is my understanding that the restricted problem is <u>not</u> applicable to the Neptune-Triton-Sun system.

Message: I agree.

Marsden: Would you care to make any comment about the alleged tenth and eleventh satellites of Saturn?

Message: I understand that their existence has not yet been confirmed.

Kozai: Regarding the question by Prof. Szebehely, the three-body problem of satellite case is more than the three-body problem as we cannot assume that the planet is a point mass.

AN IMPROVEMENT OF THE ORBITAL ELEMENTS OF HYPERION

Yoshizumi Hatanaka
Tokyo Astronomical Observatory
Mitaka, Tokyo 181, Japan

For the improvement of the orbital elements of Hyperion, a program of photographic observations of Saturnian satellites has been made since the opposition of 1966. In this paper are presented the reduction method and results derived from the observations during the oppositions of 1968 and 1977.

The photographic observation has been carried out with the 65-cm refractor of which focal length is about ten meters at the Tokyo Astronomical Observatory, Mitaka. In the earlier stage Fuji FLO plates (16.4 X21.4cm) were used without filter. Kodak 103aO plates (16X16cm) have been used since the opposition of 1970. All of the plates have been taken on or near the meridian. The several minute exposure is necessary to obtain a measurable image of Hyperion, but it is excessively long for those of the inner satellites. In order to obtain the best images of as many satellites as possible, the short time exposure ranging from twenty to forty seconds is made at the middle of the exposure under consideration, on the central area of the plate where the images of Saturn and the inner satellites are produced, though the long time exposure is made on the other parts. On the other hand, a few plates were exposed for scores of seconds on each observable night for the five inner satellites (Enceladus, Tethys, Dione, Rhea and Titan) and Iapetus that are brighter than Hyperion.

The positions of the images of the satellites were measured by the author on the Mann Type 422 comparator of the Tokyo Astronomical Observatory. Each image was bisected five times for the X and Y coordinates in one position of the plate. In order to obtain the Y position with the same accuracy as the X, the Y coordinate was measured in such an artificial posture that the measurer inclined his head by 90°. This method was able to be substituted for the measure in another position of the plate differing by 90°. All bisections were made so carefully that five measured values converged within a range of 7 microns. That is, the internal error of the measurements does not exceed 3.5 microns in the standard deviation, corresponding to 0.07 seconds of arc. When they did not converge due to the badness of the image, it was bisected again until the

R. L. Duncombe (ed.), Dynamics of the Solar System, 167–170.

number of the measured values which converged within the range amounted to five. The measured position was obtained as the average of the five measurements.

The calculated positions of the seven satellites were obtained from the orbital elements which were adopted in the Astronomical Ephemeris. But the elements of Hyperion, given by Woltjer (1928), were used without omitting the smaller perturbations. For their computation, the following original expression (Hatanaka, 1977) was used.

$$\rho\begin{pmatrix}\xi\\ \eta\\ \zeta\end{pmatrix} = R_1(90°-\delta)\ R_3(\alpha+90°)\ R_1(-\epsilon)\ R_3(-\Omega)\ R_1(-i)\ R_3(-f-\Pi+\Omega)\begin{pmatrix}r\\ 0\\ 0\end{pmatrix} \quad (1)$$

where α, δ, ρ, ϵ, r, f, Π, Ω and i are the right ascension, declination and geocentric distance of Saturn, the obliquity of the ecliptic, the distance from Saturn, true anomaly, longitudes of the perisaturnium and node, and inclination of the orbit to the ecliptic of the satellite, respectively. If necessary, the differential refraction correction was made only for the η values of Titan, Hyperion and Iapetus. The apparent Saturnicentric position (ξ, η) with respect to the equator of the Earth is referred to as the C-value.

The reduction of the observed position of Hyperion was as follows. The plate scale was determined from the measures of the angular distance between Rhea and Titan, since our analysis disclosed that the currently adopted orbital elements of Rhea and Titan seemed to be accurate for the purpose. The scale was adopted on all the plates during the opposition. Making use of the plates of which exposures were made on the satellites except for Hyperion, the observed position of Iapetus was obtained from the plate orientation and Saturn's position which were determined by the method of least squares mainly from the positions of Titan and Rhea, and by making complementary use of Enceladus, Tethys and Dione. The orbital elements of Iapetus at each opposition were improved by least squares of the equations described later, from the O-C differences of its position during the opposition. To obtain the observed position of Hyperion from the measures, Titan, Rhea and the improved Iapetus were useful as the reference satellites, that is, the orientation of the plate was derived from the positions of Rhea and Titan and the center of Saturn was determined from the improved position of Iapetus. The reason of the complicated procedure is that there is a discrepancy in the telescope guiding between the partial exposure for the inner satellites and the total for Hyperion and Iapetus. The observed position (X,Y) relative to Saturn, positive towards the east and north of the Earth, is referred to as the O-value. Figure 1 shows the O-C value with the observation date in UTC for all the observations used. An arrow directs from C to O, multiplied 25-fold in size.

The equations of condition for the improvement of the orbital elements are given by the differentiation of (1) and written in the following general forms (Hatanaka, 1977).

$$X - \xi = A da + B d\varepsilon + C d(e \sin\Pi) + D d(e \cos\Pi) + E d\Omega + F di \quad (2)$$

$$Y - \eta = G da + H d\varepsilon + I d(e \sin\Pi) + J d(e \cos\Pi) + K d\Omega + L di \quad (3)$$

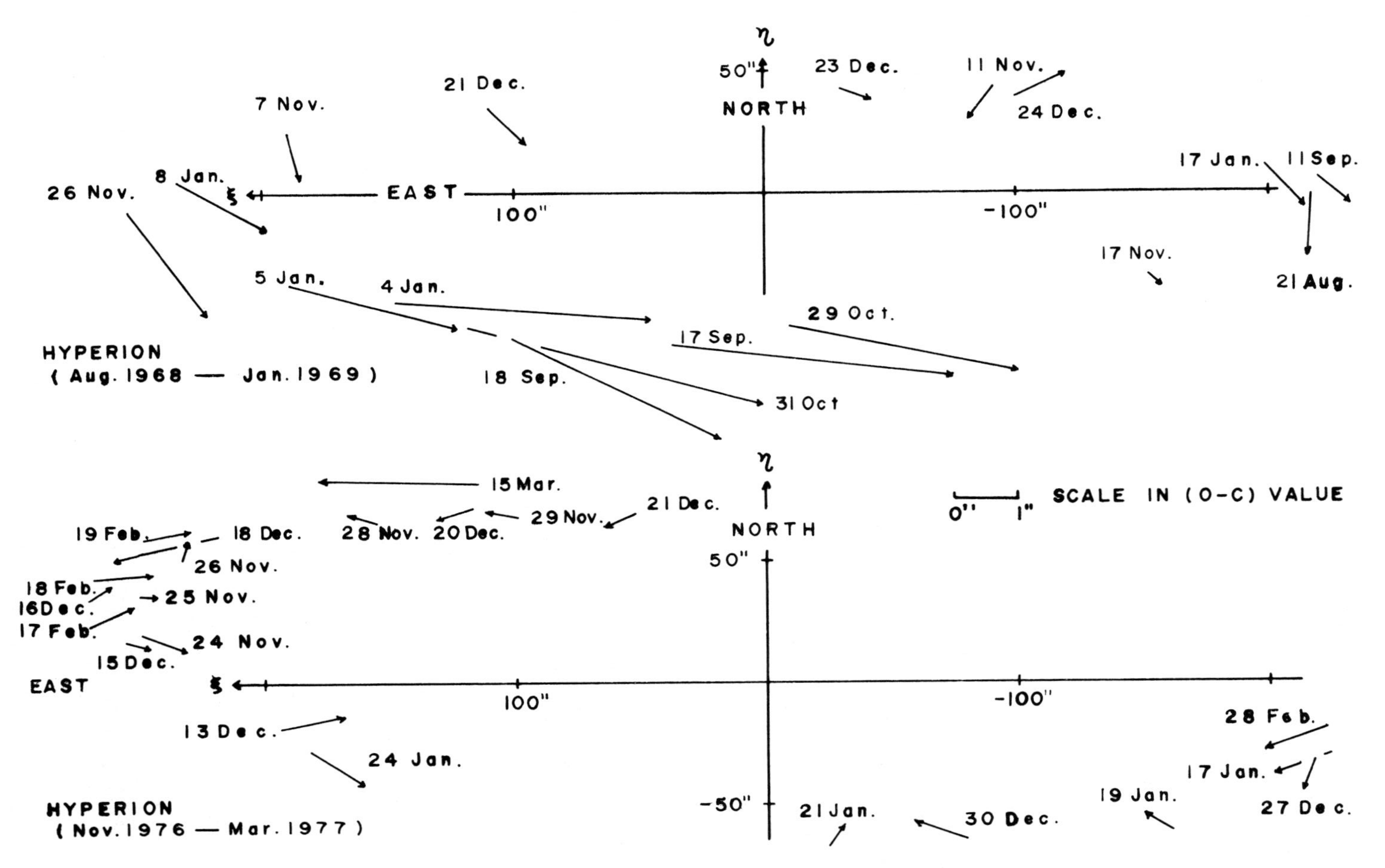

Figure 1. O–C value of the position of Hyperion. An arrow directs from C to O.

in which the twelve coefficients are expressible in terms of the orbital elements of the satellite to be improved and α, δ, ρ, ϵ. A simultaneous least-squares solution was made for corrections to the orbit during each opposition. Thus, the results obtained for Hyperion with the mean error are as follows.

1968 (17 nights)	1977 (22 nights)	
da = - 6".8 ± 1".7	da = - 5".4 ± 1".2	(at unit distance)
dε = - 0°.42 ± 0°.05	dε = - 0°.22 ± 0°.05	
de = 0.0038 ± 0.0011	de = 0.0006 ± 0.0010	
dΠ = - 4°.41 ± 0°.69	dΠ = - 1°.36 ± 0°.58	
dΩ = - 0°.16 ± 0°.11	dΩ = - 0°.14 ± 0°.10	
di = 0°.07 ± 0°.05	di = 0°.10 ± 0°.03	

The standard deviations of residuals are 0".50 and 0".45, respectively. The corrections of the positions of Iapetus determined by the method of least squares are given below. They were used for obtaining the position of Hyperion mentioned above.

1968 (37 nights)	1977 (32 nights)	
da = - 0".3 ± 1".1	da = 5".1 ± 0".6	(at unit distance)
dε = - 0°.040 ± 0°.013	dε = - 0°.181 ± 0°.007	
de = -0.00093 ± 0.00030	de = 0.00151 ± 0.00020	
dΠ = - 0°.49 ± 0°.34	dΠ = - 4°.19 ± 0°.38	
dΩ = 0°.22 ± 0°.04	dΩ = - 0°.10 ± 0°.02	
di = - 0°.122 ± 0°.012	di = - 0°.059 ± 0°.007	

The standard deviations of residuals are 0".49 and 0".26, respectively.

The computations were made with the FACOM 230-58 computor at the Tokyo Astronomical Observatory.

References

Hatanaka, Y., 1977. Proceedings of a Symposium on Celestial Mechanics, 1977, p. 42. (Edited by Hori and Yuasa, held in Tokyo)

Woltjer, J., 1928. Annalen van de Sterrewacht te Leiden, 16, part 3.

THEORIES OF RESONANT SATELLITE PAIRS IN SATURN'S SYSTEM

W. H. Jefferys and L. M. Ries
University of Texas at Austin

INTRODUCTION

Upcoming missions to the outer planets have made evident the need for better theories of their satellites. We have been involved in an effort to provide better satellite theories through new observations (Abbot, Mulholland and Shelus 1975; Mulholland, Shelus and Abbot 1976; Mulholland and Shelus 1977; and Benedict, Shelus and Mulholland 1978), and new analytical theories (Jefferys and Ries 1975; Jefferys 1976. Hereinafter these are denoted Paper I and Paper II, respectively). In this paper we report on our incorporation of new and old observations into our theories, and on our progress on the theoretical front.

REVIEW OF EARLIER WORK

Our goals for the theoretical work have been to calculate new analytical theories for the three resonant pairs in Saturn's system: Enceladus-Dione, Titan-Hyperion and Mimas-Tethys, using the same software for all theories. We are attempting to calculate all terms down to the level of a few kilometers--potentially observable from space, and a great improvement in the accuracy of currently used theories (Struve 1930, 1933; Woltjer 1928). For example, observations of the satellites of Saturn with the Space Telescope can be expected to have errors of under 10 km.

We have employed the algebraic manipulation language TRIGMAN (Jefferys 1970, 1972) to calculate our theories, using the canonically invariant Hori-Lie theory (Hori 1966) in noncanonical variables. Because of the advantages of the Hori-Lie theory, it is possible to calculate the perturbations in any desired quantity (e.g., longitude, latitude, radius vector) directly, and it is not necessary to employ canonical variables, as long as routines for calculating Poisson brackets are available.

Because the theories are being developed automatically, they will

R. L. Duncombe (ed.), Dynamics of the Solar System, 171–176.

be made available in the form of Fortran subroutines for direct calculation of the perturbations (and their partial derivatives) by machine. Since the constants of integration appear in literal form, it can be expected that the theories will be useful for a substantial period of time.

As explained in Paper I, we have developed the Hamiltonian in the variables

$$\begin{aligned} a &= \bar{a} + f_A A \\ e &= \bar{e} + f_E E \\ i &= \bar{i} + f_I I \end{aligned} \qquad (1)$$

where $\bar{a}$, $\bar{e}$, $\bar{i}$ are nominal values of the semimajor axis, eccentricity and inclination, respectively; f_A, f_E, f_I are numerical factors for controlling truncation; and A, E and I are the variables carried in the series expansions. Similar expressions involving truncation factors are used to express other quantities, e.g., the satellite masses and the dynamical form factors of Saturn.

By making use of the relations between the Delaunay variables and the usual elliptic variables, one can easily write down expressions for the partials of the variables A, E, and I with respect to the Delaunay variables, and hence of the Poisson bracket of any two functions expressed in terms of A, E, I and the Delaunay angles.

NEW THEORETICAL WORK

In Paper I a two-step method of eliminating first short-period and then long-period (i.e., resonant) arguments was described. Paper II showed how the elimination of the resonant arguments could be simplified by the introduction of a novel set of arguments, at the cost of an increase in the number of degrees of freedom of the system. This new set of arguments makes possible the treatment of the resonance completely independently of all other variables, and does not require the introduction of unnatural and awkward combinations of variables.

Our most recent work has simplified this procedure even more. The substitution of variables given in Paper II, in fact, can be made at any time, even prior to the elimination of the short-period terms. (Indeed, if there are two resonances, the same transformation can be applied to each, increasing the number of degrees of freedom by two. Under certain circumstances, such as small oscillations, the resulting equations can be solved. This approach may well provide a method for handling two simultaneous critical arguments under some circumstances). As a result, we have elected to make this transformation at the outset, and to eliminate both short and long with a single canonical transformation.

It has also become evident, since the publications of Papers I and II, that there are important terms at the 1-100 km level in the theories of these satellites when second order terms are computed. For example, the combination of the critical argument with short-periodic terms in the oblateness will, in the second-order theory of Enceladus and Dione produce terms in $(\ell \pm q)$, where q is the libration of the critical argument, having amplitudes of nearly 100 kilometers (about 0".01 when observed from Earth). In addition, there are significant terms in the mean motions, such as those found by Kozai (1957), although whether they appear in first or second order depends upon how the calculation is done.

The theory of Mimas-Tethys, which we have computed in nearly final form (except for a few terms arising from the resonance) provides yet another reason to go to higher than the first order. In this resonant pair, the libration argument is quite large (amounting to over 43° in the case of Mimas' longitude). This in turn means that the small-oscillation approximation which could be applied for Enceladus and Dione is no longer valid. Other workers (e.g. Kozai 1957) have used the exact solution in elliptic functions, but this has two drawbacks in the present theory; first, TRIGMAN being a Poisson series processor, there is no easy way to incorporate elliptic functions; moreover, the possibility of second-order contributions from other terms in combination with the libration needs to be taken into account.

Our approach has been to develop the solution of the large-amplitude oscillation as a power series in the amplitude parameter, using the Hori-Lie method to obtain as many terms as are needed. According to the prescription of Paper II for handling the critical argument, the relevant part of the Hamiltonian is

$$F = F_o + n_o\Theta + \frac{A}{2}\Theta^2 - B\cos\theta, \qquad (2)$$

where $n_o \equiv 0$ is the constraint imposed at the resonance by the procedure of Paper II, although the partials of n_o do not vanish; A and B are constants depending on the initial conditions; and (Θ,θ) are canonically conjugate variables (θ being the libration argument).

Expanding the cosine in powers of θ yields in the lowest order a harmonic oscillator:

$$F = F_o + n_o\Theta - B + \frac{1}{2}(A\Theta^2 + B\theta^2) - \frac{B}{24}\theta^4 + \ldots \qquad (3)$$

By the substitution of the canonical pair

$$\Theta = (2p\gamma)^{1/2}\cos q, \qquad \theta = (2p/\gamma)^{1/2}\sin q, \qquad (4)$$

with $\gamma = (B/A)^{1/2}$ and

$$n_q = (AB)^{1/2}$$

this can be brought into the form

$$F = F_o + n_o(2p\gamma)^{1/2}\cos q \quad - \quad B + n_q(p - \frac{1}{16\gamma}p^2 + \ldots)$$
$$+Ap^2[\frac{1}{12}\cos 2q - \frac{1}{48}\cos 4q] + \ldots$$

By considering $n_q(p - \frac{1}{16\gamma}p^2 + \ldots)$ to be the zero-order Hamiltonian insofar as (p,q) are concerned, the Hori-Lie procedure allows the elimination of the variable q through the use of a determining function S(p,q,...), where the ellipsis represents other variables.

If we consider now the effect of the libration on the longitude, the leading term in S is found to be

$$-\frac{n_o}{n_q}(2p\gamma)^{1/2}\sin q,$$

which contributes to the longitude (in first order) the term

$$\{\ell,S\} = -(2p\gamma)^{1/2}/n_q\{\ell, n_o\}\sin q$$

$$= (2p\gamma)^{1/2}/n_q \frac{\partial n_o}{\partial L}\sin q$$

Note that since $n_o \equiv 0$ because of the constraint condition, other partials do not contribute.

In second order, the elimination of terms involving p^2 yields terms in S arising from the Poisson brackets of $(2p\gamma)^{1/2}\cos q$ and $(2p\gamma)^{1/2}\sin q$ with terms in 2q and 4q, which are of the form

$$n_o p^{3/2}\sin q \quad \text{and} \quad n_o p^{3/2}\sin 3q.$$

These in turn give rise to terms in the longitude in q and in 3q, factored by $p^{3/2}$. Similarly, the third order theory provides terms in q, 3q and 5q, factored by $p^{5/2}$, and so on.

The theory of Mimas-Tethys is being extended to the level of accuracy which is the goal of this work. Judging by the ease with which the Enceladus-Dione programs were modified for the Mimas-Tethys case, we anticipate no great difficulties in this, nor in the extension to the Titan-Hyperion case.

COMPARISON OF THEORY AND OBSERVATION

We have applied our theory of Enceladus and Dione to the observations, including new observations made at McDonald Observatory over the past few years. Rather than to fit observations in each coordinate or quantity separately, we have attempted to make a single least-squares solution for all parameters. This has involved some difficulties and delays, partly because of the treatment of the critical argument that is described in Paper II. In that method, the number of degrees of freedom in the problem is increased by one through the introduction of the variables (Θ,θ), or equivalently, (p,q). This in turn means that two conditions of constraint must be imposed and consistently applied throughout the solution. There are methods of applying such constraints which are particularly elegant from a mathematical point of view (e.g., Brown, 1955). However, since these procedures were not available to us as programs, and other Least-Squares programs were available, it was decided to adapt the equations to the available programs rather than vice versa. In retrospect, this may not have been the best choice, although we have finally obtained satisfactory results.

The values of the parameters obtained from the theory do not differ greatly from those of Kozai (1957), although we have chosen to work in the equatorial system of 1950.0 rather than Struve's ecliptic system of 1889.25. The advantages of the present theory, therefore, lie not so much in the improvements that can be obtained from groundbased observations, but rather on their potential for improvement from space observations. Nevertheless, we do find that the recent observations made at McDonald Observatory are quite good, yielding mean residuals on the order of about 1 arcsecond or less in both right ascension and declination. The older observations are of variable quality, some yielding residuals as much as 5-10 times as large. Others among the older observations are of excellent quality.

We find for the forced librations of Enceladus and Dione the coefficients 12°.18 and 0°.66, respectively, and for the free librations 15°.54 and 0°.84. The value of J_2 that we obtain is +0.01666 ± 0.00001; however, we have not been able to solve for J_4 separately, and have therefore adopted the value -0.001 in our solutions.

ACKNOWLEDGEMENTS

We would like to express our appreciation for the help of Drs. J. Derral Mulholland and Peter J. Shelus, who have been most supportive of these efforts. The support of the United States National Aeronautics and Space Administration, under grants NSG 44-012-282 and NSG 7408, is gratefully acknowledged.

REFERENCES

Abbot, R. I., Mulholland, J. D., and Shelus, P. J.: 1977, "Astron. J." 80, pp. 723-728.

Benedict, G. F., Shelus, P. J., and Mulholland, J. D.: 1978, to appear in "Astron. J.".

Brown, D.: 1955, "A Matrix Treatment of the General Problem of Least Squares Considering Correlated Observations." Ballistic Research Laboratories, Report No. 937.

Hori, G.-I.: 1966, "Publ. Astron. Soc. Japan" 18, pp. 287-296.

Jefferys, W. H.: 1970, "Celest. Mech." 2, pp. 467-

Jefferys, W. H.: 1972, "Celest. Mech." 6, pp. 117-

Jefferys, W. H.: 1976, "Astron. J." 81, pp. 132-134.

Jefferys, W. H., and Ries, L. M.: 1975, "Astron. J." 80, pp. 876-884.

Kozai, Y.: 1957, "Ann. Tokyo Astron. Obs." Ser. II, 5, pp. 73-127.

Mulholland, J. D., and Shelus, P. J.: 1977, "Astron. J." 82, p. 238.

Mulholland, J. D., Shelus, P. J., and Abbot, R. I.: 1976, "Astron. J." 81, pp. 1007-1009.

Struve, G.: 1930, "Veröff. Univ. Berlin-Babelsberg" 6, part 4.

Struve, G.: 1933, "Veröff. Univ. Berlin-Babelsberg" 6, part 5.

Woltjer, J.: 1928, "Ann. Sterrewacht Leiden" 16, part 3.

DISCUSSION

Garfinkel: Does not your F_o, expanded in powers of the momentum θ, correspond to the Ideal Resonance Problem, rather than the Simple Pendulum?

Jefferys: Yes, if we take into account the powers of θ beyond the second. We have not done that in our first approximation.

THE MOTIONS OF URANUS' SATELLITES: THEORY AND APPLICATION

Richard Greenberg
Planetary Science Institute, Tucson, Arizona, U.S.A.

As spacecraft and sophisticated ground-based observations measure physical properties of many planets and satellites, dynamical theory and astrometry remain a principal source of such knowledge of the Uranian system. Study of the motions of Uranus' satellites thus has broad application to planetary studies as well as to celestial mechanics. Moreover, the structure and dynamics of the system provide important cosmogonical constraints; any theory of solar system origin and evolution must account for the formation within it of analogous systems of regular satellites.

The five known satellites of the Uranian system have nearly circular, regularly spaced, coplanar orbits. To remain coplanar, they must lie near Uranus' equatorial plane, consistent with spectroscopic measurements of Uranus' rotation, which show rotation in the same direction as the satellites' motion. Small, irregular satellites, like Jupiter and Saturn's, have yet to be discovered. Despite the internal regularity of the known system, it has a highly irregular, 98° obliquity, the origin of which is an important question for dynamical astronomy.

The rings of Uranus, recently discovered interior to the other satellites' orbits (Elliot *et al*. 1977), have properties which raise interesting dynamical problems; they are apparently narrow, in contrast to the broad rings of Saturn, and in conflict with theories of collisional evolution (Goldreich and Nicholson 1977); they are not all circularly symmetric, again in conflict with theory; and, since they are not visible in reflected solar light, the ring material must be very dark (albedo $< 5\%$) (Smith 1977), surprising for outer solar system material. Moreover, dynamical constraints on masses strongly suggest that Miranda and Ariel are icy in contrast to the nearby ring material.

Masses are determined from satellites' mutual perturbations. One diagnostic effect is precession. Dunham (1971) found Titania's precession rate to be $2^\circ.9 \pm 1^\circ.5$/yr. Assuming this behavior to be dominated by Oberon (UIV), he found the latter's mass to be $(0.8 \pm 0.6) \times 10^{-4}$ (herein Uranus' mass $\equiv 1$). Dunham's precession rate for Oberon is too

R. L. Duncombe (ed.), Dynamics of the Solar System, 177–180.

imprecise for any useful mass constraint.

Other mutual perturbations are the enhanced variations in longitude due to the near-commensurability amongst the inner three satellites, Miranda (UV), Ariel (UI) and Umbriel (UII). Their mean motions nearly obey the Laplace relation, $n_V - 3n_I + 2n_{II} = 0$. The relation implies that the combination of orbital longitudes $\Theta \equiv \lambda_V - 3\lambda_I + 2\lambda_{II}$ varies slowly. Θ can be interpreted geometrically as the angle between the longitudes of UV and UI when UI and UII are in conjunction (i.e. when $\lambda_I = \lambda_{II}$). Θ circulates through 360° in 12.5 yr, slow compared with orbital periods, so geometrical configurations of the three satellites repeat periodically, enhancing perturbations. Earlier workers (Harris 1949, Dunham 1971) assumed such effects were negligible, because the relation amongst mean motions is only approximate, not exact as for three Galilean satellites. As I shall show, their assumption was not justified *a priori* for plausible satellite masses.

Analysis of the Laplace relation is more complicated than common two-satellite commensurabilities; the critical argument, Θ, does not appear in the Fourier expanded disburbing function. Only when the perturbation theory is extended to second order in satellite masses do combinations of terms appear which have Θ, or multiples of Θ, as arguments of sines and cosines. Then, but not before, other terms with short-periods can be neglected to study the effects of the commensurability. In the Galilean case, first order terms with arguments $2\lambda_2 - \lambda_1$ and $2\lambda_3 - \lambda_2$ have long periods. They dominate and thus simplify the theory [cf. Professor Hagihara's (1972) lucid account of Souillart's theory].

In the case of the Uranian satellites, many other first-order terms have significant contributions to the second-order long-period terms. Development of the theory is a much more tedious procedure. The first partial treatment was made by Sinclair (1975). I have described in previous publications (Greenberg 1975a, 1976) an approach to the theory that uses some numerical shortcuts. Applying my methods, I find

$$dn_V/dt = \mu_I\mu_{II}n_V^2 \left[-11.5 \sin \Theta + 5.8 \sin 2\Theta + 1.1 \sin 3\Theta + \ldots\right]$$

$$dn_I/dt = \mu_V\mu_{II}n_I^2 \left[83.9 \sin \Theta - 42.5 \sin 2\Theta - 8.3 \sin 3\Theta + \ldots\right]$$

$$dn_{II}/dt = \mu_V\mu_I n_{II}^2 \left[-274 \sin \Theta + 36.2 \sin 2\Theta + 9.9 \sin 3\Theta + \ldots\right]$$

where μ's are satellites' masses. Based on known visual magnitudes, and the assumption of albedos and densities identical to Oberon's, Dunham estimated $\mu_V = 3 \times 10^{-6}$, $\mu_I = 6 \times 10^{-5}$, and $\mu_{II} = 2 \times 10^{-5}$. For these values, and the known behavior of Θ, integration gives the following amplitudes of longitude variations: $\Delta\lambda_V \sim 15^o$, $\Delta\lambda_I \sim 0\overset{\circ}{.}9$, $\Delta\lambda_{II} \sim 3\overset{\circ}{.}5$. In fact, the observed amplitudes are $< 5^o$ for UV (Greenberg 1976) and $< 0.1^o$ for UI and UII (Dunham 1971). From these we derive the following limits on mass products: $\mu_I\mu_{II} \lesssim 10^{-9}$, $\mu_V\mu_I \lesssim 5 \times 10^{-12}$ and $\mu_V\mu_{II} \lesssim 6 \times 10^{-12}$.

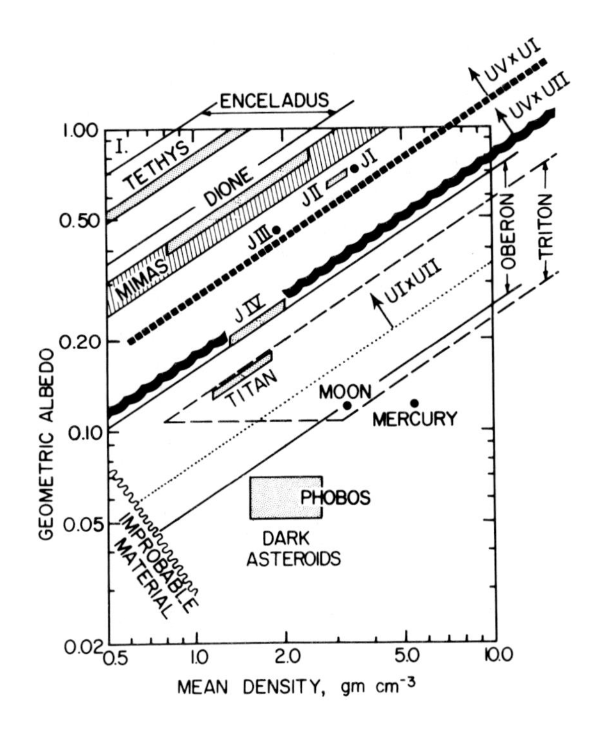

Figure 1. Density vs. Albedo plot showing limits for Uranian satellites and other planetary bodies.

These limits combined with visual magnitudes, place constraints on bulk physical properties of the satellites. Consider a plot of density versus albedo (Figure 1) on which a body's known mass and magnitude define a line of slope 2/3. For comparison limits for other planets and satellites are shown. For UV and UI the limit is shown assuming each has the same density and albedo. One might lie below this limit, but the other must then lie at least as far above it to satisfy the mass product constraint. Similar limits are shown for the pairs UI-UII and UV-UII. These limits support Dunham's assumption that Titania's precession is dominated by Oberon, whose boundaries are also shown. UI and UII are not both dark or carbonaceous material, although one of them might be. Also UV or UI must lie in the region of the plot suggestive of icy material. Although the region admits bright-rock-like albedos, this would require a low bulk density such as that of ice. Conversely, rock-like densities would imply high albedos such as a surface of ice or of evaporite salts, as has been suggested for Io (Fanale et al. 1977). The inner satellites are much brighter material than the rings and significantly brighter and/or less dense than Oberon.

Another application of Uranian Satellite dynamics is in the determination of Uranus' oblateness, a basic constraint on interior structure models. Measurements of the optical oblateness, ε, have been made by Dollfus (1970) and by Danielson et al. (1972), who found $.03 \pm .008$ and

.01 ± .01, respectively. From the apsidal precession rate of Ariel's nearly circular orbit, based on the apparent orientation of the apsides at two epochs. Dunham found J_2 = 0.012. From the apsidal and nodal precession of Miranda, Whitaker and Greenberg (1973) obtained a lower value, J_2 = 0.005. For a uniform fluid model and the long-accepted rotation period 10.8 hours (Moore and Menzel 1930), both values of J_2 correspond to ε values larger than observed optically. Greenberg (1975b) suggested that the rotation should be remeasured using modern equipment. Recent measurements show the period to be 24 ± 3 hr (Hayes and Belton 1977, Trafton 1977). Thus Dunham's J_2 corresponds to $\varepsilon \approx 0.025$ and Whitaker and Greenberg's to $\varepsilon \approx 0.013$, both in agreement with optical values. However, Podolak (1976) suggested that Dunham's J_2 is not consistent with realistic interior models, while J_2 = .005 is quite plausible.

Whitaker and Greenberg did note explicitly that their determination depended on the assumption that the Laplace relation had negligible effects on Miranda's longitude. Yet we have seen that the Laplace relation may be much stronger than previously thought. The next step will be to fit the complete theory of motion to the full set of observations, including the significant data base acquired in the last few years. It should be possible to separate the effects of planetary oblateness and satellite masses and to solve for these parameters. A great deal remains to be learned about the physical properties of the Uranian system by the methods of celestial mechanics.

REFERENCES

Danielson, R.E., Tomasko, M.G., and Savage, B.D.: 1972, "Astrophys. J." 178, pp. 887-900.
Dollfus, A.: 1970, "Icarus" 12, pp. 101-117.
Dunham, D.W.: 1971, 'The Motions of the Satellites of Uranus', Yale Univ. (Ph.D. Thesis).
Elliot, J.L., Dunham, E., and Mink, D.: 1977, "Nature" 267,pp.328-329.
Fanale, F.P., Johnson, T.V., and Matson, D.C.: 1977, in J.A. Burns (ed.), "Planetary Satellites", Univ. of Arizona Press, Tucson, pp. 379-405.
Goldreich, P. and Nicholson, P.: 1977, "Nature" 269, pp. 783-784.
Greenberg, R.: 1975a, "Monthly Notices Roy. Astron. Soc." 173, pp.121-129.
Greenberg, R.: 1975b, Icarus" 24, pp. 325-332.
Greenberg, R.: 1976, "Icarus" 29, pp. 427-433.
Hagihara, Y.: 1972, "Celestial Mechanics", Vol II, CH. 9.10, MIT Press, Cambridge, Massachusetts.
Harris, D.L.: 1949, 'The Satellite System of Uranus', Univ. of Chicago (Ph.D. Thesis).
Hayes, S.H. and Belton, M.J.S.: 1977, "Icarus"32, pp. 383-401.
Moore, J.H. and Menzel, D.H.:1930,"Publ.Astron.Soc.Pacific" 42, pp.330-335.
Podolak, M.: 1976, "Icarus" 27, pp. 473-478.
Sinclair, A.T.: 1975, "Monthly Notices Roy. Astron.Soc." 171, pp. 59-72.
Smith, B.A.: 1977, "Nature" 268, pp. 32-33.
Trafton, L.: 1977, "Icarus" 32, pp. 402-412.
Whitaker, E.A. and Greenberg, R.J.: 1973, "Monthly Notices Roy. Astron. Soc" 155, pp. 15p-18p.

METHOD OF SURFACE OF SECTION APPLIED TO A POSSIBLE CAPTURE ORIGIN OF JUPITER'S SATELLITES

K. Tanikawa
International Latitude Observatory, Mizusawa, Iwate, 023 Japan

1. INTRODUCTION

The capture of satellites is considered to be a slow process, in which particles approaching the planet lose their energy gradually and fall down into the stable orbits around it. Therefore, in order to investigate the capture problem, it seems to be necessary to clarify the behaviors of orbits near the planet. Recently Hayashi et al. (1977) and Heppenheimer and Porco (1977) found that in the restricted three body problem, particles incident through Lagrange points L_1 and L_2 revolve always directly around Jupiter as long as the potential windows at L_1 and L_2 are small. This suggests the impossibility of the capture of retrograde satellites when the potential windows are small.

In the present paper, we investigate the behaviors of the ensemble of retrograde orbits for which the potential windows are large.

2. THE EQUATIONS OF MOTION

We adopt the following formulation of the restricted three body problem. The frame of reference is the rotating rectangular coordinates (x, y) with the angular velocity of the Sun around Jupiter. Both the Sun and Jupiter are at rest on the x-axis, Jupiter being at the origin. The direction of the x-axis is from the Sun to Jupiter. The system of units is such that the distance between the Sun and Jupiter, the angular velocity of their revolution, and the sum of their masses are all unities, respectively. Denoting the mass of Jupiter by μ, we have the equations of motion

$$\ddot{x} = 2\dot{y} + x + 1 - \mu - (1-\mu)\frac{x+1}{r_1^3} - \mu\frac{x}{r_2^3} \quad ,$$

$$\ddot{y} = -2\dot{x} + y - (1-\mu)\frac{y}{r_1^3} - \mu\frac{y}{r_2^3} \quad ,$$

where

R. L. Duncombe (ed.), Dynamics of the Solar System, 181–184.

$$r_1 = \sqrt{(x+1)^2 + y^2} \quad , \qquad r_2 = \sqrt{x^2 + y^2} \quad .$$

The equations of motion admit the Jacobi integral

$$C = (x+1-\mu)^2 + y^2 + \frac{2(1-\mu)}{r_1} + \frac{2\mu}{r_2} - \dot{x}^2 - \dot{y}^2 \ .$$

3. THE BEHAVIORS OF THE RETROGRADE ORBITS IN THE SURFACE OF SECTION

The solution of the equations of motion can be represented as a trajectory in the 4-dimensional space (x,y,v,θ), where $v=\sqrt{\dot{x}^2+\dot{y}^2}$ and $\theta=\arctan(\dot{y}/\dot{x})$. If we fix the value of Jacobi constant C and put $y=0$, the trajectory can be represented as points in the surface of section (x,θ), where the corresponding orbits cross the x-axis (see Figure 1).

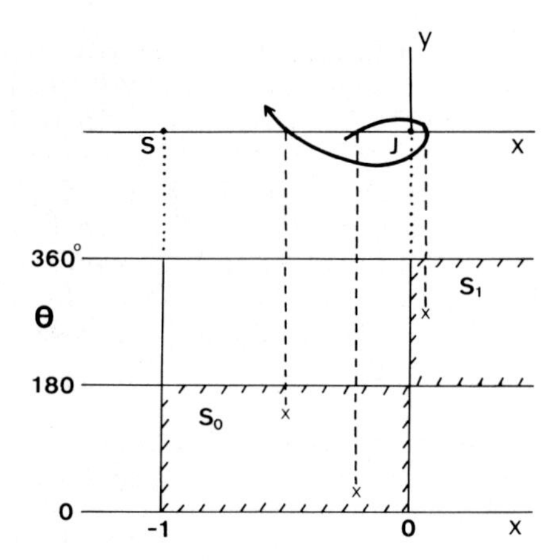

Figure 1. An orbit and the surface of section (x, θ)

Let Σ_0 and Σ_1 stand for the upper and lower halves of the surface of section respectively. In general, an orbit which crosses Σ_0 at a point P_0 crosses successively Σ_1 at P_1 and again Σ_0 at P_2. The transitions from P_0 to P_1 and from P_1 to P_2 are considered to be induced by mappings. Denoting the mapping by T, we have $P_1=TP_0$, $P_2=TP_1$, hence $P_2=T^2P_0$.

Let S_0 and S_1 be the region defined by $-1\leq x\leq 0$, $0°\leq\theta\leq 180°$ in Σ_0 and that defined by $x\geq 0$, $180°\leq\theta\leq 360°$ in Σ_1, respectively. As long as the orbit around Jupiter is retrograde, the corresponding points in the surface of section are always in S_0 and S_1.

Let the point set W_1 in S_0 be such that $TP\in S_1$ and $T^2P\in S_0$ for any $P\in W_1$ and either $TP'\notin S_1$ or $T^2P'\notin S_0$ for $P'\notin W_1$. We assume that TW_1 is compact and that the points $x=0$, $\theta=180°$ and $x=0$, $\theta=360°$ are not included in TW_1. Then W_1 and T^2W_1 are also compact. Consequently T is topological on

W_1 and TW_1. Besides the assumptions above, W_1 is assumed to include the points invariant under T^2. Then it can be shown that the boundaries of W_1 and T^2W_1 intersect each other at least at two points.

Define W_n for n=2,3,... by $W_n=T^{-2(n-1)}(W_1 \cap T^{2(n-1)}W_{n-1})$. Evidently, the relations $W_1 \supset W_2 \supset ... \supset W_n \supset ...$ hold, and W_n shrinks to the invariant set W. If we define U_n by $U_n=W_n-W_{n+1}$ (n=1,2,...), it is obvious that U_n tends to a null set as $n \to \infty$. The orbits having the initial conditions in U_n will revolve retrogradely just n times around Jupiter in the future.

Because of the invariance of the equations of motion under the transformation $x \to x$, $y \to -y$, and $t \to -t$, the past behaviors of the orbit corresponding to the point (x,θ) in S_0 is simulated by the future behaviors of the orbit corresponding to the point $(x, 180°-\theta)$ in S_0. The invariant set W is symmetric with respect to the line θ=90°. Therefore, if P is not included in W, its mirror image with respect to the line θ=90°, is not included in W. This means the impossibility of the capture of the retrograde satellites in the restricted problem.

We denote the mirror image of U_n with respect to the line θ=90° by U_n' (n=1,2,...), and define the sets U_0 and U_0' as $U_0=S_0-W_1$ and $U_0'=S_0-T^2W_1$. Then, the point set in S_0 corresponding to the ensemble of the orbits which revolve retrogradely just n times is expressed as

$$V_n = \bigcup_{\substack{k+l=n \\ k,l \geqq 0}} (U_k \cap U_l')$$

4. NUMERICAL RESULTS

The equations of motion with μ=0.001 have been integrated numerically for various values of C. It is shown that the sets W_1, TW_1, and T^2W_1 are compact at least when $C \gtrsim 2.5$. The remarkable result is that for $C \gtrsim 2.95$, the boundaries of the above sets, consequently those of W_n(n=2,3,...)too, are constituted only of the points corresponding to collision orbits. Figure 2 shows the structure of the ensemble of retrograde orbits in the region S_0 for C=3.0. In the figure, W_1 is the region inside the bold solid curve. T^2W_1 is inside the broken curve. The regions filled with circles, shaded horizontally and shaded vertically are U_1, U_2 and U_3, respectively. W_2, W_3 and W_4 are the regions obtained by subtracting U_1, U_2 and U_3 from W_1, W_2 and W_3 respectively. W_1 and T^2W_1 are symmetric with each other with respect to the line θ=90°. The approximate invariant region around the invariant point f corresponding to the class f periodic orbit (see e.g. Hénon 1965) is also shown by the dotted curve. Three crosses are the invariant points under T^6.

It has been shown numerically that most of the orbits corresponding to the points in U_1 escape from Jupiter for C=3.0. Therefore in this case V_n defined in the previous section corresponds to the orbits which approach Jupiter and escape after revolving retrogradely just n times around it. A preliminary estimate shows that the amount of such orbits decreases rather rapidly with the increase of n. Therefore the occurrence of the capture seems to be rare for the value of C examined.

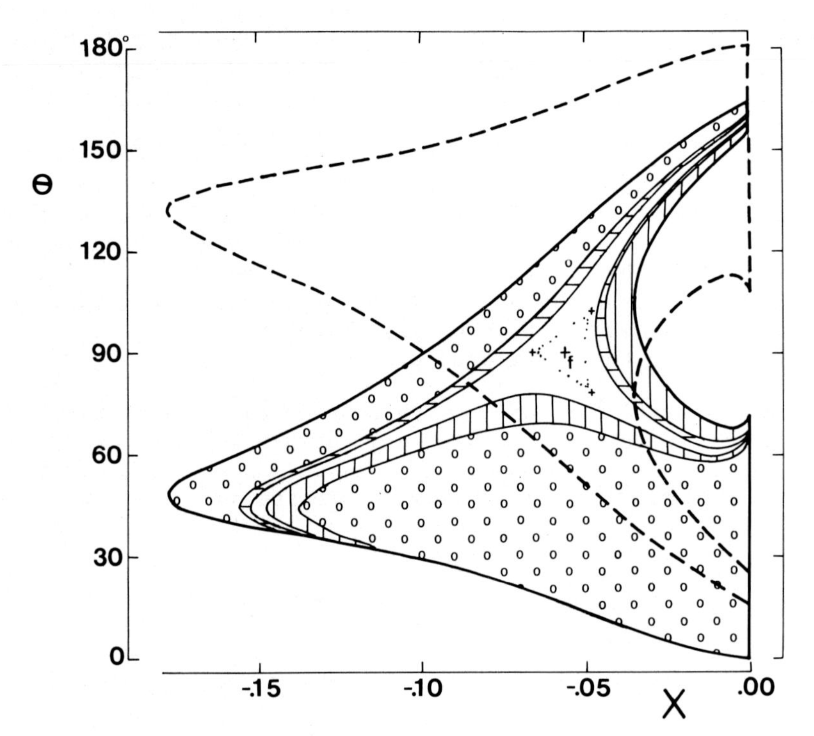

Figure 2. The region S_0 in the Surface of Section for μ=0.001, C=3.0

REFERENCES

Hayashi, C., Nakazawa, K., and Adachi, I. : 1977, *Publ. Astron. Soc. Japan*, **29**, 163.
Hénon, M. : 1965, *Ann. Astrophys.*, **28**, 449.
Heppenheimer, T. A. and Porco, C. : 1977, *Icarus*, **30**, 385.

DISCUSSION

Dvorak: How can you explain in the simple model of the cirular problem the capture of Jupiter's satellites? In this case the Hill's curves are closed!

Tanikawa: In this case, Hill's curves are not closed around Jupiter; and, for our result, shows that the capture is impossible in the restricted problem. The other forces (like dissipation) are needed.

PLANETARY CLOSE ENCOUNTERS: AN INVESTIGATION ON TEMPORARY SATELLITE-CAPTURE PHENOMENA

A. Carusi
Laboratorio di Astrofisica Spaziale,CNR,Frascati,Italy
F. Pozzi and G. Valsecchi
University of Rome,Italy

This research is part of a wider one (Carusi & Pozzi,1978a,b) concerning a detailed study of the dynamics of close encounters between a giant planet and a minor object. A special result of that investigation was the recognization of some satellite-capture events,already found by Everhart (1973). An important remark about this previous work is that all satellite-captures occurred with low inclination objects which orbits were initially near-tangent to the Jupiter's one. Starting from this consideration,a hundred fictitious orbits have been generated in order to study the phenomenon in greater detail. Their initial distribution is shown in fig.1. The initial angular parameters i,ω,Ω,were chosen to be equal to those of the most interesting case of the previous research. Eccentricities were selected regularly in the range .01-.5,with a step of .01, giving the same value to an object of the upper band and to the next of the lower. The semimajor axes were chosen at random between limits computed so that the aphelion for the lower band,or the perihelion for the upper band,would lie within a distance of 10^8km from Jupiter's orbit. As the orbital planes do not coincide with that of Jupiter,the minimum distance point between the two orbits does never coincide with the object perihelion or aphelion,but is always close to them. Fig.2 shows the final situation of this population:we note that in no case a permanent binding occurred. We can do some remarks on this picture. First of all we note that 56% of objects experienced a temporary binding to the planet. Secondly,67% of objects,bounded or not,had a final orbit lying, on the a-e diagram,on the opposite band with respect to their initial one. This kind of transition is especially significant if compared with the case of observed comets,because it gives a simple mechanism to transform long-period comets in short-period ones, and to transfer comets from one family to another. Actually,on the basis of the computations of Kazimirchak-Polonskaya (1972),we can say that a similar process has been experienced, for example,by comets Whipple,Oterma,Brooks 2,Lexell,Kearns-Kwee and o-

R. L. Duncombe (ed.), Dynamics of the Solar System, 185–189.

thers. A third remarkis that,between the initial and final situations, small eccentricities are increased,as a consequence of the encounter, whilst the great ones are decreased. This phenomenon leads to a clustering of final orbits in the eccentricity range .1-.2.

It is quite interesting to analyze,just to give an example, the path of object 24 in a jovicentric rotating frame,as shown in fig.3 This object binds itself to Jupiter at point a,then becomes temporary unbounded from the Sun between the points b and c,and finally it unbinds itself from the planet at point d. The maxima and minima of semiaxis occur in correspondance of the conjunctions. We call inner conjunction the one in which the object is located between Jupiter and the Sun,the other situation representing an outer conjunction. Then,we note that a maximum of semiaxis always occurs during an inner conjunction on a retrograde planetocentric orbit,or during an outer conjunction on a direct planetocentric orbit. The minima of semiaxis occur in the remaining two cases. In order to get a better understanding of these occurrencies we can use the formulas for heliocentric energy and angular momentum:

$$E = mv^2/2 - GmM/r = - GmM/2a$$

$$|\underline{L}| = mvr \sin\varphi = m\sqrt{GM}\sqrt{a(1-e^2)}$$

It is easy to see that we have relative maxima of E and $|\underline{L}|$,and then of a, in a direct outer conjunction,and relative minima in a direct inner conjunction. From an exam of our objects we have seen that,for a retrograde planetocentric orbit,things go the opposite way. It follows,for instance,that an object can unbind itself from the Sun only in inner retrograde or in outer direct conjunction. Fig.4 clearly explains what we said. In this picture the energy and angular momentum with respect to the Sun are plotted. The abscissa gives the number,to be multiplied by 50,of the integration time steps,and so it is a not linear time scale. We can note that the positions of maxima and minima are in good agreement with what we said. Moreover,in these points the object is always near to its osculating perihelion or aphelion. Referring to the "mirror theorem" demonstrated by Roy & Ovenden (1955),we note that in the case of number 24 we have three instants in which a configuration of this kind is quite well verified,that is in the 3rd,5th and 6th conjunctions. The mirror theorem,however,is not completely satisfied,because in none of these conjunctions Jupiter is located on its aphelion or perihelion,and the lines of nodes are not aligned. Now a comparison with a really observed case is quite interesting:that is the case of comet Oterma,which orbital evolution has been analyzed by Kazimirchak-Polonskaya (1967). An inspection of the orbital history of this comet shows that,with respect to July 1950, the evolution was almost symmetric for a period of about 17 years forwards and backwards. Fig.5 shows this symmetry for the semiaxis and the eccentricity. Let's now spend some words about another quite interesting experimental evidence,that is the case of the objects 25 and 28. The trajectories of these two bodies are shown in figs.6 and 7:they can overlap by a rotation of π about z-axis. In fact,the maxima and the minima of E and $|\underline{L}|$ of number 25 coincide with the minima and the maxima respectively for number 28. Similar cases occur even when the objects,although bounded to Jupiter, do not close any orbit about it.

Owing to the scarceness of the allowed space,we have limited ourselves to a few comments. A more complete discussion of our results will be published elsewhere.

REFERENCES

Carusi, A., and Pozzi, F.:1978a, submitted to "Astrophys. Space Sci."
Carusi, A., and Pozzi, F.:1978b, submitted to "Astrophys. Space Sci."
Everhart, E.: 1973, "Astron. J." 78, pp.316-329.
Kazimirchak-Polonskaya,E.I.:1967, "Sov. Astr.-AJ" 11, pp.349-365.
Kazimirchak-Polonskaya,E.I.:1972, IAU Symp. 45, pp. 373-397.
Roy,A.E.,and Ovenden,M.W.:1955,"Mon. Not. R. Astr. Soc." 115,pp.296-309.

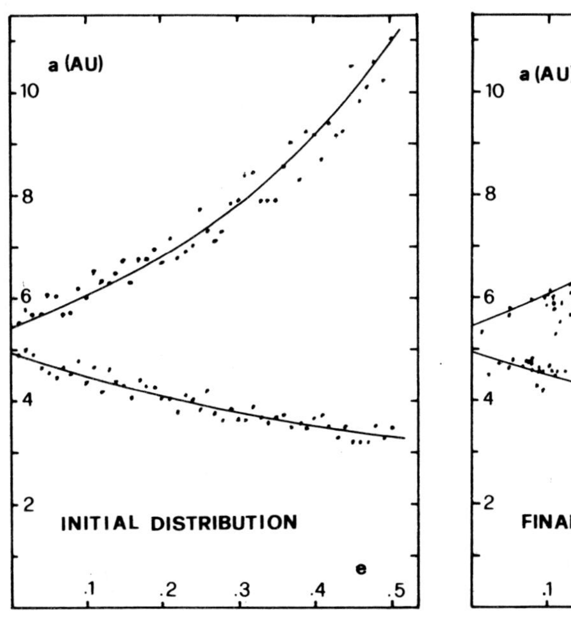

Fig. 1

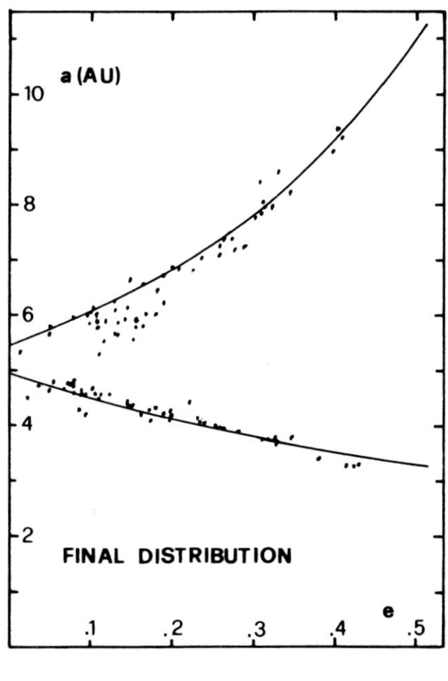

Fig. 2

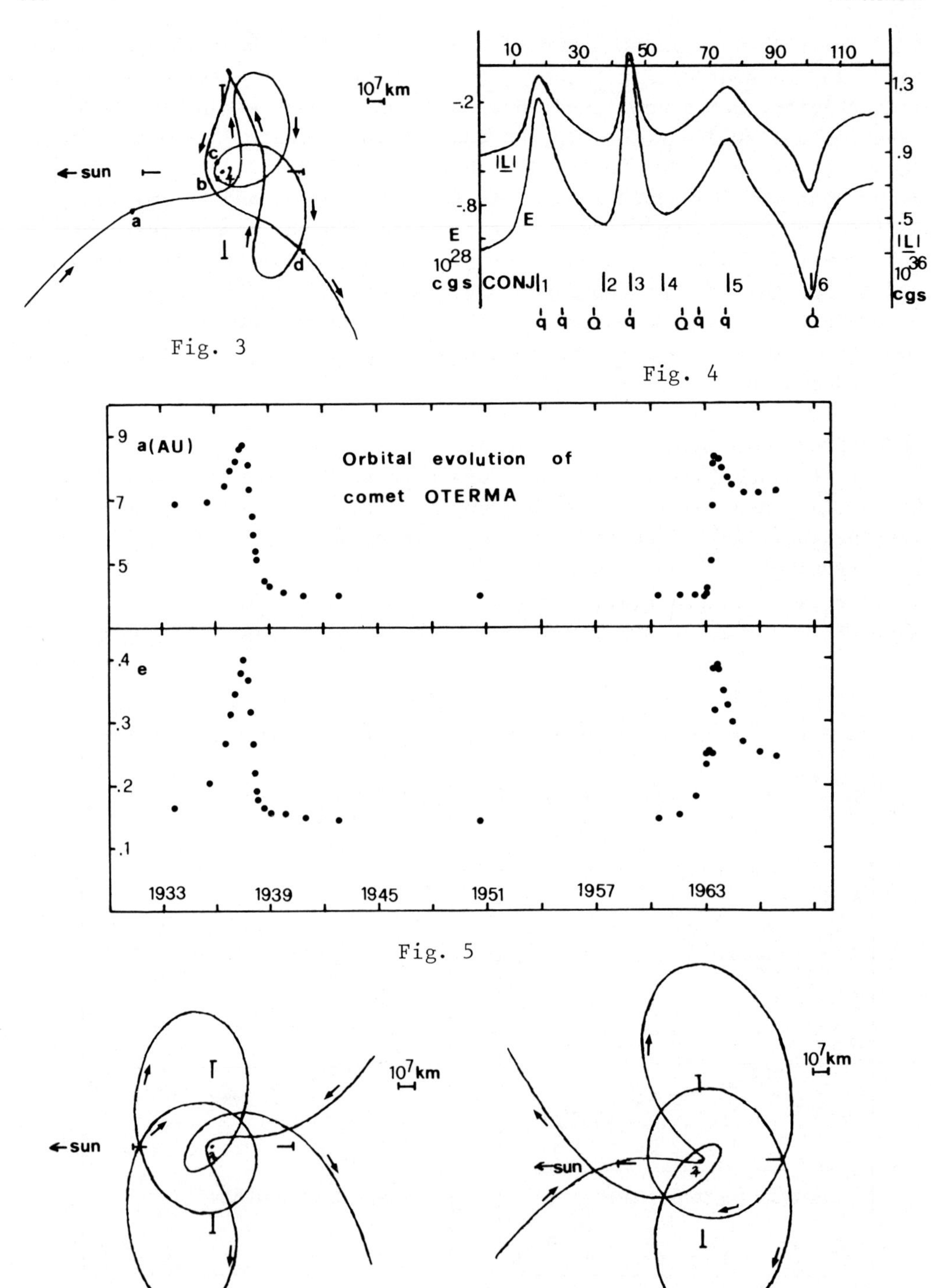

Fig. 3

Fig. 4

Fig. 5

Fig. 6

Fig. 7

DISCUSSION

Szebehely: What equations did you integrate, the circular or the elliptic restricted problem? In the first case, did the Jacobian constant have a constant value? Did you use regularization at the close approaches?

Carusi: I have used the Greenspan's "discrete mechanics" (LeBudde & Greenspan:1976,Numer.Meth.25,323; 1976,Numer.Meth.26,1) which consists in recursive formulas giving positions and velocity components for all bodies. The main feature of that method is the exact conservation of the invariants of motion for any n-body system, without restrictions on the initial mass distribution. During the close approach the integration time step is automatically scaled, proportionally to the relative distance of the two bodies, which cannot become less than 7×10^5 km ($\sim$ Jupiter's radius). In this case a collision occurs. Positions and velocities are computed in a heliocentric ecliptical reference frame; in its last version the computer program supplies, at every time step, the osculating heliocentric and jovicentric parameters, together with other useful quantities, such as heliocentric and jovicentric energies, angular and linear momenta, components of the Landau vector, Tisserand and Collenbreau-Motukume invariants, and so on. The computation is fast: 100 close encounters (mean time length about three years) are computed in about ten minutes of a UNIVAC 1106.

Kresak: This is just a nice example of the difference in the behavior of the Jacobi integral and Tisserand invariant. While both of them remain approximately constant before and after the perturbation, the Tisserand value may change appreciably during the approach. This is due to the neglected term containing the reciprocal distance from Jupiter. However, even the Tisserand criterion resumes its original value after the encounter, except for a small deviation produced by Jupiter's orbital eccentricity.

Carusi: This is true: the Tisserand constant may change even on the first decimal digit, during the close encounter. It must be noted, however, that the variations between initial and final values are not so small. We have found variations of the order of $\pm$ 1% or 2% with respect to the initial value.

Dvorak: Did you use Tisserand's criterion in your numerical calculations?

Carusi: My numerical method consists in solving the Newtonian equations of motion directly, by means of the "discrete mechanics" method of Greenspan (LeBudde & Greenspan,1976: Numer.Meth.25, 323; Numer.Meth.26,1), so I can use the actual positions and velocities of all objects. I can compute the Tisserand invariant, but I've found that it is not invariant when the objects are very close ($\sim 10^8$ km).

THE RINGS OF SATURN AND URANUS

Peter Goldreich
California Institute of Technology

1. INTRODUCTION

Nature has presented us with two systems of planetary rings. They are not at all similar. Saturn has bright, broad rings separated by narrow gaps. The rings of Uranus are dark, narrow and widely spaced. Presumably both sets of rings lie inside the Roche limit and this is why the ring material has not condensed into satellites.

I will briefly review what is known about each ring system with particular emphasis on properties of significance to dynamical astronomy.

2. SATURN'S RINGS

2.1 Observations

Observations at visual wavelengths reveal two bright broad rings separated by a narrow gap known as the Cassini division. The optical depth normal to the ring plane is of order several tenths (1,2). The thickness of the rings is less than 3 km (1).

Near infrared reflectance spectra show absorption features due to water frost (3). Thus the ring particles are at least coated with ice.

Measurements of the thermal emission at far infrared wavelengths yield brightness temperatures of order 90°K (4). These values are consistent with the equilibrium kinetic temperature for highly reflecting particles at Saturn's distance from the sun.

Interferometric observations at centimeter wavelengths show that the rings are very weak radio emitters, $T_B < 10^{\circ}$K (5,6). The rings partially block radio emission from Saturn's disk. The optical depth at radio wavelengths is comparable to that at visual wavelengths (6).

R. L. Duncombe (ed.), Dynamics of the Solar System, 191–196.

The rings are excellent reflectors of centimeter wavelength radar waves (7). This implies that the ring particles are greater than a few centimeters in size. Incident polarized radar signals are largely depolarized after reflection, presumably the result of multiple reflections (8).

2.2 Basic deductions

The ring particles are probably largely composed of water ice. Consideration of cosmic abundances, the low densities of the inner satellites of Saturn and, above all, the near infrared spectral data all support this view. If ice, the low radio emission implies that the mean particle radius is < 0.1 absorption lengths or perhaps < 10 m. The composition independent lower limit of 1 cm radius is set by the high radar reflectivity.

Silicate particles may be ruled out because the ratio of their emissivity to reflectivity at centimeter wavelengths is much too high to satisfy the radio and radar data.

Metallic particles coated with ice cannot be excluded. They could be of any size greater than 1 cm in radius. However, iron particles would be outside the Roche limit and should have collected into satellites.

2.3 Dynamical problems

The velocity dispersion of the ring particles is determined by the details of their mutual collisions. Collisions give rise to a viscous stress that converts orbital energy into random motions. Since the collisions are not perfectly elastic, the energy in random motions is dissipated as heat. The velocity dispersion adjusts so that the effects of these two processes balance (9,10). For ice particles collisions are likely to be quite inelastic even for impact velocities as low as 10^{-3} cm s^{-1}. This implies that a ring of ice particles would be a monolayer (10).

The outer or A ring shows an azimuthal brightness variation which is asymmetric with respect to the Earth-Saturn center line. Brightness maxima and minima precede and follow conjunctions of the ring particles with this line (11,12). No convincing explanation of this phenomenon exists. Suggestions include albedo variations on synchronously rotating particles and gravitationally enhanced, elongated, particle density fluctuations inclined to the radial direction (13).

The inner edge of the Cassini division is near the position of the 2:1 orbital resonance with Mimas. The problem is to explain how tiny Mimas has managed to clear such a large gap and why the resonance radius lies close to its inner edge. A new mechanism, based on the collective response of the ring particles to the resonant perturbations by Mimas, has been proposed (14). It accounts in a natural way for the size and location of the gap. The principal hypothesis is that Mimas excites a trailing spiral density wave at the position of the 2:1 resonance. The wave carries negative energy and angular momentum and propagates outward. The wave is damped by viscosity (due to collisions) and its negative energy and angular momentum are transferred to the ring particles. Consequently, the particles just outside the 2:1 resonance move inward opening a gap.

3. THE RINGS OF URANUS

3.1 Observations

Uranus is encircled by at least 9 narrow rings. The 5 most prominent rings were discovered during an occultation observation in March 1977 (15,16). Of these, the inner 4 are of order 10 km wide and have eccentricities $< 10^{-3}$. The outermost or ϵ ring is of variable width (20 to 100 km) and has an eccentricity of order 10^{-2}. An additional 4 rings were recognized in a more careful analysis of the March 1977 data and were also detected during an occultation observation in April 1978 (17,18).

Optical detection of the rings is difficult because of their proximity to the much brighter planet. Observations made in methane bands shortward of 1 μm indicate that the albedo of the rings is $\leqslant 0.05$ (19,20). One marginal detection has been reported (21). The stratoscope pictures of Uranus reveal a faint shadow, apparently cast by the rings (19,22).

In May 1978 the rings were successfully mapped at 2.2 μm (23). At this wavelength the rings are brighter than Uranus even though their albedo is only of order 0.05. These measurements cannot resolve the individual rings but they do reveal an azimuthal brightness variation which is plausibly attributed to the variable width of the ϵ ring.

3.2 The ϵ ring

An analysis of all available occultation data reveals the following (24). The width of the ϵ ring varies linearly with radius. Its optical depth profile is remarkably similar at different locations. The edges of the ring are abrupt.

The positions of all the ϵ ring crossings are well fit by a precessing Keplerian ellipse with $a = 51{,}284$ km, $e = 7.8 \times 10^{-3}$ and $\dot{\tilde{\omega}} = 1\overset{\circ}{.}37\ \text{day}^{-1}$. The linear relation between ring width and radius is consistent with a small spread in the semimajor axes and eccentricities of the ring particles.

3.3 Dynamical problems

Narrow rings tend to spread due to particle collisions. The lifetimes of the Uranian rings are probably comparable to the age of the solar system because diffuse rings are not seen and there is very little inter-ring material. In the absence of confining forces, these considerations would imply that the sizes of the ring particles were less than one centimeter. However, over the age of the solar system the Poynting-Robertson effect would produce a substantial decay of the orbits of sub-centimeter size particles. Thus, the presence of confining forces seems likely (25,26).

Attempts to relate the 5 original rings to a series of orbital resonances with the Uranian satellites have not been successful (27,28, 29). The discovery of 4 additional rings makes this approach less attractive.

My guess is that each ring contains a small satellite and that the ring particles are debris from its surface. However, I have yet to investigate the stability of such a configuration.

4. ACKNOWLEDGMENTS

I thank G. Colombo and P. Nicholson for many informative discussions.

This work was supported by NASA Grants NGL-05-002-003 and NGL-05-002-140 and by NSF Grant AST 76-24281.

BIBLIOGRAPHY

1. Bobrov, M.S.: 1970, in A. Dollfus (ed.), "Surfaces and Interiors of Planets and Satellites," Academic Press, New York, p. 377.
2. Cook, A.A. and Franklin, F.A.: 1958, "Smithsonian Contrib. Astrophys." 2, p. 377.
3. Pilcher, C.B., Chapman, C.R., Lebofsky, L.A. and Kieffer, H.H.: 1970, "Science" 167, p. 1372.
4. Murphy, R.E.: 1974, in F.D. Palluconi (ed.), "The Rings of Saturn," NASA, Washington, D.C., p. 65.
5. Berge, G.L. and Muhleman, D.O.: 1973, "Astrophys. J." 185, p. 373.
6. Briggs, F.H.: 1974, "Astrophys. J." 189, p. 367.
7. Goldstein, R.M. and Morris, G.A.: 1973, "Icarus" 20, p. 260.
8. Goldstein, R.M., Green, R.R., Pettengill, G.H. and Campbell, D.B.: 1977, "Icarus" 30, p. 104.
9. Brahic, A.: 1977, "Astron. Astrophys." 54, p. 895.
10. Goldreich, P. and Tremaine, S.: 1978, "Icarus" in press.
11. Reitsema, H.J., Beebe, R.F. and Smith, B.A.: 1976, "Astron. J." 81, p. 209.

12. Lumme, K. and Irvine, W.M.: 1976, "Astrophys. J. Lett." 204, p. L55.
13. Colombo, G., Goldreich, P. and Harris, A.W.: 1976, "Nature" 264, p. 344.
14. Goldreich, P. and Tremaine, S.: 1978, "Icarus" in press.
15. Elliot, J.L., Dunham, E. and Mink, D.: 1977, "Nature" 267, p. 328.
16. Millis, R.L., Wasserman, L.H. and Birch, P.: 1977, "Nature" 267, p. 330.
17. Elliot, J.L., Dunham, E., Wasserman, L.H., Millis, R.L. and Churms, J.: 1978, submitted to Astron. J.
18. Persson, E., Nicholson, P., Matthews, K., Goldreich, P. and Neugebauer, G.: 1978, "IAUC" No. 3125.
19. Sinton, W.M.: 1977, "Science" 198, p. 503.
20. Baum, W.A., Thomsen, B. and Morgan, B.L.: 1977, "Bull. Am. Astron. Soc." 9, p. 499.
21. Smith, B.A. and Reitsema, H.J.: 1977, "Bull. Am. Astron. Soc." 9, p. 499.
22. Colombo, G.: 1977, "Sky Telesc." 54, p. 188.
23. Matthews, K. and Neugebauer, G.: 1978, private communication.
24. Nicholson, P., Persson, E., Matthews, K., Goldreich, P. and Neugebauer, G.: 1978, submitted to Astron. J.
25. Brahic, A.: 1978, paper presented at IAU Symp. No. 81., Tokyo.
26. Goldreich, P. and Nicholson, P.: 1978, in preparation.
27. Dermott, S.F. and Gold, T.: 1977, "Nature" 267, p. 590.
28. Aksnes, K.: 1977, "Nature" 269, p. 783.
29. Goldreich, P. and Nicholson, P.: 1977, "Nature" 269, p. 783.

DISCUSSION

Van Flandern: Elsewhere in the solar system, we tend to find other satellites, rather than gaps, at resonance positions. Would you comment on the possibility that the Cassini gap is actually cleared by a small Saturnian satellite?

Goldreich: The Cassini gap may be cleared by a small satellite. However, our mechanism would lead to an accumulation of material just inside the 2:1 resonance with Mimas. Thus, if a satellite could form in the rings that is where we would expect it to.

Van Flandern: Isn't it possible to explain the unusual optical and dynamical properties of the Uranian rings, particularly the "precession," if each ring is caused by a single satellite diffusing material, similar to the sodium-hydrogen torus in the orbit of Io?

Goldreich: The ring particles must be at least a few microns in size. However, they might have come from a small satellite.

Marchal: How do you explain the stability of the asymmetric ring since the precession rate decreases with increasing radius?

Goldreich: We are looking for the explanation; the motions are likely forced with numerous collisions.

Kozai: What do you mean when you say that ε-ring of Uranus is precessing? Do the apsides move as if the ring is a solid body?
Goldreich: We think so.

Brahic: In your model of Cassini's division, are the density waves trailing or leading?
Goldreich: Only a trailing wave is excited.

Scholl: Can your theory about the formation of the Cassini division be applied to the Kirkwood gaps? Why?
Goldreich: It cannot because the density in the belt is too low for cooperative gravitational effects to be of importance.

DYNAMICS OF GRAVITATING SYSTEMS OF COLLIDING PARTICLES IN PLANETARY DISCS

André Brahic
Observatoire de Paris, 92190 Meudon et
Université Paris VII, 75005 Paris
France

1. INTRODUCTION

During this symposium on the dynamics of the solar system, we have mainly studied the movements of the bodies of the solar system submitted to gravitational perturbations. The next step is to take into account the physical collisions. Indeed, there can be little doubt that collisions between "macroscopic bodies" are of frequent occurence in the Universe. All kinds of quite different objects undergo such collisions: these may range from large interstellar clouds to small solid bodies in the solar system. Collisions have surely played an important role in the formation of planets and satellites and continue to play a central role in the behaviour of the planetary discs. For example for Saturn's rings, one can see intuitively that until the optical depth drops much below unity, the rings are still evolving. Each orbiting particle can be taken as occupying a kind of torus, and collisions will continue until there is only one particle in each such "orbital tube"; this corresponds to a very small optical depth.

Since the time of Poincaré (1911), it is known that inelastic collisions tend to flatten any system; inelastic collisions tend to damp out the motions perpendicular to the plane of the disc as well as the radial motions, so that the orbits become more and more circular and coplanar. With the help of Hénon, I am studying systematically three-dimensional gravitating systems of colliding particles by numerical simulation and analytically. This work finds an immediate application to the dynamics of planetary discs for example. These simulations lend particularly well to astrophysical problems in which the mean free path of a given particle is of the order of or larger than the dimensions of the system.

In order to understand first the basic mechanics of the process, I have considered the simplest models in which attraction between particles has been neglected (and so particles orbits are keplerian around a central mass point), and in which particles have the same masses and radii. In a collision, the grazing component of velocity is conserved and the perpendicular component is multiplied by a coefficient k which lies between 0 and -1. The evolution of these first models have been

R. L. Duncombe (ed.), Dynamics of the Solar System, 197–202.

already published (Brahic, 1977a) and can be briefly summarized in the following way: after a very fast flattening of the order of twenty collisions per particle, the system reaches a quasi-equilibrium state in which the thickness of the newly formed disc is finite and in which collisions still occur. Under the combined effect of differential rotation and inelastic collisions, the disc spreads very slowly, particles move both inwards and outwards carrying out some angular momentum. For Saturn's rings, the time scale of flattening is of the order of a few weeks, the time scale of the quasi-equilibrium state is of the order of 10^{14} years and the system reaches a third hypothetical collisionless state in a time scale of the order of 10^{21} years. This result disagrees strongly with the time scale obtained by Jeffreys (1947). His paper contains an erroneous calculation as noted by Hénon (1975).

Now, I have considered different collision laws, particles of different masses and radii,..... I have no place here to give all the results which will be published in the near future. During the few minutes of this communication, I shall give you just the results concerning the existence of a new transition in the (k,k') diagram and the results concerning the mutual evolution of two different types of particles under various conditions.

2. A NEW TRANSITION IN THE (k,k') DIAGRAM

If the grazing component of the relative velocity of two colliding particles is reduced after each collision by a factor k' which lies between 0 and +1, an instability phenomenon appears. For very inelastic collisions and contrary to the behaviour of the first models (Figure 1, zone B), the disc is completely flattened and the system become a two-dimensional configuration in which collisions continue to occur (Figure 1, zone A). A sharp transition L_1 separates zones A and B. A linear stability analysis of the observed behaviour has been made: differential rotation constantly feeds energy into horizontal motions and thus maintains a finite dispersion of horizontal velocities, whilst vertical motions are only a by-product of the horizontal activity and do not necessarily arise (Brahic, 1977b).

There exists a second sharp transition L_2 between zone B and zone C (Figure 1). In zone B, the thickness of the disc decreases and reaches a value of the order of r/R, where r is the radius of a particle and R some characteristic dimension of the system. In zone C, the thickness of the disc increases. On the one hand, keplerian motion tends to establish an anisotropic distribution of velocities with a velocity two times larger in the radial direction than in the transverse direction. On the other hand, collisions tend to establish an isotropic distribution of velocities. If the degree of inelasticity is small (Figure 1, zone C), a large quantity of energy is extracted from the ordered motion and expended in the collisions. If the degree of inelasticity is larger (Figure 1, zones A and B), a smaller quantity of energy is extracted from the ordered motion and the orbits become more and more circular and coplanar.

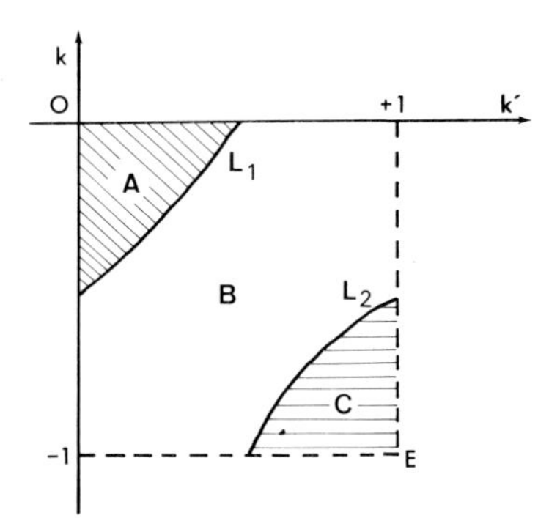

Figure 1. The flattening of the system is a function of the rebound coefficients k and k'. In zone A, the system is completely flattened. In zone B, the thickness of the disc decreases towards a value of the order of a few times the size of a particle. In zone C, the thickness of the disc increases. There exist two sharp limits L_1 and L_2 between the three zones A, B and C.

3. PARTICLES OF DIFFERENT MASSES AND RADII

In a real system, most collisions would be between particles of very unequal size. I shall give here just the results concerning the mutual evolution of two different types of particles. Generally speaking one can say that the two groups of particles evolve in essentially the same way. There is no real equipartition of energy, but rather a kind of separation. The mean inclination and the mean eccentricity of the large particles is smaller than the mean inclination and the mean eccentricity of the small particles (Figure 2). The evolution is essentially similar to that of the first models. The more elastic the collisions, the bigger the separation; the larger the mass ratio, the bigger the separation; there would not seem to be any separation in the plane of the flattening.

Generally speaking, the massive particles evolve faster than the light ones. The relation ($a r^2 < 3.10^{10}$) between the size r of the particles and the age a of the system (resp. in meters and years), proved for systems of identical particles (Brahic, 1977a), remains valid. To the extend that Saturn's rings may be modelled in this way, then if the rings are as old as the solar system, the maximum size of the particles is 2.5 meters. A system, which to day is essentially made of large particles, must presumably have been created much later.

4. CONCLUSIONS

The rings of Uranus could be made of very small colliding particles. If the above relation is applied to Uranus'rings, rings having the age of the solar system would be composed of particles whose size is less

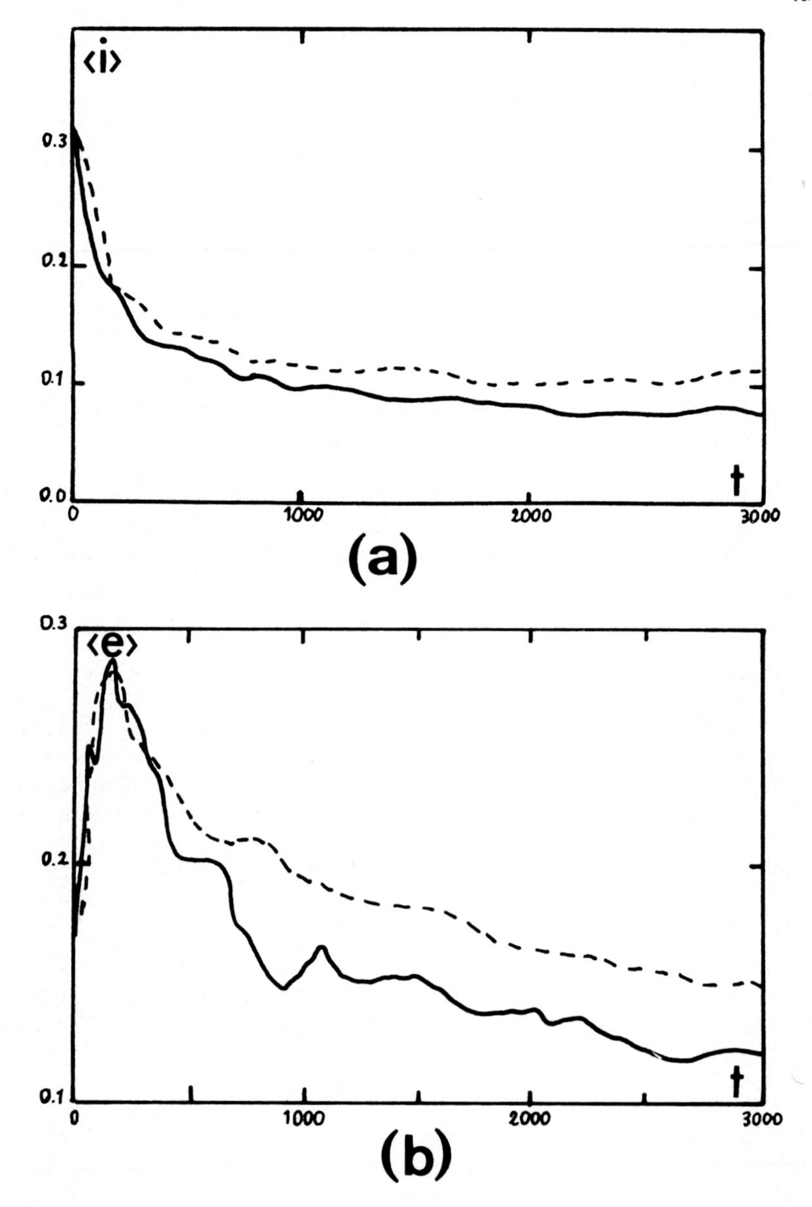

Figure 2 (a). Variations as a function of time of the mean inclination of the particles (which is a good measure of the flattening of the system). The large particles are in full line and the small particles are in dashed line. The particles have the same density, the large particles are four times more massive and about 1.5 larger than the small particles.
(b). Variations as a function of time of the mean eccentricity of the same system.

than five millimeters. The limiting size given by the Poynting-Robertson effect is curiously of the same order of magnitude: smaller particles have fallen onto Uranus. The rings could be very young, allowing particles of larger size.

In the near future, new data (observations of Saturn's rings from spacecrafts and observations of some occultations of stars by Uranus' rings) will be particularly helpful to test theories. Note that planetary rings is a case where on the one hand the input of the theory is relatively free of uncertainties, and of the other hand detailed observations will be soon available, so that observations and theories can be compared fruitfully.

REFERENCES

Brahic, A.: 1977a, *Astron. Astrophys.* 54, pp. 895-907.
Brahic, A.: 1977b, *Astron. Astrophys.* 59, pp. 1-7.
Hénon, M.: 1975, private communication.
Jeffreys, H.: 1947, *Mon. Not. R. Astron. Soc.* 107, pp. 263-268.
Poincaré, H.: 1911, *Leçons sur les hypothèses cosmogoniques*, Hermann, Paris, pp. 86-89.

DISCUSSION

Carusi: Have you taken into account the possibility of adhesion or fragmentation of particles in the ring? Which is the initial velocity distribution function?

Brahic: The particles are indestructible spheres, and so there is no fragmentation after a collision. This would seem reasonable; relative velocities are small in real physical cases as Saturn's rings, and also for the major part of the evolution of our models. Coalescence is not considered here. Nevertheless, we can apply the above results; in this case, the total volume $N(r/R)^3$ of the particles is conserved, and $N(r/R)^2$ varies as R/r. Thus evolution will be accelerated by fragmentation in the proportion $1/r$. The initial conditions are set up by selecting at random the six elements of the Keplerian orbit of each particle in such a way that the initial trajectories are all ellipses, lying between two spheres of radius R_1 and R_2, respectively. In order to start from a system already a little flattened, the initial inclinations i of the orbits are taken between zero and some maximal value i_{max}.

Marchal: Do some bodies escape to infinity?

Brahic: I have assumed that particles on hyperbolic trajectories escape at once. Even for very inelastic collisions one of the particles involved can acquire a hyperbolic velocity. For typical values of the parameters, this happens on the average of once every 1000 collisions, and twenty-two particles have fallen onto the central body.

Marchal: What is the importance of the oblateness of the planet?
Brahic: The oblateness of the central planet has no very important effect on the evolution of the system. Even with an oblateness of the central body ten times larger than in the case of Saturn, I have not been able to see any difference.

THE RINGS OF PLANETS AND COSMOGONY OF THE SOLAR SYSTEM

S. K. Vsekhsvyatskij
Kiev State University, Kiev, U.S.S.R.

SUMMARY

It is shown that the hypothesis that comets originate in the vicinity of each of the giant planets is consistent with the idea that these planets possess rings. It is suggested that the rings of Uranus, the probable gaseous ring around Jupiter and the clouds surrounding Io are rapidly evolving phenomena - the same may also even be true of Saturn's rings - which is to be expected of agglomerations of cometary meteoroids losing their icy constituents by sublimation.

R. L. Duncombe (ed.), Dynamics of the Solar System, 203.

PART V

MINOR PLANETS

ASTEROIDAL MOTION AT COMMENSURABILITIES TREATED IN THREE DIMENSIONS

Joachim Schubart
Astronomisches Rechen-Institut, Heidelberg, Germany, F.R.

1. SOME METHODS APPLIED TO CASES OF COMMENSURABLE MEAN MOTIONS

This paper consists of a review about work done on three-dimensional motion at commensurabilities of either the mean motions, or of secular periods, and of a report on the author's recent results on some special cases. Real and fictitious asteroidal orbits and the corresponding long-period effects are the main subject of interest. At first, methods are listed.

Let n and n_J be the mean motions of an asteroid and Jupiter (the index J relates to Jupiter) and let n/n_J be close to a rational number $(p+q)/p$, using the author's preferred notation (Schubart, 1964, 1968). The motion of the asteroid is described by osculating elements with respect to the sun, a, e, i, Ω , $\tilde{\omega} = \omega + \Omega$, and by the mean longitude, ℓ. Periods of interest are given by the circulation or libration of ω, $\tilde{\omega} - \tilde{\omega}_J$, $\Omega - \tilde{\omega}_J$, and of a critical argument σ, where

$$q\sigma = (p+q)\ \ell_J - p\ell - q\tilde{\omega} \quad .$$

Numerical integration applied to the equations of motion of a suitable model for the acting forces of attraction has become a reliable means to study the changes in an orbit over an extended period. If the model fits well to the conditions of nature, the results will be applicable to real cases. The sun, the major planets, and an asteroid can be integrated on a computer as an n-body-system (Schubart and Stumpff, 1966), and ephemerides of high accuracy result. If, however, the interval of interest is very large, it will be necessary to reduce the number of attracting planets to two, or to one (sun-Jupiter-asteroid problem, $e_J \neq 0$ or $e_J = 0$).

Hagihara (1961) published an extensive listing of references of former work about commensurabilities. In many cases the work was based on analytical theories. A method by Poincaré (1902) suggests to drop the short-period terms in a development of the disturbing function of the sun-Jupiter-asteroid problem, but to retain terms depending on σ and

R. L. Duncombe (ed.), Dynamics of the Solar System, 207–215.

other slow arguments. To-day, it is no longer necessary to use a series development for this simplification, since numerical averaging of a Hamiltonian function (Schubart, 1964, 1966) or of a suitable set of differential equations (Schubart, 1968, 1978) allows an equivalent elimination of the short-period terms and a much wider range of applications. Numerical integration is used in the latter case. Care is necessary with respect to such an approximate method, since it is certainly not suitable for the study of orbits that allow close approaches to Jupiter (compare Schubart, 1970).

A further simplification of a corresponding model is possible, if the critical argument is eliminated during a suitable formal process of canonical transformation (Giacaglia, 1969). This method was applied to several special cases of commensurability.

2. METHODS GIVING EVIDENCE FOR SECULAR RESONANCES

If the orbital period of an asteroid is not close to a commensurability, methods are available to study the secular variations of e, $\tilde{\omega}$, i, Ω. Methods using numerical averaging allow an individual treatment of most asteroids (compare Smith, 1964). The usual analytical method (Brouwer and van Woerkom, 1950) is not applicable to large values of e or i, but it allows an explicit solution of the differential equations corresponding to an asteroid, in general. The constants of integration, called proper elements, gave rise to the discovery of families of asteroids by Hirayama. A basic period appearing in the solution depends on the element a. For certain values of a this period equals one of the fixed secular periods that are present in the differential equations, if the variations of e_J, i_J, and of analogous quantities are considered. Such a secular resonance needs a special treatment (Hagihara, 1972), since the difference of two respective frequencies causes a small divisor otherwise.

J.G. Williams (1969, 1971) proposed a new theory that allows to derive proper elements for most asteroids without a limitation in e or i. He starts with a zero-order approximation in neglecting the eccentricities and inclinations of the perturbing planets. These quantities are later considered by first-order terms, which introduce secular periods and can cause secular resonances again. Now the resonances are no longer restricted to specific values of a. They appear on surfaces in the space given by a and by proper elements corresponding to e and i. Comparatively few asteroids are close to these surfaces, but Williams (1969, 1973) performed a general study of such cases. There is evidence for a special type of libration of corresponding critical arguments. Strong variations of e or i can occur in case of a secular resonance, compare also the reviews by Peale (1976) and Greenberg (1977).

3. FORMER WORK ON SPECIAL ASTEROIDS WITH COMMENSURABLE MEAN MOTIONS

As demonstrated by Peale's (1976) review, asteroid-Jupiter commensurabilities contribute many interesting cases to the wider field of orbital resonances in the solar system. At first commensurable asteroids were studied with methods based on a development of the disturbing function, but later, especially in Japan, astronomers started to replace the approximate analytical theories by computations of orbital elements with numerical integration (Hagihara, 1961). This was most important for asteroids with n/n_J close to 3/2, or 4/3, compare the papers by Akiyama (1962) and Takenouchi (1962), who studied the motion of (153) Hilda and (279) Thule more recently. They found evidence for libration of σ in both cases. Belyaev and Chebotarev (1968), and Chebotarev et al. (1974) applied numerical integration over an interval of 400 yr to many interesting asteroidal cases of resonance, especially to Trojan asteroids (1/1 - case). For some other cases compare Froeschlé and Scholl (1978). Schweizer (1969) used integration for the asteroids corresponding to the vicinity of the 2/1 - case. This type of resonance was studied in detail by Franklin et al. (1975), after Marsden (1970) had treated this and other types as given by minor planets and comets. The 2/1 - case corresponds to a typical Kirkwood gap (Hagihara, 1961), but (1362) Griqua and two other numbered asteroids show an oscillation of the mean motion about the exact value of resonance with respect to Jupiter. These three numbered objects have a large eccentricity, but there is some indication for the existence of Griqua-type objects with lower e as well (Franklin et al., 1975).

Numerical integration of averaged differential equations was mainly applied to the asteroids of the Hilda group (3/2 - case). Planar models gave first results and demonstrated, that almost all asteroids of the group show libration of σ about 0° (Schubart, 1968; Ip, 1976), compare the next chapter. Direct numerical integration gave evidence about resonances and librations of Apollo and Amor asteroids with respect to earth and Venus (Janiczek et al., 1972; Ip and Mehra, 1973). The librations with respect to inner planets are temporary in most, and perhaps in all cases.

4. RECENT WORK ON GRIQUA- AND HILDA-TYPE ASTEROIDS

The author's results presented in this and the following chapter depend on three models for the acting forces. Model 1 is a generalization to three dimensions of the former computer program (Schubart, 1968), that allowed the elimination of short-period terms from the differential equations of the planar, elliptic restricted three-body problem by averaging, and then the numerical integration of these equations (compare Schubart, 1978). Model 2 is the corresponding rigorous elliptic sun-Jupiter-asteroid problem. Model 3 approximates the real variations of e_J and $\tilde{\omega}_J$, since Saturn is added to the bodies of the second model on an appropriate orbit that is turned into the orbital plane of Jupiter, so that this fixed plane can be used as the plane of reference in all three models ($i_J = 0$). The masses of Jupiter and Saturn correspond to the IAU (1976) System of Constants, while all other perturbing masses are entirely neglected. The two last models are realized by an n-body program (Schubart and Stumpff, 1966). Further definitions are $a_J = 1$, $e_J = 0.048$, $\tilde{\omega}_J = 0^o$; they are related to a moment in 1975 August in case of model 3, together with $\ell_J = 0^o$. The time is counted from this moment, or otherwise from the date of osculation of the asteroidal orbit to be studied. The elements of this orbit are transformed to the new scale, plane of reference, and zero-direction of longitude.

The three numbered Griqua-type objects ($p = q = 1$) and ten members of the Hilda group ($p = 2$, $q = 1$) have been studied over periods that range from 20 000 yr to 140 000 yr in total and cover both past and future motion in most cases. Results by model 1 are available in all cases. They depend on orbital elements that were averaged over the interval of an appropriate short-period integration corresponding to model 2, before the long-term integration by model 1 started. This numerical integration allowed a step-length of a half to one year for twelve of the asteroids, but 0.25 yr was needed for (1921) Pala, a special case of 2/1 - libration. From former experience, and from recent comparisons it is known that model 1 gives a good approximation to the conditions of model 2 except in special cases, compare chapter 1. Model 2 is expected to give a qualitatively correct picture of the real motion, but frequencies of very long periods and related amplitudes will differ from values found by more accurate models (Froeschlé and Scholl, 1978). Model 3 is important for tests about this, and for special cases with a strong influence of the motion of $\tilde{\omega}_J$ and of the long-term oscillation of e_J between 0.03 and 0.06 (compare Cohen et al., 1973). A comparison of results by model 1 for (1748) Mauderli with n-body results by Froeschlé and Scholl (1978), that correspond to model 3, appears satisfactory.

Orbital elements from Ephemerides of Minor Planets for 1978 are the basis of study for twelve of the asteroids under consideration. In case of (153) Hilda, the starting values used before (Schubart, 1978) were retained. Table 1 shows results obtained by model 1 for all 13 objects. The results on (1746), (1748), and (1921) refer only to future motion, although a backward computation is available for (1921), see below. During the intervals related to Table 1, all 13 asteroids show stable libration of σ about a mean value of 0° with a mean period T_L. The periods of retrograde revolution of perihelion (T_P) and node (T_N) are comparatively large. They are given in millennia and represent mean values in many cases. If (1921) Pala is excluded, the variations of the orbits appear to be quasi-periodic, and it is expected that this will show up in more extended computations by model 1 as well. It appears from more detailed studies of four objects, that mean frequencies corresponding to T_L, T_P, and T_N, and combinations of them, as the mean angular velocity of $\omega = \tilde{\omega} - \Omega$, determine the changes in the elements of the orbits. Since it turned out that the former planar model (Schubart, 1968) gave a good description of basic effects for Hilda (Schubart 1978), the present study includes mainly objects with larger inclination and the more recently numbered Hilda-asteroids (compare Marsden, 1970). Table 1 shows for the

Table 1. Periods and extreme values for three 2/1 - librators followed by ten Hilda-type asteroids (3/2 - case). Results from model 1.

No.	Name	T_L	T_P	T_N	σ_M	e_M	i_M	Δ_m
1362	Griqua	396	10.6	36	122°	0.37	23°.6	2.11
1921	Pala	307	29	22	140	.56	20.2	1.97
1922	Zulu	422	26	132	98	.53	39.5	2.65
153	Hilda	275	2.7	22	43	0.24	9.3	1.89
361	Bononia	270	4.1	26	66	.29	12.6	1.94
1345	Potomac	273	3.8	27	50	.28	11.2	1.98
1746	Brouwer	278	1.8	18	54	.21	9.6	1.76
1748	Mauderli	253	2.7	17	87	.26	2.3	1.73
1754	Cunningham	269	3.5	24	70	.27	11.7	1.88
1877	Marsden	283	4.2	30	60	.29	18.2	1.98
1902	Shaposhnikov	277	3.3	26	35	.26	11.7	1.96
1911	Schubart	267	3.1	23	47	.26	3.1	1.93
1941	Wild	258	4.2	24	68	.30	3.2	1.95

Notes. T_L (yr): mean period of σ - libration; T_P, T_N (10^3 yr): mean or approximate periods of retrograde revolution of $\tilde{\omega}$ and Ω; The subscript M indicates the maximum - value of $|\sigma| \leqq 180°$, e, or i found ($i_J = 0$); Δ_m is the smallest distance Jupiter-asteroid in AU. For (1921) Pala the backward computation is not considered.

ten objects of this type a sufficiently large minimum of distance to Jupiter, Δ_m, and a similar amount in each of the three periods. Differences in amount of T_P or T_N are correlated with Δ_m. The upper limit for e is about 0.3. (1877) Marsden has the largest inclination of the group.

The three 2/1 - librators reach much larger values in e or i, and the amplitude of libration in σ is comparatively large, as found before by Franklin et al. (1975). Δ_m is large, since a is smaller than in the Hilda-case, and due to the libration. T_P and T_N are rather large, especially for (1922). They have a comparable amount in case of (1921). Although this turns out to hold only for a limited interval, it gives rise to slow variations of ω during this interval, and ω can remain between 0° and 90° for more than 20 000 yr. This seems to cause an exchange between the mean amounts of i and e. A backward computation leads to small values of e, large amplitudes in σ, and indicates the possibility of a temporary circulation of σ. A computation on (1921) Pala by model 2 gives analogous results. They are shown in Fig. 1 and Fig. 2, left curve. All these curves are smoothed with respect to effects by T_L. In Fig. 2 the solid part of the left curve extends backward in time for 30 000 yr. It shows temporary libration of $\tilde{\omega}$. A little open dot close to the curve indicates a time 14 500 yr ago, where one single revolution of σ is observed. Libration of σ takes place at earlier and later times. More typical examples of change between libration and circulation of σ were described by Froeschlé and Scholl (1977). The right curve in Fig. 2, smoothed like the others, represents a solution of the planar elliptic restricted problem, that is not periodic. It has the quality of simultaneous libration of $\tilde{\omega}$ and σ (2/1 - case), with large amplitudes about 0° in both cases. The two respective mean periods are about 18 000 and 340 yr. The amplitude and period for σ vary along the curve shown. Simultaneous small - amplitude oscillation of $\tilde{\omega}$ and σ may occur near stable equilibrium points, given by a Hamiltonian that corresponds to model 1. Libration of $\tilde{\omega}$ can also occur in nearly resonant orbits, if e is small, but in some cases studied, σ circulates, if $\tilde{\omega}$ librates. The plane orbit of Fig. 2 corresponds to starting values $a = 0.63$, $e = 0.37$, $\sigma = 133^{\circ}$ $(p = q = 1)$, $\tilde{\omega} = \ell_J = 0^{\circ}$. Model 2 was used with $i = 0^{\circ}$.

The solid part of the left curve in Fig. 2 resembles the right one. The approximate symmetry of these curves with respect to the direction $\tilde{\omega}_J = 0^{\circ}$ suggests, that the amount of e_J is important for the effects of libration in $\tilde{\omega}$. Probably due to the comparatively small amount of e_J in the respective interval, a backward computation for (1921) Pala by model 3 did not confirm the temporary libration of $\tilde{\omega}$ and circulation of σ. However, a corresponding forward computation agrees to the prediction of a slowly changing ω, and of big changes in e and i, and it may be expected, that libration of $\tilde{\omega}$ and circulation of σ will occur at more remote times, when the phase of oscillation of e_J is more favorable for this.

5. HILDA- AND THULE - TYPE ORBITS OF HIGH INCLINATION

The known asteroids of Hilda- and Thule - type have a small or moderate inclination. However, there is theoretical evidence for the possibility of long-term motion on highly inclined orbits at the 3/2 and 4/3 commensurabilities as well. This appears from the author's recent studies (compare Schubart, 1978) by means of models 1 - 3. Orbits were found that give evidence for a permanent simultaneous libration of ω about 90° and of σ about 0°, with different main periods of libration, and with a sufficiently large minimum distance Δ_m. The following two sets of starting values refer to typical orbits of this kind with small-amplitude libration.

$$p = 2 \ , \ q = 1 \ , \ a = 0.7630 \ , \ e = 0.54 \ , \ i = 41^{\circ} \ , \ \omega = 90^{\circ} \ ;$$

$$p = 3 \ , \ q = 1 \ , \ a = 0.8277 \ , \ e = 0.38 \ , \ i = 30^{\circ} \ , \ \omega = 90^{\circ} .$$

These values were used together with $\sigma = \tilde{\omega} = \ell_J = 0^{\circ}$. Studies by both models 2 and 3 cover more than 30 000 yr in each case. The observed amplitudes of ω-libration do not exceed 10° and 22° in the two cases, respectively. The amplitudes in σ are small too. Since high-inclination orbits occur in other regions of the asteroid belt, there may be small unknown objects on orbits of the above types as well.

I thank Mrs. E. Miltenberger and Mrs. I. Seckel for aid in typing and drawing. I used the IBM 360-44 and 370-168 computers at the University of Heidelberg's Rechenzentrum.

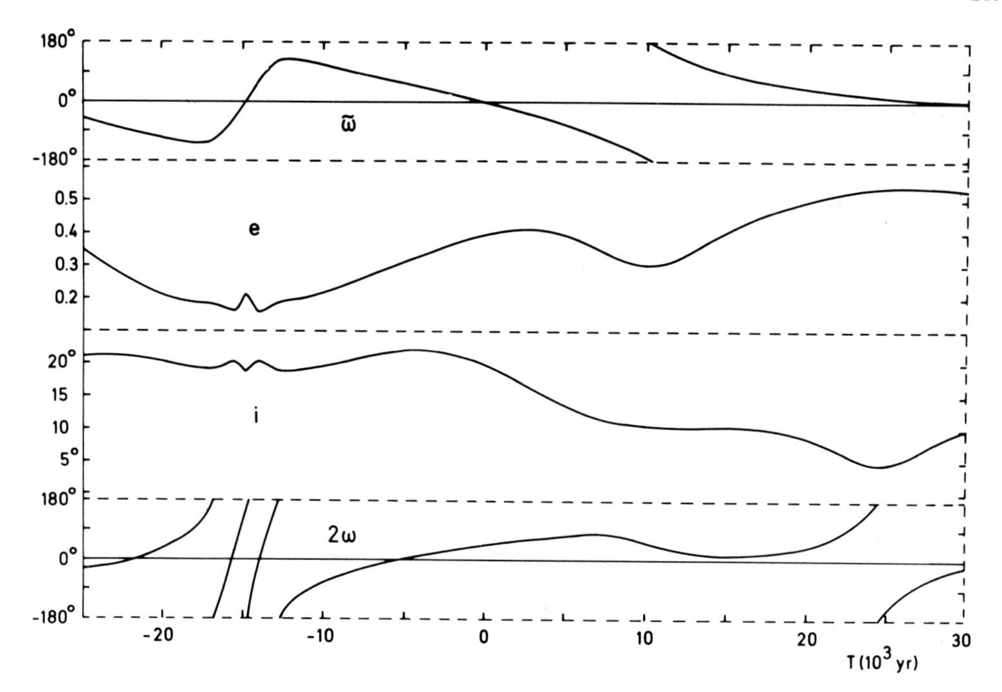

Fig. 1. Smoothed curves representing a solution of the elliptic restricted problem (model 2) at the 2/1 - resonance. $\tilde{\omega}$, e, i, 2ω are plotted against time. The starting values correspond to (1921) Pala.

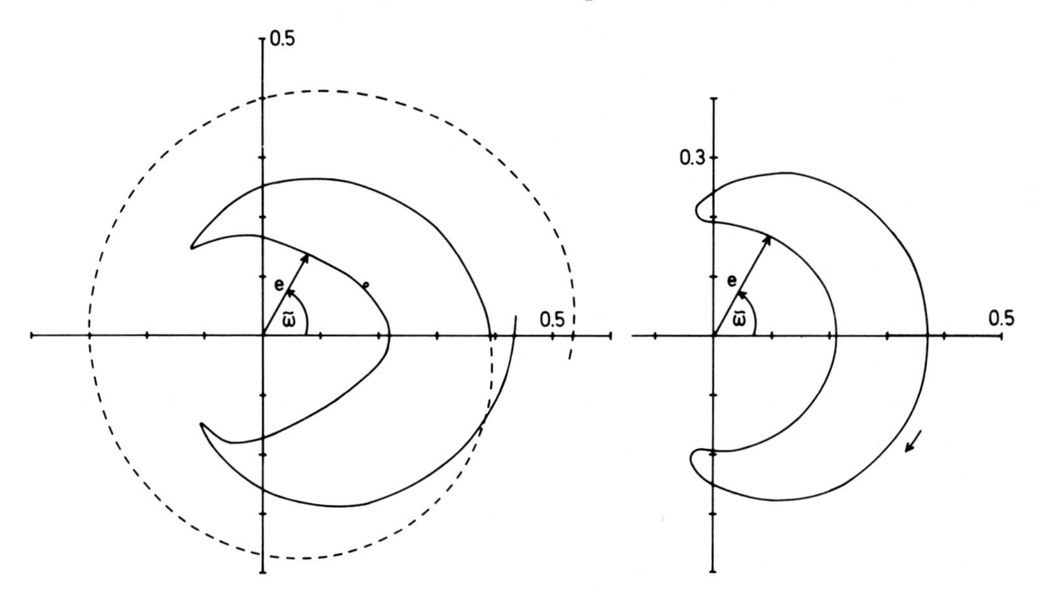

Fig. 2. Smoothed curves showing e and $\tilde{\omega}$ in polar coordinates. The left curve corresponds to the upper part of Fig. 1 (dashed line gives future motion), the right one to a case of planar motion (see text).

REFERENCES

Akiyama,K. : 1962, "Publ.Astron.Soc. Japan" 14, pp.164-197.

Belyaev,N.A. and Chebotarev,G.A. : 1968, Astron.Circ. Moscow No.480, Aug.22, 1968.

Brouwer,D. and van Woerkom,A.J.J. : 1950, Astron. Papers American Ephemeris Washington, Vol.13, part 2.

Chebotarev,G.A., Belyaev,N.A., and Eremenko,R.P. : 1974, IAU Symposium No.62, pp.63-69.

Cohen,C.J., Hubbard,E.C., and Oesterwinter,C. : 1973, Astron. Papers American Ephemeris Washington, Vol.22, part 1.

Franklin,F.A., Marsden,B.G., Williams,J.G., and Bardwell,C.M. : 1975, "Astron.J." 80, pp.729-746.

Froeschlé,C. and Scholl,H. : 1977, "Astron.Astrophys." 57, pp.33-39.

Froeschlé,C. and Scholl,H. : 1978, "Astron.Astrophys.", to be published.

Giacaglia,G.E.O. : 1969, "Astron.J." 74, pp.1254-1261.

Greenberg,R. : 1977, "Vistas Astron." 21, pp.209-239.

Hagihara,Y. : 1961, "Smithsonian Contrib.Astrophys." 5, pp.59-67.

Hagihara,Y. : 1972, Celestial Mechanics Vol.2, The MIT Press, pp.187-199.

Ip,W.-H. : 1976, "Astrophys. Space Sci." 44, pp.373-383.

Ip,W.-H. and Mehra,R. : 1973, "Astron.J." 78, pp.142-147.

Janiczek,P.M., Seidelmann,P.K., and Duncombe,R.L. : 1972, "Astron.J." 77, pp.764-773.

Marsden,B.G. : 1970, "Astron.J." 75, pp.206-217.

Peale,S.J. : 1976, "Ann.Rev.Astron.Astrophys." 14, pp.215-246.

Poincaré,H. : 1902, "Bull.Astron." 19, pp.289-310.

Schubart,J. : 1964, Smithsonian Astrophys.Obs.Spec.Rept. No.149.

Schubart,J. : 1966, IAU Symposium No.25, pp.187-193.

Schubart,J. : 1968, "Astron.J." 73, pp.99-103.

Schubart,J. : 1970, Periodic Orbits, Stability and Resonances, Ed. : G.E.O. Giacaglia, D. Reidel, Dordrecht, pp.45-52.

Schubart,J. : 1978, Dynamics of Planets and Satellites and Theories of their Motion, IAU Coll. No.41, Ed. : V. Szebehely, D. Reidel, pp.137-143.

Schubart,J. and Stumpff,P. : 1966, "Veroeffentl.Astron. Rechen-Inst." Heidelberg No.18.

Smith,Jr.,A.J. : 1964, A Discussion of Halphen's Method, part 2, NASA Techn.Rept., TR R-194.

Schweizer,F. : 1969, "Astron.J." 74, pp.779-788.

Takenouchi,T. : 1962, "Ann. Tokyo Astron.Obs." 7, pp.191-208.

Williams,J.G. : 1969, Secular Perturbations in the Solar System, Ph.D. Dissertation, Univ. of California, Los Angeles, pp.1-270.

Williams,J.G. : 1971, Physical Studies of Minor Planets, IAU Coll. No.12, Ed. : T. Gehrels, NASA SP-267, pp.177-181.

Williams,J.G. : 1973, "Bull.Am.Astron.Soc." 5, p.363 (Abstract).

DISCUSSION

Garfinkel: What is the number of secondary resonances that you mentioned in your talk?

Schubart: I have only mentioned cases of simultaneous libration. I did not treat resonances of σ with respect to short-period arguments.

Garfinkel: How do you treat the critical divisors?

Schubart: I have done numerical integration, where such divisors do not arise.

Marsden: In the model giving circulation of σ for (1921) Pala, what was the minimum approach distance to Jupiter?

Schubart: In the backward computation by model 2 the minimum approach distance was about 1.3 AU.

RECENT WORK ON THE ORIGIN OF THE KIRKWOOD GAPS

H. Scholl
Astronomisches Rechen-Institut, Heidelberg, F.R.G.

The frequency distribution of the asteroidal mean motions shows the well known Kirkwood gaps which are found near the commensurabilities 3/1, 5/2, 7/3, and 2/1. The number of asteroids in the gaps is significantly smaller than those in the corresponding regions around these commensurabilities. In the outer part of the asteroidal belt, the frequency distribution is reversed: The number of asteroids at the commensurabilities 3/2, 4/3, and 1/1 is much larger than in the region around these commensurabilities.

This reversal of the frequency distribution is one of the main problems of any hypothesis that tries to explain the Kirkwood gaps, since it is not at all obvious how asteroids can stay close to the commensurabilities 3/2 and 4/3 but avoid the commensurabilities 3/1, 5/2, 7/3, and 2/1. Different attempts have been made to explain the Kirkwood gaps based on statistical, gravitational, collisional or cosmogonic mechanisms.

Before looking for a sophisticated explanation for the Kirkwood gaps, one has to be sure that the gaps really exist and are not simply statistically underpopulated regions. According to the statistical hypothesis, asteroids librate through the gaps moving fastest at the center of the gap and remaining outside of the gaps over a comparatively long time span. The probability of observing an asteroid inside of a gap, is therefore very low.

This statistical hypothesis was tested by Schweizer (1969) and by Wiesel (1976). In order to find librators, Schweizer calculated the orbits of numbered minor planets which surround the gaps. He thus obtained a time - averaged frequency distribution of mean motion which did not differ strongly from the non-averaged distribution. A large majority of minor planets remained on one side of the gap whereas only a few asteroids were found to librate through the gaps. A list of the few librators is given by Williams (1977).

Franklin et al. (1975) investigated 30 orbits of unnumbered minor planets and of P-L objects which might librate about the 2 : 1 resonance.

R. L. Duncombe (ed.), Dynamics of the Solar System, 217–222.

However, as these orbits are very uncertain, one has to wait for more observations of these objects in order to determine the number of minor planets librating about the 2 : 1 resonance.

Another statistical attempt was made by Wiesel (1976). He used an analytical method in order to investigate the evolution of an initially uniform distribution of fictitious asteroids in the Kirkwood gaps. His method is based on Poincaré's integrals for resonant motion. According to Wiesel's calculations, the statistical behaviour of the orbits does not indicate a depletion of the gaps for $t \to \infty$. The density of asteroids in the gaps never becomes significantly lower than in the surrounding regions. The emptiness of the gaps in Wiesel's model is not as pronounced as indicated by observations.

Both, Wiesel's and Schweizer's work, negate the validity of the statistical hypothesis for the Kirkwood gaps. Only if further observations yield many more minor planets librating about the commensurabilities, should this hypothesis be reexamined.

The second hypothesis, the gravitational hypothesis, has been investigated ever since the discovery of the gaps by Kirkwood (1867). A lot of effort was spent on analytical approaches, since the classical first order perturbation theory fails for cases of near resonant motion. In the classical theory, the disturbing function is expanded in trigonometric series of the osculating orbital elements. The occurence of small denominators destroys the formal convergence of these series for resonant motion and the osculating elements become infinitely large. This fact provides the basis for the following gravitational hypothesis: An asteroid situated in a gap will suffer especially strong perturbations by Jupiter and therefore will leave the gap, because very small denominators occur in the series development for the semi-major axis.

Recently, this gravitational hypothesis was tested by analytical and numerical methods. The analytical approaches use either Poincaré's or related variables which are better suited than the classical orbital elements since these variables vary slowly with time for resonant motion. A simplification of the problem was achieved by discarding the short periods and by investigating only the long period commensurable terms. Then, for further simplification, the resonant motion of an asteroid was investigated in the restricted Sun-Jupiter-Asteroid problem. Different models are now available for the two-dimensional or three-dimensional, circular or elliptic case.

For the simplest of these gravitational models, the circular planar model, Schubart (1964) published a systematic survey of orbital types for several commensurabilities. His dynamical system has one degree of freedom. The possible orbits can be classified in three main types: periodic orbits, librators or circulators with respect to the critical argument σ. According to Schubart's survey, an asteroid which is situated in a gap does not leave the gap.

Obviously, this model represents only a rough approximation to reality as Jupiter's eccentricity is neglected and because the perturbations of the other planets are not taken into account. Therefore, Scholl and Froeschlé (1974, 1975) investigated numerically fictitious asteroidal orbits at the low order commensurabilities 3/1, 5/2, 7/3, and 2/1 in Schubart's elliptic planar Sun-Jupiter-Asteroid model. As in the circular model, no asteroid was found to leave a gap. Most of the orbits exhibit a quasi periodic behaviour with the exception of peculiar orbits which alternate between circulation and libration.

Two objections were raised against these calculations:

1.) We calculated over 100 000 years only, which is a short period compared to the age of the solar system of some 10^9 years.
2.) Schubart's model uses an averaging method in order to discard the short periodic effects, thus obtaining an integral of motion. That integral, which does not exist in the unaveraged model, might stabilize the calculations.

However, a few calculations in the unaveraged Sun-Jupiter-Saturn-Asteroid model over 100 000 years also failed to yield an orbit that left a gap. We have to concede that this result is not necessarily valid for much longer periods than 10^5 years. What happens with these orbits for $t \rightarrow \infty$, remains an open question.

From all our numerical experiments we can say that Schubart's figures obtained for the averaged circular case are a good approximation for the unaveraged Sun-Jupiter-Saturn model for 10^5 years.

Another approach to the gravitational hypothesis came from the technique of surface of sections which Hénon and Heiles (1964) applied to the restricted Three-Body-Problem. The application of this technique indicates what parts of an appropriately defined phase space an asteroid for instance can cover. Especially this method enables us to say whether the motion of an asteroid is limited to a small portion of the phase space (integrable case) or whether it can fill the whole phase space (ergodic case). Applying this technique to the Hecuba gap and the Hilda group, Giffen (1973) found an ergodic behaviour for orbits starting with low eccentricities in the Hecuba gap. On the other hand, he did not find any ergodic behaviour for fictitious Hildas. We, therefore, speculated that asteroids starting in the Hecuba gap might drift out of the gap while such a drift is not possible for the Hildas.

Froeschlé and Scholl (1976) tested this hypothesis in more detail applying the surface of section technique to the 3/1, 5/2, 7/3 and 2/1 commensurability. The calculations were based on Schubart's averaged planar elliptic model. Summarizing, we can say that we found a few quasi ergodic zones. However, these zones are surrounded by forbidden regions or invariant curves. An asteroid, therefore, situated in a quasi ergodic zone can not drift out of the gap. Here, we have to emphasize that this result is only valid for the applied model and for time spans up to 10^5 years. We cannot exclude the possibility that the invariant curves will dissolve on a much longer time scale and that an asteroid, therefore, might leave a gap following Arnold's diffusion process. This problem is still open.

A third approach to the gravitational hypothesis was made recently by Kiang (1978). He investigated the stability of orbits in the Hecuba gap and for the Hilda group. The stability of an orbit is determined by Hill's exponent. For his calculations, Kiang used Schubart's averaged planar circular model. According to his astonishing results, motion in the Hecuba gap is unstable while it is stable for the Hilda group in the sense of Hill's exponent. Therefore, Kiang supposes that an asteroid which started in the Hecuba gap will drift out of the gap. This supposition however is not supported by numerical calculations nor by theoretical considerations of Schubart's circular model.

Besides the gravitational hypothesis for the origin of the Kirkwood gaps, the collisional hypothesis has been discussed by numerous authors (Jefferys, 1967; Sinclair, 1969; Lecar and Franklin, 1973; Giffen, 1973; Scholl and Froeschlé, 1974; Wiesel, 1976). This hypothesis is very attractive as it seems to solve the problem why we observe so many Hildas at the 3/2 commensurability while we observe so few asteroids in the Kirkwood gaps. A collision between two asteroids can only occur when their orbits intersect. Therefore, the collision probability for an asteroid depends on the numbers of orbits it intersects. Thus, the collision probability depends on the variation of the orbital eccentricity. The stronger the eccentricity varies, the larger the collision probability.
Collision here means both destructive collision or a close approach which changes the semi-major axes of the orbits.

It is known, that asteroids close to a low order commensurability show strong variations in eccentricity. Therefore, these asteroids have a comparatively larger collision probability. Since there are almost no asteroids around the Hilda group, the collision probability for the Hilda family is very low.

In order to test the collisional hypothesis which needs large variations in eccentricity, Scholl and Froeschlé (1974, 1975) calculated a large number of orbits with Schubart's averaged planar elliptic model. The main purpose of these calculations was to find out how strong and how fast the orbits, i.e. their eccentricities, vary. According to these calculations the following orbits do not show strong variations in eccentricity and therefore do not support the collisional hypothesis:

a) almost circular orbits close to the 3/1, 5/2, and 2/1 commensurabilities;
b) orbits started at the borders of the corresponding observed gaps;
c) all the orbits close to the 7/3 commensurability.

These orbits behave like non-resonant orbits and therefore represent the problematic cases for the collisional hypothesis.

Severe objections to the collisional hypothesis came from more detailed calculations of the collisional probability of asteroids and from calculations of the kinetic energy required for a destructive collision. Recently, Ip (1977) investigated the probability of collision for main belt asteroids. According to his results, the concept of collisional probability has to be revised. The probability of colliding with another asteroid is not only a function of orbits intersected but it is also a function of the time the asteroid remains outside the main belt. Therefore, an asteroid with a low eccentricity may have a higher collision probability than an asteroid with a large eccentricity. A realistic model which takes into account the variation of orbits and the structure of the asteroidal belt might answer the question if an asteroid in a gap has a significantly higher collision probability than an asteroid outside of a gap.

Heppenheimer (1975) calculated the kinetic energy gained by an asteroid which increases its eccentricity due to resonant motion. According to his results, this energy is not sufficient to destroy the asteroid by a collision or by a sequence of collisions. He therefore concludes that the collision theory has to be abandoned. Especially, large asteroids with some 100 kms in diameter cannot be destroyed completely within the age of the solar system. In addition, it seems to be difficult to remove such large objects from the gaps by close approaches.

Because of these objections, the number of adherents of the collisional hypothesis is decreasing. All the three hypotheses mentioned above, the statistical, gravitational, and collisional assume that the Kirkwood gaps formed after the formation of the solar system and that the gaps were originally filled with asteroids.

Recently, two different hypotheses were presented which propose the formation of the Kirkwood gaps at the same time when the planetary system formed. Greenberg (1978) proposes a mechanism which produces the Kirkwood gaps within a few thousand years by a drag effect. The larger objects of 1 km size which formed at the commensurabilities collide frequently with the smaller particles. These small particles form the drag effect which removes the larger objects to the inner edge of the gap. Because of the drag effect, the eccentricities of the larger objects are dampened and as a result their semi-major axes experience a secular decrease, and the orbital energy is dissipated. This mechanism works especially well for resonant motion because the eccentricity is more strongly increased than for nonresonant motion. This energy dissipating hypothesis is very appealing as it seems to explain also other resonance phenomena in the solar system. Those orbits which do not vary their eccentricities strongly are the same problematic cases for the energy dissipating hypothesis as for the collisional hypothesis. According to the energy dissipating hypothesis, the members of the Hilda group formed close to the 3/2 commensurability and were captured in the 3/2 resonance by the same drag effect.

Another hypothesis which will be called the cosmogonic hypothesis has been proposed by Heppenheimer (1978). He shows that planetesimals could not have formed in the gaps. The strong variations in eccentricity of a growing particle produce strong variations of the particles orbital velocity which in turn prevents the planetesimal's formation. The cosmogonic hypothesis is a completely different approach to the problem of the Kirkwood gaps because it does not presuppose asteroids or planetesimals in the gaps like the other hypotheses.

The cosmogonic hypothesis has at the moment three weaknesses:

1.) It depends strongly on the formation mechanism of planetesimals.
2.) It cannot produce the Hilda family which must be formed somewhere else in the solar system and captured into resonance, because the observed Hildas show strong variations in eccentricity.
3.) It does not explain the absence of objects in the gaps which do not vary their eccentricities strongly.

More extended investigations of the cosmogonic hypothesis might solve these problems.

Of all the extant hypotheses for the origin of the Kirkwood gaps, the statistical hypothesis appears to have been sufficiently refuted. The gaps are not a statistical phenomenon. The gravitational hypothesis is still open, since we do not know how the dynamical system Sun-All the planets-Asteroid does evolve for $t \to \infty$. If the gravitational hypothesis is correct, then the formation of the Kirkwood gaps needed much longer than 10^5 years. The collisional hypothesis can not explain why we do not observe large asteroids of 100 km size in the gaps, as it is a problem to destroy or remove them because of energetic reasons. The collisional, energy dissipating and cosmogonic hypotheses still have to explain the absence of objects with almost circular orbits and the formation of the gap at the 7/3 commensurability.

REFERENCES

Franklin,F.A., Marsden,B.G., Williams,J.G., Bardwell,C.M. : 1975, "Astron.J." 80, pp.729-746

Froeschlé,C., Scholl,H. : 1976, "Astron.Astrophys." 48, 389-393

Giffen,R. : 1973, "Astron.Astrophys." 23, pp.387-403

Greenberg,R. : 1978, "Icarus" 33, pp.62-73

Hénon,M., Heiles,C. : 1964, "Astron.J." 69, pp.73-79

Heppenheimer,T.A. : 1975, "Icarus" 26, pp.367-376

Heppenheimer,T.A. : 1978, "Astron.Astrophys." to be published

Ip,W.-H. : 1977, "Icarus" 32, pp.378-381

Jefferys,W.H. : 1967, "Astron.J." 72, pp.872-875

Kiang,T. : 1978, "Nature" to be published

Kirkwood,D. : 1867, "Meteoric Astron." Ch.13

Lecar,M., Franklin,F.A. : 1973, "Icarus" 20, pp.422-436

Scholl,H., Froeschlé,C. : 1974, "Astron.Astrophys." 33, 455-458

Scholl,H., Froeschlé,C. : 1975, "Astron.Astrophys." 42, 457-463

Schubart,J. : 1964, "SAO Special Report" 149

Schweizer,F. : 1969, "Astron.J." 74, pp.779-788

Sinclair,A.T. : 1969, "Mon.Not.R.Astron.Soc." 142, pp.289-294

Wiesel,W.E. : 1976, "Celestial Mechanics" 13, pp.3-37

Williams,J.G. : 1977, "IAU Comm. No.20" Circular No.6

EVOLUTION OF ORBITS IN THE OUTER PART OF THE ASTEROIDAL BELT AND IN THE KIRKWOOD GAPS

C. Froeschlé and H. Scholl
Observatoire de Nice, Astronomisches Rechen-Institut,
Heidelberg, F.R.G.

It is well known that the semi-major axes of the asteroids between the orbits of Mars and Jupiter are not uniformly distributed. The depleted regions in the outer part and the Kirkwood gaps in the inner part of the belt represent singularities in the frequency distribution of the asteroidal semi-major axes. The question whether these depleted regions and the Kirkwood gaps are due to gravitational perturbations of Jupiter or due to cosmogonic effects has not yet been solved and the discussion is still continuing.

In a series of numerical experiments (1974, 1975), we have tried to depopulate the Kirkwood gaps at the 2/1, 3/1, 5/2 and 7/3 commensurability by Jupiter's gravitational action on fictitious asteroids starting in a gap. In no case did an asteroid leave any of the investigated gaps and remain outside the gap. All the fictitious asteroids librate through the gaps or mostly within the gaps with quite different amplitudes and frequencies depending on starting values. In addition, no fictitious asteroids approached Jupiter closely and therefore none escaped out of the gap. According to our numerical experiments, the Kirkwood gaps cannot be depopulated by Jupiter's perturbations.

We therefore tested the collisional hypothesis for the origin of the Kirkwood gaps, which assumes originally existing asteroids or planetesimals in the gaps. Because of the especially strong perturbations, these objects vary their eccentricities strongly, and therefore can cover a large portion of the asteroidal belt thus increasing the probability of a collision with a belt asteroid. The larger the variation in eccentricity, the higher the collision probability. In our test for the collisional hypothesis, we used a planar Sun-Jupiter-Asteroid model averaged by Schubart's method (1964). Most of the calculated orbits support the collisional hypothesis. The few problematic cases which do not support the collisional hypothesis are the almost circular orbits, the orbits starting at the edges of the observed gaps and the orbits at the 7/3 commensurability. These problematic cases do not vary their eccentricities strongly and therefore cannot be explained by the collision hypothesis.

R. L Duncombe (ed.), Dynamics of the Solar System, 223–226.

Compared to the depletion of the Kirkwood gaps, the gravitational explanation for the depletion of the outer part of the belt between the 2/1 commensurability and the Hilda family seems to be even more difficult according to Lecar and Franklin's (1973) numerical experiment. Lecar and Franklin calculated orbits of fictitious objects over a few thousand years using the elliptic planar Sun-Jupiter-Asteroid model. The expected mechanism which depletes that region is based on perturbations in the semi-major axis and eccentricity of an asteroidal orbit which results in an Jupiter crossing orbit. After a close approach to Jupiter, the asteroid escapes from the considered region.

In Franklin and Lecar's experiment, the objects with higher eccentric orbits, $e > 0.25$, escaped, while for objects with $e < 0.25$ only a few had close encounters with Jupiter.

The existence of these problematic cases in both experiments, for the Kirkwood gaps and for the outer part of the belt, suggests that the corresponding hypotheses which are mainly based on gravitational affects are false. However, before looking for different hypotheses which would need more sophisticated physics, one has to refine the gravitational models with respect to the number of perturbing bodies and with respect to the periods covered. A calculation including Saturn over much longer time spans might reduce the number of problematic cases considerably.

Our calculations were based on Schubart and Stumpff's N-Body Program (1966). Over 100 000 years, we computed the orbits of fictitious asteroids which represent the problematic cases for the collision hypothesis of the Kirkwood gaps and for the ejection hypothesis of the outer belt. Jupiter and Saturn were included as perturbing bodies.

For the Kirkwood gaps, our new numerical experiment yielded the same negative result as before. The number of problematic cases could not be reduced.

The depopulated region, $3.6 < a < 3.9$ AU, in the outer belt was depleted by our experiment (Froeschlé and Scholl, 1978) to a larger extent than by Lecar and Franklin's experiment. The average time scale for the excitation of an orbit in order that it crosses Jupiter's orbit is somewhat larger than the period covered by these authors. After 2 000 years, 25% of all the escapers were obtained in our experiment, 75% after 15 000 years and 100% after 60 000 years. However, our experiment did not depopulate the region $3.6 < a < 3.9$ AU completely. Most of the objects with starting eccentricities of $e < 0.10$ remained in that region and had no close approach to Jupiter. In addition, another family of objects which minimize their eccentricities if their aphelia are precessing through Jupiter's orbital plane, avoided a catastrophic encounter with Jupiter.

For both types of families, observed asteroids are known in the range under consideration but they exist in a much lower abundance than should be expected from our experiment. The three asteroids (721)

Tabora, (522) Helga, and PL - 4164 show a coupling between the precession of perihelia and the long period in eccentricity. The eccentricity becomes smallest when the perihelion and therefore also the aphelion lies in Jupiter's orbital plane. That yields the largest possible distance to Jupiter when the asteroid is at its aphelion. On the other hand, when the eccentricity reaches its maximum and therefore the aphelion distance is largest, thus yielding the smallest possible distance to Jupiter, the argument of perihelion is close to 90° or 270°. Therefore, the asteroid passes through its aphelion when it is high above or below Jupiter's orbital plane. This mechanism prevents close encounters with Jupiter.

We suppose that more objects of that kind might exist but have not yet been detected as observers ordinarily detect minor planets close to the ecliptic. Objects of the kind described above, however, can be detected best at high ecliptic latitudes when they pass through their perihelia. The discrepancy between expected and observed asteroids might be reduced by observations.

For the first mentioned type of asteroids having $e < 0.1$, more observations with very powerful instruments might reduce that discrepancy. It is an interesting problem why we observe so few objects with almost circular orbits in the region 3.6 - 3.9 AU as well as in the Kirkwood gaps, because gravitational models including collisions do not remove such objects to a large extent. For the gap at the 2/1 resonance, Franklin et al. (1975) list objects with low eccentricities which seem to librate in the gap. However, most of these minor planets have very uncertain orbits. Further observations in order to find these objects might reduce the discrepancy between observation and calculation.

Such long runs over 100 000 years are rather expensive. Therefore, several authors (e.g. Nacozy 1976) propose to increase the masses of the perturbing bodies in order to shorten the time scale and consequently the computing time and in order to obtain perturbations with larger amplitudes. The intrinsic problem for such numerical experiments with larger masses consists in the equivalence of a Sun-Jupiter-Asteroid and a Sun -"Super Jupiter"- Asteroid model. It is not obvious that the latter model on a shorter time scale yields the same orbits as the first one. The perturbing mass may not be increased too greatly.

For our purposes, and especially for the depopulation of the outer belt, it is necessary to keep the Hilda family stable in an experiment with a "Super Jupiter", since the Hilda family is actually observed. In order to determine a limiting Jupiter mass up to which the Hilda family remains stable, we computed orbits around the 3/2 commensurability with different Jupiter masses.

According to our calculations, the model with a Jovian mass of 0.007 solar masses yields no stable orbits. All the objects escape. Therefore, a value of 0.005 solar masses might be used for a Super Jupiter without significantly destroying the topology of the model. We repeated the experiment for the region $3.6 < a < 3.9$ AU with a value of

5 times Jupiter's mass ($\gamma = 5$) in order to compare the depletion curves for the Sun-Jupiter-Asteroid and the Sun-Super Jupiter-Asteroid model. The orbits of 47 fictitious objects were calculated. The following Table 1 shows the total number of escapers after certain time intervals given in years.

Table 1

Sun-Jupiter ($\gamma = 1$)		Sun-Super Jupiter ($\gamma = 5$)	
Time	Total Number of Escapers	Time	Total Number of Escapers
0	–	0	–
1 000	3	100	7
2 000	5	200	12
3 000	6	500	17
10 000	10	1 000	26
15 000	16	2 500	33
60 000	19		
100 000	19		

Obviously, the depletion of both models is different. Most of the escapers in the model with $\gamma = 1$ were found after 15 000 years, while in the model with $\gamma = 5$, the 2 500 years do not seem to be sufficient to obtain all the escapers. The use of a Super Jupiter destroys the protection mechanisms which in the model with $\gamma = 1$ avoid close encounters with Jupiter. Therefore, the model with a Super Jupiter yields too many escapers and is therefore not equivalent to the original model ($\gamma = 1$).

ACKNOWLEDGMENTS

This work was partially supported by grant No.3705 of the ATP Planetologie (France).

REFERENCES

Franklin,F.A., Marsden,B.G., Williams,J.G., Bardwell,C.M. : 1975, "Astron.Journ." 80, pp. 729-746
Froeschlé,C. and Scholl,H. : 1978, submitted to "Astron. and Astrophys."
Lecar,M. and Franklin,F.A. : 1973, "Icarus" 20, pp. 422-436
Nacozy,P.E. : 1976, "Astron.Journ." 81, pp. 787-791
Scholl,H. and Froeschlé,C. : 1974, "Astron. and Astrophys." 33, pp. 455-458
Scholl,H. and Froeschlé,C. : 1975, "Astron. and Astrophys." 42, pp. 457-463
Schubart,J. : 1964, "SAO Special Report" No. 149
Schubart,J. and Stumpff,P. : 1966, "Veröff.d.Astron. Rechen-Inst." No. 18

HYPERPERIODS, ORBITAL STABILITY, AND SOLUTION OF THE PROBLEM OF KIRKWOOD GAPS

T. Kiang
Dunsink Observatory, Castleknock, County Dublin, Ireland

I believe I have solved, at least in principle, the long-standing problem of the Kirkwood gaps, and have incidentally initiated a new approach to questions of orbital stability. I shall begin with the concept I call hyperperiod. A given periodic dynamical system S with period P may or may not have a latent long period - the hyperperiod P*. If P* exists, then any small displacement or variation, actual or virtual, once-for-all or recurrent, will induce a displacement y which will be periodic with period P* and will be of bounded amplitudes. We can then say that S is stable. If P* dose not exist, then y will eventually become indefinitely large - and we say that S is unstable.

Example 1. An idealised Sun-Jupiter-Halley system was idealised to be periodic with P = 154.2 yr. It was found (refs 1,3) that, for some initial configurations, P* exists and is about 600 yr; while for others, P* does not exist.

Example 2. An upright rod 13.3 cm long has its lower end moved up and down at 50 Hz (ref.1). Here, P = 0.02 sec. If the stroke exceeds 0.45 cm, then P* exists and is about 1.25 sec, and, given any small push, the rod will simply sway with that period (stable !). If the stroke is less than 0.45 cm, then P* does not exist and the rod falls over (unstable !).

The technique of finding P* is this: we first derive a Hill's equation for y (or a linear function thereof):

$$\frac{d^2y}{dt^2} + G(t)\,y = 0 \qquad (1)$$

where G(t) is a known period function of period P. The solution of (1) is of the form

$$y = e^{icx} \sum b_k e^{ikx}, \qquad (x = 2\pi t/P_0) \qquad (2)$$

where c is a latent frequency of the system and the b's integration constants. I shall call c the Hill exponent. The evaluation of c was first given in ref. 4; but see my remarks in ref. 3. If c is real, then it defines a hyperperiod:

R. L. Duncombe (ed.), Dynamics of the Solar System, 227–230.

$$P^* = P/c \tag{3}$$

If c is imaginary, then y will eventually be dominated by the term $b_0 \exp(icx)$, and P^* does not exist.

Thus, if, for a given periodic dynamic system S, we can write down a Hill's equatinn for the displacement or variation, we can then define the stability of S according as P^* exists or not, or as c is real or imaginary.

Now, each orbit in what I call the Schubart diagram (ref. 3) is a periodic dynamic system. Unlike the examples above, there are no further parameters to be specified so that any verdict we may return on the stability character of a given orbit will be an unconditional one. Of course, in order to reach a verdict, we must first write down a Hill's equation. Here, again, the case of resonant asteroids is different from the examples given above. In the examples, the derivation follows entirely the classical treatment; but if we try to do the same in the present case, we shall never succeed. This is because the classical treatment of the present case, namely, a conservative, periodic system of one degree of freedom will lead to the general assertion that <u>all</u> such systems are unstable in the sense of asymptotic stability and stable in the sense of orbital stability (for the various types of stability, see ref. 5), whereas if we could write down a Hill's equation, then the stability character of a given orbit will have to depend on the individual properties of the orbit. In other words, the classical treatment is incompatible with the possibility of writing down a Hill's equation. To do the latter we must at some stage modify the classical treatment.

The problem of the Kirkwood gaps reaches its most acute form when we compare the Hecuba and Hilda regions in the intermediate eccentricity range. Nature suggests that, in this range, the Hilda librators are orbitally stable and the Hecuba librators are orbitally unstable. The classical treatment with its sweeping statements will never be able to resolve the antimony, while the method of Hill's equation or hyperperiods offers a possibility of doing so.

Let us see what modifications are necessary. An orbit in the Schubart diagram is defined by the canonical equations

$$\frac{dx}{dt} = F_2, \quad \frac{dy}{dt} = -F_1 \tag{4}$$

where the numerical suffix (1 for x, 2 for y) denotes partial differention of the Hamiltonian F and setting the variations u and v equal to zero. (x and y are the canoncial variables here; do not confuse with their previous usage). Differentiating (4), we have

$$\frac{d^2x}{dt^2} = F_{21} F_2 - F_{22} F_1 \tag{5}$$

Now, apply to (5) the variations

$$x \longrightarrow x + u, \quad y \longrightarrow y + v. \tag{6}$$

Then, after some reduction, we have

$$\frac{d^2u}{dt^2} = \frac{dF_{21}}{dt} u + \frac{dF_{22}}{dt} v + (F_{12}^2 - F_{11} F_{22}) u \tag{7}$$

The equation of v can be obtained by interchanging u and v, and the suffixes 1 and 2. Because of the term in v, (7) is not a Hill's equation. We can therefore get a Hill's equation by dropping that term. We get another Hill's equation by dropping also the preceding term. I opt to do the latter and obtain

$$\frac{d^2u}{dt^2} + (F_{11} F_{22} - F_{12}^2) u = 0 \tag{8}$$

The reason for my choice is that the form (8) can also be derived from the following scheme:

$$\frac{d^2u}{dt^2} = \lim_{h \to 0} \frac{1}{h} \left\{ \frac{d}{dt} (x + u + h \frac{du}{dt}) - \frac{d}{dt} (x + u) \right\} \tag{9}$$

in the evaluation of which we always set

$$\frac{du}{dt} = F_{21} u + F_{22} v, \quad \frac{dv}{dt} = -F_{11} u - F_{12} v \tag{10}$$

which are the classical variation equations of (4). In this scheme, no use is made of equation (5) so that an "Enabling Principle" can be ennunciated as follows: "In order to be able to write down a Hill's equation, we must refer the variation to an <u>unaccelerated</u> frame in the phase space". The classical treatment, by contrast, refers the variation to the accelerated frame "following the natural motion".

The results of applying (8) to the orbits in the Schubart diagram are that the Hilda librators are, in general, stable, and the Hecuba librators are, in general, unstable. Details will be published elsewhere and the qualitative results have been given in ref. 3.

The "Enabling Principle" stated above offers the possibility of a new approach to a general theory of stability of dynamical systems, i.e., not only of systems of one degree of freedom.

REFERENCES

1. Kiang, T., Mon. Not. R. Astron. Soc. 162 (1973) 271-287
2. Kiang, T., Nature, 273 (1978) 734-736.
3. Kiang, T., Paper VI.8 of this Symposium.
4. Hill, G.W., Acta Math. 8 (1886) 1-36.
5. Nacozy, P.E., Paper I.2 of this Symposium.

DISCUSSION

Scholl: Schubart's model which you used does not yield orbits in the Hecuba gap which drift out of the gap. According to Schubart's figures you use, that is not possible. Apparently, your stability investigation about Schubart's dynamical system, which has one degree of freedom, predicts the behavior of a system with two or more degrees of freedom. This procedure is, however, very doubtful.

Kiang: My faith and intuition is that the stability of resonant asteroids in real life can be studied using the simplest model. It is the same kind of faith that made Poincaré concentrate so much on periodic orbits.

Message: Would each of the speakers give their views as to the extent to which the simplifications, required to reduce the problem to one degree of freedom, conceal features of the actual motion over very long periods of time?

Kiang: I have to refer you to my paper in Nature, in which I actually used the "simplifications" to sharpen my qualitative results in the Hecuba case.

Schubart: I think that one can try to use the model that is simplified to one degree of freedom, with an addition that follows from the model without difficulty. After solving the problem of one degree of freedom for the arguments of slow variation, the short-period argument follows by means of a quadrature, and this gives the period of revolution of the longitude of perihelion, $\tilde{\omega}$. If a period of libration follows from the solution of the basic problem, the ratio of the two periods may be a characteristic for the stability of the orbit under consideration against perturbations that are not considered in the simplified model.

Schubart's comments: [For references compare paper 5.1] I am not in favor of the designation "Schubart diagrams" for the figures in a former paper (Schubart, 1964) mentioned by Dr. Kiang. Poincaré (1902) drew the first diagram of this type, so it is better to use "Poincaré diagrams" or "Izsak diagrams," since the late Dr. I. Izsak had asked me to plot the diagrams in just this way. The one-degree-of-freedom problem mentioned by Dr. Kiang (Schubart, 1964) is well defined. A numerical test about the time dependence of the relative distance of two points in phase space can give evidence about "hyperperiods." Some test about Dr. Kiang's theoretical results is necessary, according to my opinion.

I like the suggestion given by Dr. Kiang's recent treatment, to think again about the differences between the Hecuba and Hilda commensurabilities, using gravitational theory alone.

SECULAR PERTURBATIONS OF ASTEROIDS AND COMETS

Yoshihide Kozai
Tokyo Astronomical Observatory, Mitaka, Tokyo 181, Japan

In 1962 the author published a paper on secular perturbations of asteroids with high eccentricity and inclination (Kozai, 1962) and found that for such asteroids the secular perturbations are more complicated than those estimated by the classical linear theory. Before that the author(Kozai, 1954) tried to extend the classical theory to include terms of squares of disturbing masses, eccentricities and inclinations in expressions of secular motions of angular variables and found that the sum of the longitudes of the perihelion and the ascending node is not as stable quantity as the classical theory predicted, and, therefore, it cannot be used to estimate the ages of asteroid families. Then Williams(1969) and Yuasa(1973) extended the classical theory further.

In this paper the author computes the secular perturbations of numbered asteroids and short-periodic comets whose semi-major axes are less than 30 a.u. by assuming that (1) the disturbing planets are Jupiter, Saturn, Uranus and Neptune, (2) their orbital semi-major axes are not much disturbed, and (3) the disturbing planets are moving along circular orbits on Jupiter's orbital plane. According to the assumption (2) any asteroid or comet with commensurable mean motion and any comet which can approach to any disturbing planet very closely should be excluded.

According to the assumption (2) the mean anomaly of the disturbed body and the mean longitude of each of the disturbing planets can be eliminated from the disturbing function by averaging it with respect to each of the angular variables independently. The averaging is carried out not by any analytical way but by a numerical harmonic analysis with 5° as one step for the variables. After the elimination the semi-major axis can be assumed to be constant. By the assumption (3) the longitude of the ascending node of the disturbed body does not appear in the averaged disturbing function when the orbital plane of Jupiter is adopted as the reference plane, and, therefore, the z-component of the angular momentum, $(a(1-e^2))^{1/2}\cos i$, is conserved, where i is the inclination with respect to Jupiter's orbital plane. And as the averaged Hamiltonian as well as the disturbing function do not depend on time explicitly, the Hamiltonian F is conserved.

R. L. Duncombe (ed.), Dynamics of the Solar System, 231–237.

Now the system of the equations of motion is reduced to that of one degree of freedom, and it is the equation for the secular perturbations. The variables are $G = (a(1-e^2))^{1/2}$, the angular momentum, and g, the argument of perihelion, and the Hamiltonian is regarded as a function of $x = (1-e^2)^{1/2}$ and 2g with a and H = x cos i as parameters. Therefore, x can be solved as a function of 2g, in fact, a periodic function of 2g, and as H is given, both the eccentricity and the inclination can be graphically expressed as functions of g. As the values of a and H are given for each of the asteroids and the comets, the values of the Hamiltonian can be computed for various sets of x and g, and then equi-F-value curves are drawn in g-x plane, where g-axis extends from 0° to 90° and x-axis extends from H to 1. There are symmetries with respect to g = 0°, 90°, 180° and 270°. Two lines, x = 1 and x = H, are always equi-F-value curves and x = 1 corresponds to a circular orbit whereas x = H corresponds to an orbit on Jupiter's orbital plane. The two orbits are not disturbed due to the secular perturbation unless the argument of perihelion is of libration. As H is decreased, the variation of x becomes wider, and if the value of H is below a certain value around 0.8, a libration region appears in the g-x diagram. If the value of F corresponding to the concerned asteroid or comet is that in the libration region, the argument of perihelion does not make a complete revolution but makes a libration around 90° or 270°.

Except for the libration case the eccentricity of any of the asteroids takes its minimum and the inclination takes its maximum at g = 0° and 180° and vice versa at g = 90° and 270°. For the libration case both maximum and minimum take place at 90° or 270°. Therefore, the orbits become the least eccentric when the perihelion is on Jupiter's orbital plane and can avoid any very close approach to Jupiter if the orbit is inside of Jupiter. For most of the asteroids it is found that x is larger than cos i and for such cases the variation of the eccentricity is wider than that of the inclination. In Figure 1 the maximum, the minimum and the present values of the eccentricities and the inclinations of eight asteroids belonging to a family, Pallas family by Hirayama(1928) and family No. 28 by Brouwer(1951), are plotted with asteroid numbers. For each of the asteroids a narrow libration region appears in the g-x diagram as the valuses of H are nearly equal to 0.8, although for non of them the argument of perihelion is of libration. In the figure the crosses in the upper left

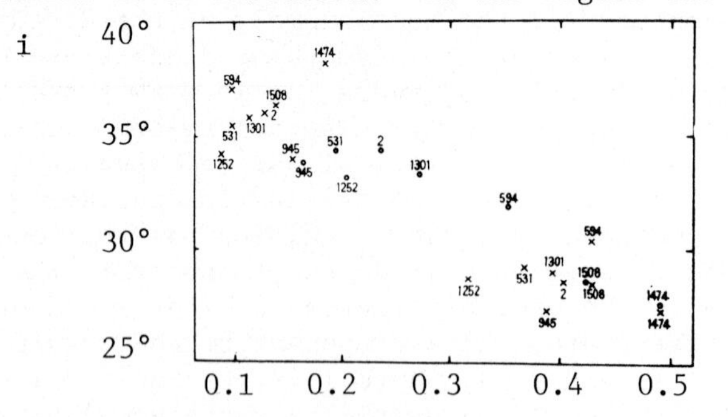

Figure 1. Variations of the eccentricities and the inclinations for asteroids belonging to Pallas family.

corner correspond to the values of the eccentricity and the inclination at g = 0° and 180° and those in the lower right corner correspond to those at g = 90° and 270°. The circles correspond to the present values. For these asteroids the eccentricities change by more than 0.3 and the inclinations change by 10° as functions of the argument of perihelion. Differences between the proper eccentricities and inclinations and osculating ones are much smaller than the variations. The asteroids 1474 and 1508 were not included in this family in Brouwer's paper, however, it is found that they are real members of the family.

There are four asteroids among the numbered ones, for which the arguments of perihelion are of libration. They are 944 Hidalgo, 1373 Cincinnati, 1866 Sisyphus and 1981. Kozai(1962) found that the argument of perihelion of Cincinnati(a = 3.411 a.u.) is of libration between 69° and 111° by a simpler computation. The eccentricity and the inclination change between 0.318 and 0.559 and between 40.°2 and 29.°2, respectively, as one sees in Figure 2,where the scales of the ordinate are not given in x but in e and i although it is scaled with equal interval of x. The present place is marked with the cross on one of the equi-F-value curves. Because of the libration both the perihelion and the aphelion of this asteroid cannot be placed on Jupiter's orbital plane, and, therefore, it cannot approach to Jupiter very closely even though the aphelion distance can become larger than the semi-major axis of Jupiter. There are three asteroids, for which the minimum aphelion distances are larger than 4.4 a.u. One of them is this asteroid, and the other two belong to Hecuba group. For the latter two the critical arguments are of libration around 0°(Franklin et al., 1975) and, therefore, they can avoid any close approach to Jupiter.

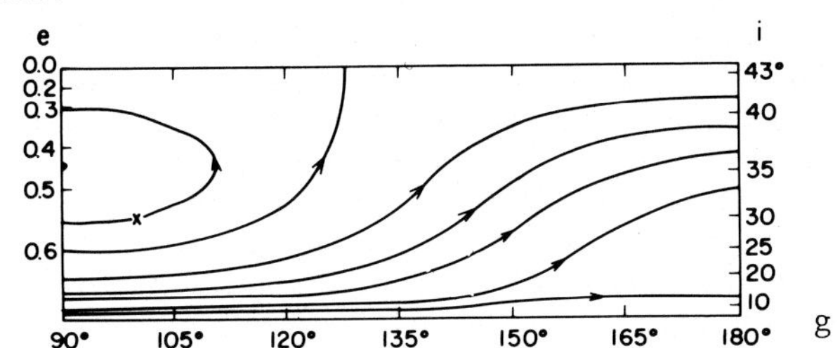

Figure 2. The g-x diagram for 1373 Cincinnati.

According to the present computation the argument of perihelion is of libration for a peculiar asteroid, Hidalgo, (a = 5.82 a.u.) between

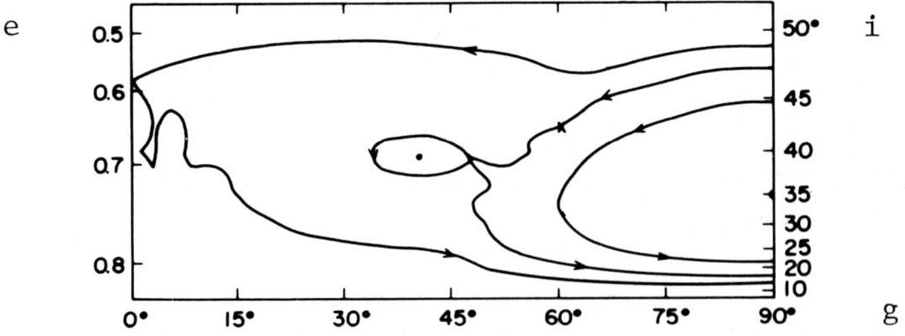

Figure 3. A part of the g-x diagram for 944 Hidalgo.

48° and 132°, and the eccentricity and the inclination change between 0.568 and 0.811 and between 47.°3 and 17.°2, respectively, whereas the present values are e = 0.658 and i = 42.°26. Though the argument of perihelion is now decreasing from 59.°5, after reaching to 48° it will increase and when it will become 90° the perihelion distance will be as small as 1.1 a.u. Then the orbit will be quite similar to that of a comet. As Hidalgo has a possibility to approach very closely to Jupiter, the present computations might have some errors. However, Nakano's numerical integration results(1977) suggest that the errors are not so large. For the other two asteroids, for which the arguments of perihelion are of libration, the semi-major axes are 1.89 and 1.78 a.u. and their orbits are quite similar to those of comets and can avoid any very close approach to Mars. Besides the four asteroids there appear librations regions in g-x diagrams for 23 asteroids.

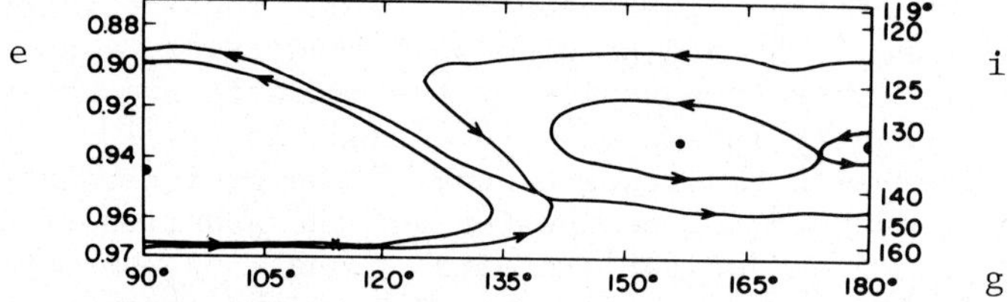

Figure 4. A part of the g-x diagram for Halley's comet. The present place is marked by the cross which is on one of the equ-F-value curves.

The situation is quite contrary for short-periodic comets. Of 104 comets treated here 78 comets have their semi-major axes less than that of Jupiter. Of them 68 comets have libration regions in their g-x diagrams because the values of H are usually small. However, for none of them the argument of perihelion is of libration. For most of them x is less than cos i and the variation of the inclination is wider than that of the eccentricity. However, of the short-periodic comets the semi-major axes of 26 comets are between that of Jupiter and 30 a.u. Of the 26 comets 25 have libration regions in g-x diagrams, the only one exception being P/Schwassma-n-Wachmann I. And the arguments of perihelion are of libration for 15 comets. However, even for the libration cases the eccentricities do not change so much as those of some asteroids. For Halley's comet, for an example, the argument of perihelion is of libration between 47° and 133° and the eccentricity and the inclination change between 0.899 and 0.968 and between 123.°3 and 164.°4, respectively. The present values are e = 0.967, i = 161.°2 and g = 114.°4 which is increasing. Therefore, the eccentricity of this comet is now nearly maximum and the perihelion distance will not decrease in near future. However, for this short range of the variation the perihelion distance is changed by 1.2 a.u. which is large enough to affect the evolution of the comet. Also there are a few comets, for which the eccentricities are changed in wider ranges like Hidalgo.

Since the eccentricity and the inclination do change very much for asteroids with small values of H, the proper eccentricity and inclination which are derived by the classical linear theory are not adequate to clas-

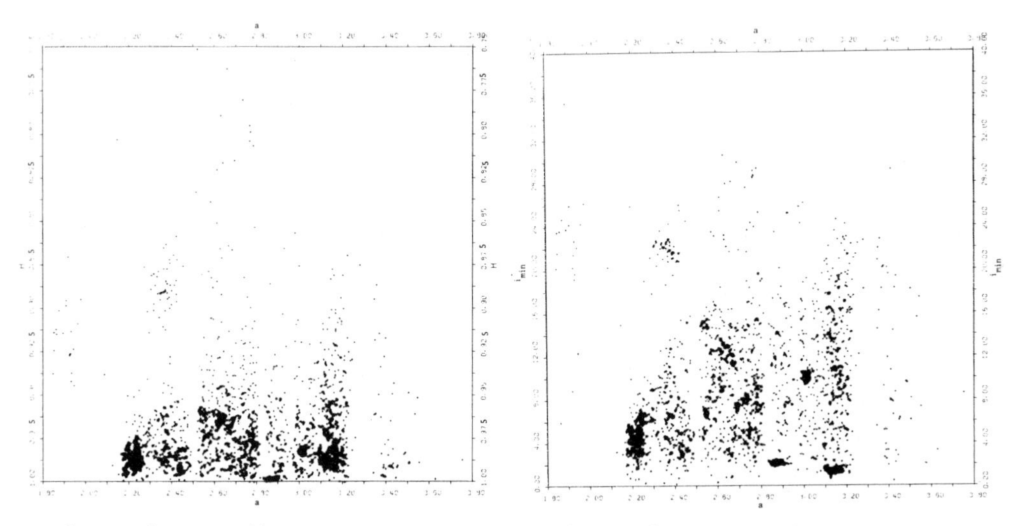

Figure 4. a-H diagram. Figure 5. a-i_{min} diagram.

sify such asteroids into families. The author believes that the minimum value of the inclination(or the maximum value or the eccentricity) and H as well as the semi-major axis are more adequate and stable quantities which can be used for classification of the asteroids with not only small values of H but also H nearly equal to 1. In Figures 4 and 5 the numbered asteroids are plotted in a-H plane and a-i_{min} plane, respectively. Even in these figures 6 famous families discovered by Hirayama(1923 and 1928) as well as the family in Figure 1 clearly show up. More clear pictures can be obtained when the asteroids are plotted in a-H-i_{min} space. By this way the seven families in the following table are derived. In the table the names of the families after Hirayama(1923 and 1928), the numbers of their members by the present analysis, Hirayama(1928) and Brouwer(1951) in this order, and the mean values of the semi-major axes, H and the minimum inclinations as well as their mean deviations are given. Also the mean values of the maximum eccentricities are given for a reference.

Name	Numbers			a	H	i_{min}	e_{max}
Themis	77	32	53	3.140± 0.024	0.9881± 0.0033	1.°40± 0.°21	0.108
Eos	89	27	58	3.015± 0.007	0.9821± 0.0013	10. 01± 0. 27	0.052
Coronis	48	20	33	2.875± 0.018	0.9984± 0.0007	2. 14± 0. 14	0.030
Maria	20	14	17	2.543± 0.014	0.9609± 0.0023	14. 93± 0. 30	0.074
Flora	228	63	125	2.229± 0.032	0.9860± 0.0060	4. 58± 1. 40	0.104
Phocaea	34	11	21	2.371± 0.033	0.8882± 0.0093	21. 85± 0. 60	0.207
Pallas	8	3	6	2.723± 0.057	0.8023± 0.0167	32. 16± 2. 43	0.229

As one can see, some families are very compact whereas others are not. The scattering of H is small for the families whose eccentricities are small. This may suggest that the neglecting the eccentricities of the disturbing planets gives stronger effects on more eccentric asteroids such as those belonging to Phocaea and Pallas families. The effects are very small for those belonging to Eos, Coronis and Maria families. Even though these effects are taken into account, there are still some loose families such as Flora and Pallas. Therefore, the author has an idea that these loose families were not originated from single mother asteroids as the other compact ones were probably originated from single ones. Probably many asteroids were pushed out from regions where secular commensurabilities exist and gathered there as loose families as Williams(1969) pointed out. In fact several gaps seen in Figures 4 and 5 are situated near regions of secular commensurablity.

The computations were made with the computor FACOM 230-58 at the Computing Center of the Tokyo Astronomical Observatory by the orbital elements of the asteroids and comets provided by Dr. K. Hurukawa.

REFERENCES.

Brouwer, D.: 1951, "Astron. J." 56, pp.9-32.
Franklin, F. A. et al.: 1975, "Astron. J." 80, pp.729-746.
Hirayama, K.: 1918, "Astron. J." 31, pp. 185-188.
Hirayama, K.: 1923, "Japan J. Astron. Geophys." 1, pp. 55-93.
Hirayama, K.: 1928, "Japan J. Astron. Geophys." 5, pp. 137-162.
Kozai, Y.: 1954, "Publ. Astron. Soc. Japan" 6, pp. 41-66.
Kozai, Y.: 1962, "Astron. J." 67, pp.591-598.
Nakano, K.: 1977,"Private Communication".
Williams, J. G.: 1969, "Secular Perturbations in the Solar System", Dissertation, University of California, Los Angeles, U.S.A.
Yuasa, M.: 1973, "Publ. Astron. Soc. Japan" 25, pp. 399-446.

DISCUSSION

Kiang: 1. Did I understand that comet Halley is a librator?

2. Hirayama and later Brouwer use proper e, i and a to find families. What are your defining parameters? Do you get better defined families? Have you found any new families?

3. Would not the functional relation between e and i in your simplified model not explain the elongated form of the families in the proper (e, i)-plane?

Kozai: 1. Yes, it is, according to my computation. However, as you can see in Fig. 4 the equi-F-value curve for Halley's comet is just inside a limiting curve and, therefore, if there is a small error in my computation, it is possible that the argument of perihelion is not of libration.

2. My parameters are a, $1-e^2 \cos i$, and the minimum values of i. I can get better defined families at least for the Pallas case.

3. No.

Marsden: Don't you think there is a significant difference between (1373) Cincinnati and the other objects for which you now find argument-of-perihelion libration? The libration of (1373) very successfully keeps this object away from Jupiter, and the orbit is thus rather stable. The comets and the other minor planets for which you find libration all make close approaches either to Jupiter or to the Earth, so it would seem that these librations can have little effect on the stability of the orbits.

Kozai: The two asteroids can avoid any very close approach to Mars due to librations.

THREE-DIMENSIONAL DISTRIBUTIONS OF MINOR PLANETS AND COMETS

Ľubor Kresák
Astronomical Institute, Slovak Academy of Sciences, Bratislava, Czechoslovakia

The attitude toward the systems of interplanetary objects is largely different from that accustomed in stellar astronomy. The stellar systems are viewed from the outside, with their shape, population, and internal structure apparent at first glance. The motions are difficult to recognize, if ever, and so the individual orbits are degraded to a background mechanism maintaining the observed organization and rotation of the system. In contrast to this, we dispose of copious information on the dynamics of individual interplanetary objects, but the bodies to which the orbital data refer are sampled under the influence of strong selection effects. Lacking a consistent survey of their systems, we tend to think of them in terms of the biased statistics in the phase-space of orbital elements, rather than in terms of the real three-dimensional distributions. No wonder that we often come across serious misrepresentations : the asteroid belt depicted as a plane ring divided by the Kirkwood gaps like the ring of Saturn; the Trojan clouds as small spherical systems resembling globular star clusters; or the meteor streams as elongated rings of uniform width and population all around. And we seem to receive these gross misrepresentations with surprisingly little annoyance.

The aim of the present analysis was just to reconstruct the outer aspect of the systems of minor planets and comets, using all orbital data as available by the end of 1977, and attempting to eliminate the observational biases. The main results are shown in Figures 1 and 2, in the form of smoothed equidensity plots in a plane passing through the Sun perpendicular to the ecliptic. The equidensity numbers Q are logarithms with base two of the reciprocal mean local densities. Hence, an increase of Q by one indicates a reduction of the density to one half, and an increase by three (solid lines) indicates a doubling of the mean separation of the objects. The unit density, $Q = 0$, is defined by the whole population uniformly dispersed within a volume of 1 $A.U.^3$; a north-south symmetry is assumed throughout. The cylindrical co-ordinates used are $X = R \cos B$, $Z = R \sin B$, L. Since the structure of the systems is mainly controlled by Jupiter, averaging of the equidensity contours in the X,Z - plane and referring the pattern to the mean heliocentric distance of Jupiter makes some features, depending on L, disappear. These are : 1. The effect of inclination of

R. L. Duncombe (ed.), Dynamics of the Solar System, 239–244.

Jupiter's orbit to the ecliptic, smearing out the density pattern by $\pm$ 0.023 X in Z. This can be neglected at the resolution used. 2. The effect of eccentricity of Jupiter's orbit, smearing out the pattern by $\pm$ 0.048 X in X. Since, due to the law of areas, the radius vector of an eccentric ring with one focus at the Sun is proportional to the square root of the population in L, the eccentricity of each system can be evaluated from the longitude distribution. 3. Periodic perturbations by Jupiter, making some pronounced irregularities revolve around the Sun with the planet. The extreme example, the Trojan clouds, has been omitted from the present analysis and will be discussed elsewhere.

MINOR PLANETS

The section through the asteroid belt, as plotted in Figure 1, above, is a composite picture. Its inner region, up to Q = 10, refers to the 121 largest asteroids, with individual masses presumably exceeding 3×10^{18} kg. Their selection was based on the list of compositional types and diameters by Zellner and Bowell (1977) and absolute magnitudes by Gehrels and Gehrels (1977). Carbonaceous and anomalous dark asteroids were included down to a diameter of 130 km; for silicaceous and other asteroids of higher reflectivity the limiting diameter was set 10% lower, to account for their presumably lower bulk densities. There are exactly 100 objects meeting these criteria, in the list of Zellner and Bowell. Using the observed proportion of different compositional types as a function of the revolution period, it was estimated that the objects for which no physical observations are available include about 21 additional carbonaceous objects and no silicaceous object above the adopted mass limit. Therefore, the 21 absolutely brightest asteroids of this group were added to the sample, making a total of 121 objects with a probable 4% share of wrong classifications. To map the outskirts of the belt, 76 outlying objects with masses presumably exceeding 3×10^{15} kg were extracted from a complete list of objects with known orbits (Efemeridy Malykh Planet, Minor Planet Circulars) and elaborated in a similar way. Hence, the outer equidensities ($Q > 10$) are based on the statistics of objects of continually decreasing size. Some uncertainty in the density gradient is due to an implicit assumption of the mass distribution index, adopted as a compromise between the results of Zellner and Bowell and those of the McDonald and Palomar-Leiden Surveys.

It is noteworthy that at $Q > 7$ the thickness of the belt perpendicular to the ecliptic is even a bit greater than its width. Major deformations are evidently associated with the depletion by Earth, Mars, and Jupiter. The first outer boundary (I) applies to the circulating asteroids. The second (II), stretching all around except in the vicinity of Jupiter, is formed by the librators of Hilda type. It may be imagined as a horseshoe-shaped extension of the belt near its central plane. The size of the whole system is determined by the carbonaceous objects, with the silicaceous population prevailing in a flat sunward core and essentially absent beyond X = 3.6. The belt is markedly eccentric, with the major axis aligned to that of Jupiter and with nearly the same eccentricity, 0.05. At any time, 50% of the asteroids are situated within $2.56 < X < 3.20$ and 90% within $2.15 < X < 3.66$; 50% within $-0.28 < Z < +0.28$ and 90% within $-0.91 < Z < +0.91$.

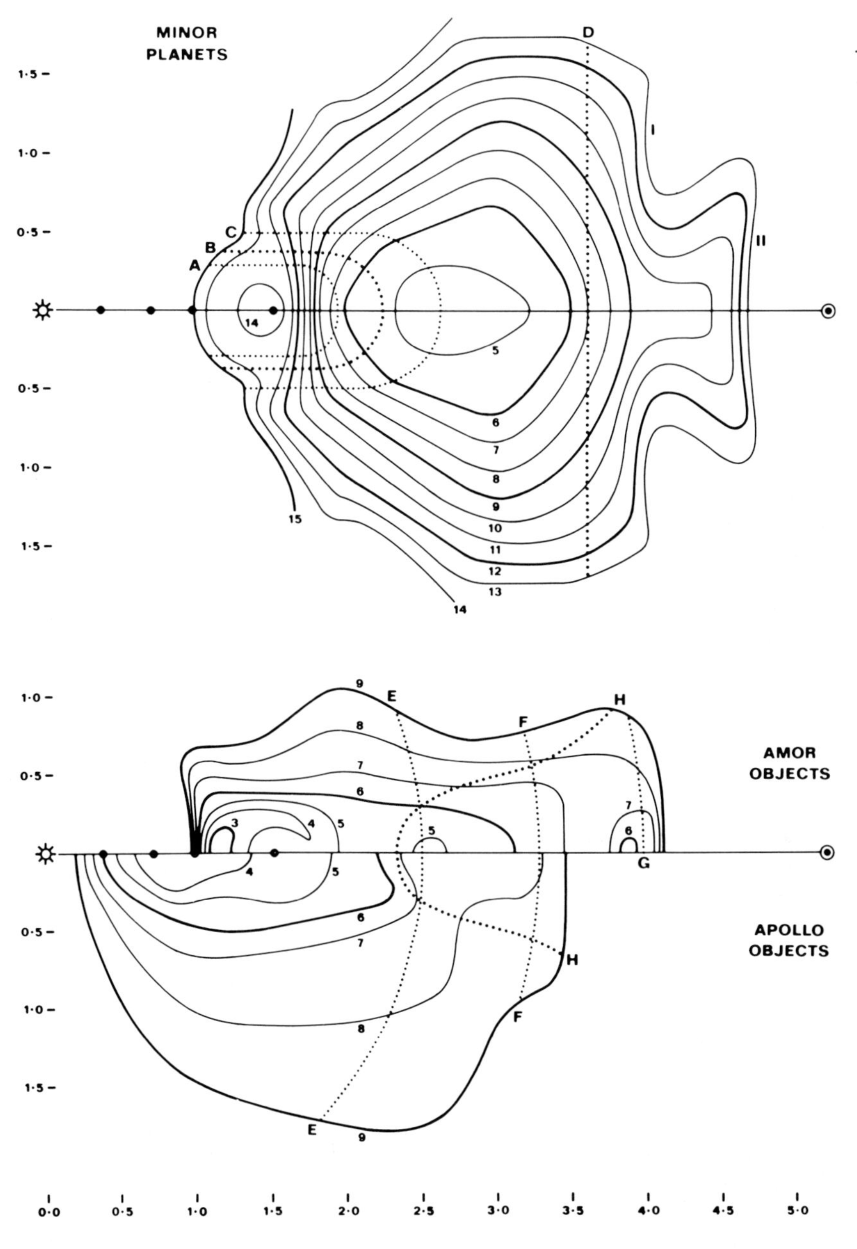

Figure 1. Equidensity contours of the main asteroid belt (above), and of the imbedded systems of the Amor and Apollo objects (below), in the plane passing through the Sun perpendicular to the ecliptic. Each step corresponds to a density ratio of 2:1; mean positions of the major planets are marked by solid dots. A, B, C, D - density ratios of 2:1, 1:1, 1:2, and 1:100 of the silicaceous to carbonaceous population. E, F, G - mean solar distances corresponding to the resonances of 1:3, 1:2, and 2:3 with Jupiter. H - density ratio of 1:1 of the Amor to Apollo objects.

AMOR AND APOLLO OBJECTS

With two exceptions, the Amor and Apollo objects fall short of the size limit adopted for determining the contours of the main asteroid belt. Therefore, a separate analysis was based on the orbits of all such objects known by the end of 1977 : 18 of Amor type ($1 < q < 1.25$) and 25 of Apollo type ($q < 1$, but including 1580 Betulia and 1917 Cuyo). As the discoveries of these objects are seriously biased by observational selection, subsamples of different absolute brightness were formed to extrapolate to complete counts.

The Amor objects exhibit a steep decline of the density towards the Earth's orbit and, on the other side of a sharp concentration, a zone of moderate depletion around the orbit of Mars. Secondary maxima appear at heliocentric distances corresponding to the 1:3 and 2:3 circular resonances with Jupiter, the latter being especially pronounced. The system is rather flattened and markedly aligned to the apsides of Jupiter, $e = 0.13$; the density gradient is similar to that of the main asteroid belt. The Apollo objects exhibit a flat maximum in the region of terrestrial planets. An abrupt change of the density gradient near $Q = 7$ suggests the presence of two components : a thinner core, akin to the Amor population but concentrated within the 1:3 resonance mark, and a thick diffuse halo. Only six of the currently known objects, including those with nearly cometary values of the Jacobian constant in the system Sun-Jupiter, can be definitely classified as belonging to the halo. Thus the three-dimensional distribution of the Apollo objects lends indirect support to the assumption that some of them may be extinct cometary nuclei. Calibration of the equidensity curves by the figures obtained for the Earth's vicinity (Kresák, 1978) suggests that there exist, in all, about 250 Apollo objects and 150 Amor objects with diameters exceeding 1 km. However, these may be slight underestimates if the system includes a significant proportion of objects of very low reflectivity (cometary nuclei ?)

COMETS

The long-period comets exhibit a mean density of about 0.09 A.U.$^{-3}$ throughout the inner solar system (Kresák and Pittich, 1978). They form a uniform background which is enhanced around the ecliptical plane and, in particular, near the orbit of Jupiter, by the comets revolving in short-period orbits. A complete list of comets of $P < 20$ years, which are currently under observation, includes 70 objects; an additional 23 objects can be regarded as lost. The available sample of orbits (Marsden, 1975, and other sources) is seriously biased by observational selection, mainly on account of the considerable brightness changes of comets with heliocentric distance. To eliminate or, at least, mitigate the bias, two lines of evidence were used : 1. Comparison of the total sample with a subsample of 27 presumably largest objects, namely those observable beyond $R = 3$ A.U. 2. An empirical relation between the perihelion distance and the discovery probability, as established earlier for the long-period comets. The degree of reliability of the density contours obtained is conceivably lower than for the asteroids; nevertheless, the general pattern should be correct.

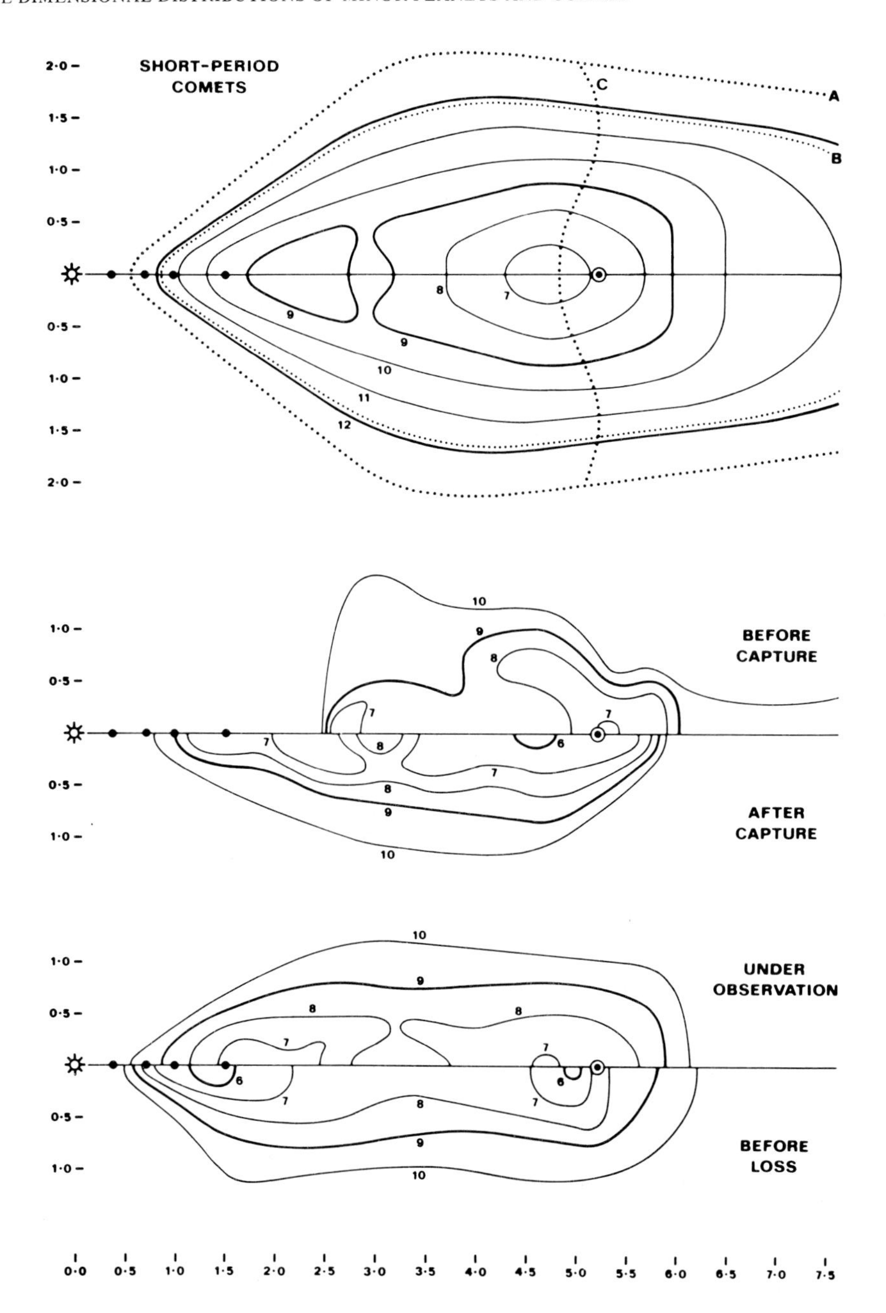

Figure 2. Equidensity contours of the system of short-period comets, P < 20 years (above), and the flow of the observed objects through the system (below). Notation is the same as in Figure 1, only the scales are reduced by a factor of 2/3. A - density equals to the uniform background of long-period comets; B - the same for nuclear diameters exceeding 1 km; C - density equal that of the main-belt asteroids with diameters exceeding 1 km.

The population of the wole system, depicted in its quasi-steady state in Figure 2 above, undergoes frequent changes. While the perturbational displacement of a given object is equally probable in both directions, towards the Sun and away from it, the presence of an outer source (Oort's cloud) and the progressive disintegration (at a rate depending on perihelion distance) tends to separate the objects by age, with physically older and deactivated comets preferring the sunward region of the belt. The only irregularity which appears statistically significant, is a moderate depletion near X = 2.8 to 3.0, Z = 0. This may be due to the fact that newly captured objects, with perihelia in this zone and aphelia close to Jupiter, enter unstable orbits of approximate resonance 2:3 and are subject to repeated strong perturbations three revolutions later.

The distribution patterns preceding and following the capture by Jupiter refer to the same 17 objects which experienced an encounter at $q > 2.5$ and were discovered, within the next two revolutions, in orbits of substantially reduced perihelion distance ($\Delta q = -1.0$, on the average). Some of the precapture orbits were kindly supplied, in advance of publication, by N.A. Belyaev, B.G. Marsden, E. Pittich, and H. Rickman. The precapture orbits obviously pass close to Jupiter, and the oblique extension protruding from the area of maximum concentration is due to their orbital inclinations. After the capture, the main concentration passes from outside to inside the orbit of Jupiter and a secondary maximum, produced by the perihelia situated near the other node, appears. At the same time, a distinct alignment to the apsides of Jupiter is built up, with an effective eccentricity of 0.20. The pattern for the 70 comets under observation, in their current osculating orbits, shows similar features; only the maximum, no longer coinciding with Jupiter in longitude, becomes less pronounced, and the effective eccentricity drops to 0.15. The last pattern refers to the 23 lost comets at their last observed apparitions. Although this does not necessarily mean the final deactivation, a shift towards the Sun is apparent. Another characteristic feature is a sharp concentration just within the orbit of Jupiter.

The relative densities calibrated by the absolute near-earth data (Kresák, 1978) again permit us to estimate the total number of the objects. The result is 300 active, or potentially active, comets with nuclear diameters exceeding 1 km, and 600 to 700 smaller ones, moving in orbits of $P < 20$ years at any time. For comparison, the total number of long-period comets situated at any time within the heliocentric distance of Jupiter is about 50. In the region of their maximum concentration, short-period comets outnumber the long-period comets by a factor of 100; but anywhere beyond 2 A.U. from the ecliptical plane the long-period comets prevail.

REFERENCES

Kresák, L.: 1978, "Bull. Astron. Inst. Czechosl." 29, pp. 114-125.
Kresák, L., Pittich, E.M.: 1978, "Bull. Astron. Inst. Czechosl.", in press.
Marsden, B.G.: 1975, "Catalogue of Cometary Orbits", SAO Cambridge.
Zellner, B., Bowell, E.: 1977, in A.H. Delsemme (ed.), "Comets, Asteroids, Meteorites", IAU Coll. 39, Univ. Toledo, pp. 185-197.

THE DISCOVERY AND ORBIT OF (2060) CHIRON

C. T. Kowal
Hale Observatories, Carnegie Institute of Washington
California Institute of Technology, Pasadena, CA 91125, U.S.A.

W. Liller and B. G. Marsden
Harvard-Smithsonian Center for Astrophysics
Cambridge, MA 02138, U.S.A.

"Slow-moving Object Kowal" was recognized by the first author on 1977 November 1 on plates obtained with the 122-cm Schmidt telescope at Palomar two weeks earlier (Kowal 1977a). The object, asteroidal in appearance and of photographic magnitude about 18, was found to have moved less than 3' between similar 75-minute exposures made on October 18 and 19. For something only 8° from opposition this retrograde motion was significantly slower than the 8'-15' daily motion of normal minor planets at opposition and immediately suggested that the object was located at almost the distance of Uranus. There was a slight possibility that the object was closer and moving almost directly toward or away from the earth, but the fact that the two trails were identical in length to ± 6 percent suggested that this was unlikely. That the object was well beyond the main belt of minor planets was confirmed when Gehrels (1977) succeeded in identifying it close to the extrapolated position near the corner of an exposure obtained with the same telescope on October 11.

From accurate measurements of the Gehrels plates of October 11 and 12, the discovery plates of October 18 and 19, and further plates obtained with the 122-cm Schmidt on November 3 and 4, the third author could demonstrate conclusively that the object was located between 14 and 17 AU from the earth (Marsden 1977a). The orbit itself was completely indeterminate, however, with the various combinations of three observations departing from great-circle motion by less than 3", and a low-inclination near-circular orbit of semimajor axis $a \simeq 16$ AU was selected as a compromise between a set of ellipses with rapidly increasing aphelion distances Q and slowly diminishing perihelion distances q (to 12.6 AU in the case of a parabola), and the complementary set with q rapidly decreasing to a limiting value of around 7 AU for an orbital eccentricity $e \simeq 0.5$.

With the availability of a pair of observations on November 9 and 10 it became clear that the more eccentric orbits of the large-Q set ($e > 0.4$, say) were no longer viable, and when further observations, extending to November 18, became available, it was apparent that only the small-q orbits could be considered. By this time it was in fact possible

R. L. Duncombe (ed.), Dynamics of the Solar System, 245–250.

to make a meaningful least-squares solution for all six orbital elements. The result from the 15 available observations gave $e = 0.35 \pm 0.04$, $q = 9.9 \pm 1.2$ AU. It seemed strange that the object, by then designated 1977 UB, had been discovered so far from perihelion, when it should have been significantly brighter. There was also the point that the object might be strongly perturbed near perihelion by Saturn, although the revolution period $P = 59.1$ years suggested that the orbit's stability could perhaps be maintained by a 1:2 resonance with Saturn.

Still, this orbit clearly indicated that 1977 UB would have been off the edge of the 1975 discovery plates of the presumed fourteenth satellite of Jupiter, excellent plates that were repeatedly examined in vain for this object. The first author had taken other plates in the general vicinity of 1977 UB on pairs of nights in July 1972, October 1970 and September 1969, however, and the new orbit suggested that there was an excellent chance the object would be on the 1969 plates. After a brief search it could be identified (Kowal 1977b) only about 55' east and 41' north of the predicted position, and an orbit fitted to the 17 observations (Marsden 1977b) showed quite definitely that $e = 0.38$, $q = 8.5$ AU (and hence $Q = 18.9$ AU, $a = 13.7$ AU, $P = 50.7$ years). A numerical integration of the orbit within ± 550 years of the present time yielded a mean value of $P \simeq 49.0$ years, suggesting that the object was in 3:5 resonance, rather than 1:2 resonance, with Saturn. Very close approaches to Uranus would be precluded by the fact that conjunction would occur when Uranus was near its aphelion distance of 20.1 AU. It was still somewhat curious that 1977 UB had not been discovered at magnitude ∿ 15 near perihelion, but perihelion passage had occurred in 1945, when few minor-planet patrols were being made, and those that were would probably have missed a very slow-moving object. The Lowell Observatory's survey for distant planets (Tombaugh 1961) would not have gone down faint enough in the 1930s and did not cover the right region of the sky in the 1940s.

After the 1969 identification had been made, additional past images of the object were uncovered in quick succession, including some on exposures in November and December 1976 with the 155-cm reflector at Harvard Observatory's Agassiz Station. Because of the small field, the chance that the object would have been recorded on plates taken with a large reflector was minute, but the field had been effectively widened with multiple exposures to follow up two minor planets discovered with this telescope some weeks earlier. A search of the Palomar Sky Survey yielded tentative images, of blue magnitude 17, on both the blue and red exposures of the appropriate field, photographed in August 1952; these images were situated about $\frac{1}{4}^{\circ}$ west of the predicted position in a very crowded area near the galactic plane in northwestern Sagittarius, but comparison with other Palomar-Schmidt exposures of the same region yielded no other moving objects within $\frac{1}{2}^{\circ}$ of the prediction (Kowal 1977c). Armed with this knowledge, the second author, who was at the time inspecting a 3-hour exposure obtained in January 1941 with the 61-cm Bruce astrograph at what was then Harvard Observatory's Boyden Station (now the Boyden Observatory), directed his attention some $\frac{1}{2}^{\circ}$ west of the predicted position and immediately noticed the trail of the object, estimated at perhaps magnitude 15 (Liller 1977). Curiously enough, this trail had been marked, but apparently never followed up, when the plate was examined in 1951 for faint galaxies. The 1941 observation clearly confirmed the 1952 identification and made it a simple matter to locate the object on similar, but inferior, Boyden plates taken in 1943 and 1948. Subsequently, images were identified on a plate taken at the Turku Observatory in 1945 (Niemi 1978) and on exposures with the Tokyo Observatory's 105-cm Schmidt telescope in 1976 (Kosai 1977).

With the help of an orbit refined to fit the 1941 and 1952 observations, it became possible to identify the object on a 60-minute exposure obtained with the Bruce astrograph near the time of the object's previous perihelion passage in 1895 (Liller and Chaisson 1977), when the instrument was being tested in Cambridge prior to its removal to the southern hemisphere. By modern standards none of these ancient plates is of particularly good quality, and this trail was so weak that it seems very unlikely that any other observations of this vintage will be found.

Using an orbit determined from observations in 1895, 1941, 1952, 1969, 1976 and 1977 (Marsden 1977c), the third author has attempted to trace the motion of 1977 UB over an extended interval of time. Perturbations by the five outer planets were considered, and some of the results are shown in Figure 1. As expected, there are several moderately close approaches to Saturn and Uranus, but there was no sign of instability in the whole 5500-year span of the orbit integration into the future. During this interval 1977 UB makes only one approach within 1 AU of Saturn, to 0.86 AU in the year 4689; the only approaches within 1.5 AU of Uranus occur in 4199 (1.10 AU) and 4871 (1.14 AU), and the only approaches within 3.0 AU of Jupiter occur in 5444 (2.86 AU) and 5932 (2.96 AU). The past motion seems to have been quite stable back to an approach within 0.7 AU of Saturn in the year -1400. The earlier motion must be considered rather uncertain, but our calculation suggests that there was an even closer approach in the year -1664, possibly to within a distance of 0.1 AU. It is of course completely meaningless to attempt to trace the motion before that time.

Figure 1 indicates that during the relatively unperturbed interval the value of a seems to vary with a period averaging about 600 years; a larger fluctuation has a period of perhaps 4000-5000 years. There is perhaps some kind of temporary libration around the 3:5 resonance with Saturn, but the long-term fluctuation tends to bring the motion more under the influence of the stronger 2:3 resonance ($P \simeq 44$ years) after about the year 6000. The oscillations in e to some extent match those in a and indicate that the changes in q must be rather small. In fact, since -1664 q varies only between 8.3 and 8.7 AU. The inclination i to the ecliptic shows a rather steady increase over the range discussed, this increase evidently being correlated with the argument of perihelion ω, which advanced by about 100° from 307° to 402°(=42°), while the longitude of the ascending node Ω shows a corresponding retrogression from 256° to 156°.

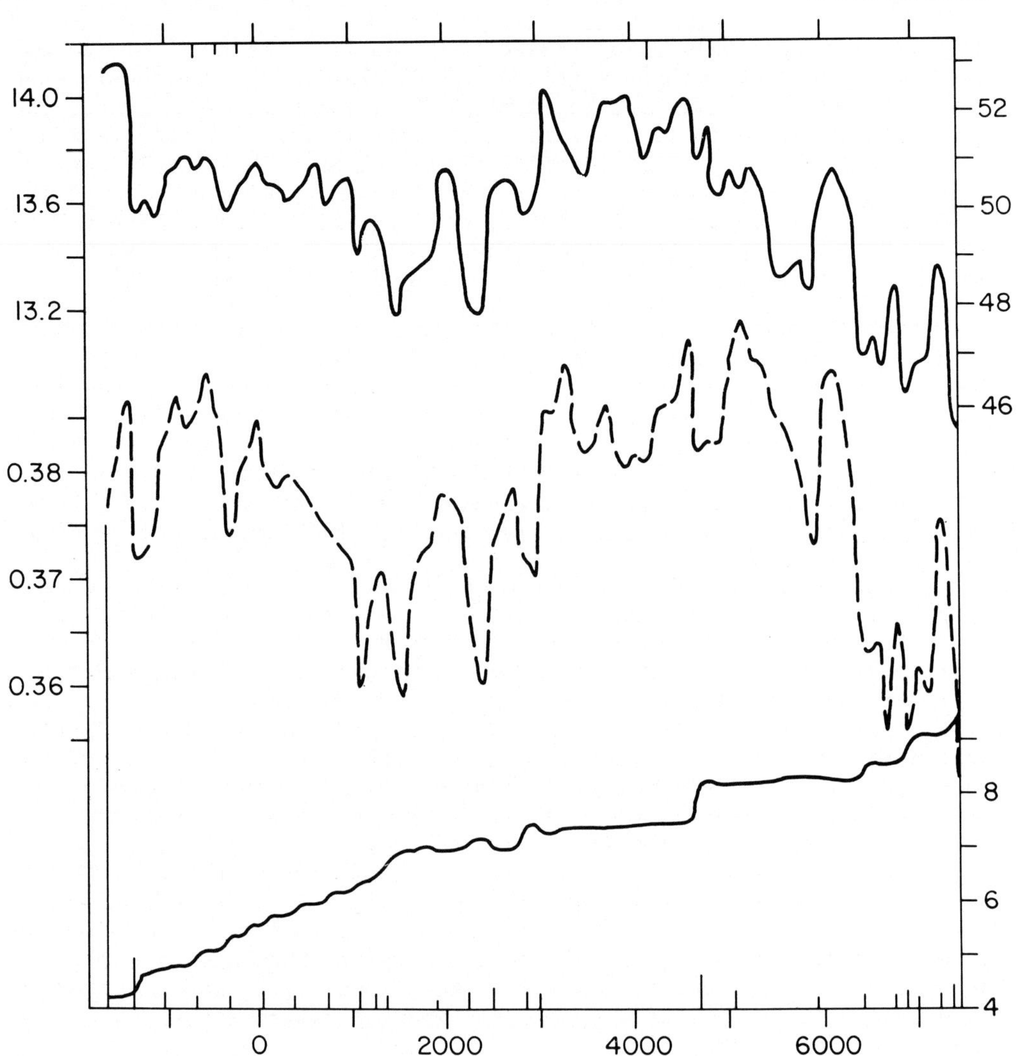

Figure 1. The variations in the orbital elements a (and P), e and i between the encounter with Saturn in -1664 and about the year 7400. The vertical lines inside the frame of the figure represent close approaches to Saturn (bottom) and Uranus (top). Only approaches Δ within 2 AU are marked, the lengths of the lines being proportional to $\Delta^{-1} - 0.5$.

The nature of 1977 UB can be only a matter of speculation. Its absolute magnitude of about 6 is comparable to the values for the brightest minor planets, but its albedo is unknown. The relatively unstable nature of its orbit, the known existence of comets at great distances from the sun and the distinct possibility that Saturn (perhaps with the help of the other outer planets) could perturb a comet with a low-inclination orbit of much longer period into one with an orbit like that of 1977 UB can of course be used as arguments for the object's cometary nature. On the other hand, 1977 UB would have to be a comet of record large size; no other comet has ever been observed beyond a heliocentric distance of 11.3 AU, and while one might guess that comet 1729, observable with the naked eye in spite of its perihelion distance of 4.05 AU, could also have had a nuclear magnitude of about 6, there is no real proof that this was the case. Even at perihelion, a cometary 1977 UB with a water-ice nuclear surface would be completely inactive. Although much smaller in size, (944) Hidalgo has also been regarded by some as an inactive comet, but with the important difference that its inactivity would be due to its water-ice content being sublimated away during its passages through perihelion: if (944) Hidalgo is a "dead" comet nucleus, 1977 UB is one that has never had a chance to live. In this connection it is useful to mention that attempts to trace back the orbit of P/Schwassmann-Wachmann 1, a large comet whose heliocentric distance currently varies between about 5 and 7 AU, lead to the result that a few thousand years ago this object, moving under the influence of both Jupiter and Saturn, had an orbit with q comparable to that of 1977 UB (although a was more like 10 AU than 13 AU).

On the other hand, both 1977 UB and (944) Hidalgo could perhaps have been perfectly ordinary minor planets, deflected into their present orbits by collisions with other minor planets and then a complex series of encounters with Jupiter (and also Saturn in the case of 1977 UB). The collisions presumably would have taken place rather recently. It can be argued that 1977 UB would be an exceptionally large collisional fragment, but it is no larger than (2) Pallas, whose highly inclined orbit also quite plausibly arose as a result of collisions.

On the whole, it seems prudent to continue to classify 1977 UB as a minor planet, and it has very recently received the permanent number (2060). The discoverer intends to give this object the name Chiron. Chiron was one of the centaurs, and it is suggested that the names of other centaurs be reserved for other objects of this same type that may be discovered in the future.

We also remark that our calculations suggest that (2060) may have approached Saturn in -1664 to about the distance of Phoebe. Chiron and Phoebe are probably of comparable size. It is therefore tempting to speculate that Phoebe and Chiron were originally members of a distinct group of related objects.

The first author gratefully acknowledges the support of NASA Grant NGL 05-002-140 and NSF Grant AST 76-81089 A01.

REFERENCES

Gehrels, T.: 1977, *IAU Circ.* No. 3129.
Kosai, H.: 1977, *IAU Circ.* No. 3156.

Kowal, C. T.: 1977a, *IAU Circ.* No. 3129.
Kowal, C. T.: 1977b, *IAU Circ.* No. 3145.
Kowal, C. T.: 1977c, *IAU Circ.* No. 3147.
Liller, W.: 1977, *IAU Circ.* No. 3147.
Liller, W. and Chaisson, L.: 1977, *IAU Circ.* No. 3151.
Marsden, B. G.: 1977a, *IAU Circ.* No. 3130.
Marsden, B. G.: 1977b, *IAU Circ.* No. 3145.
Marsden, B. G.: 1977c, *IAU Circ.* No. 3151.
Niemi, A: 1978, *IAU Circ.* No. 3215.
Tombaugh, C. W.: 1961, In G. P. Kuiper and B. M. Middlehurst (eds.), *Planets and Satellites*, Vol. III in the series: *The Solar System*, University of Chicago Press, Chicago, pp. 12-30.

DISCUSSION

Delsemme: Before the discovery of Chiron, I had predicted (A.H. Delsemme,1973,Astron.Astrophys.29,377) the existence of between 30,000 and 300,000 comets of intermediate period, in the same range of a and q as Chiron. If we use the observed brightness distribution of the 600-odd comets observed so far, normalized for an absolute magnitude of +4, to a radius of 4 km, 50 to 500 objects must have a radius larger than 100 km between Jupiter and Uranus and therefore the presence of a large object like Chiron is not surprising in the general population of comets that is likely to exist at these distances.

Marsden: It is also my personal opinion that Chiron is most likely to be a comet, but I don't think we should completely exclude other possibilities.

RECENT PROGRESS IN THE THEORY OF TROJAN ASTEROIDS

Boris Garfinkel
Yale University Observatory
New Haven, Connecticut, U.S.A.

§1 SUMMARY OF PART I

In order to provide the necessary background for this report, this section summarizes the previously published[5] "Theory of the Trojan Asteroids, Part I". Treating the system as the case of 1:1 resonance in the restricted problem of three bodies, the author constructs a formal long-periodic solution of $O(m)$, where m is the mass-parameter of the system, assumed to be sufficiently small.

The variables of the problem are the angular momentum G, the conjugate mean synodic longitude λ measured from the line of syzygies in a rotating coordinate system, and the complex Poincaré eccentric variable,

$$z = \xi + i\eta = \sqrt{2\Gamma} \quad \exp i\ell,$$

where ℓ is the mean anomaly and Γ is defined in terms of the Delaunay variables by

$$\Gamma \equiv L - G.$$

The Hamiltonian of the system is written as

$$F = \frac{1}{2}(1-m)^2 L^{-2} + G + mR\,, \tag{1}$$

and the heliocentric disturbing function,

$$R = (1 + r^2 - 2r \cos \Theta)^{-1/2} - r \cos \Theta,$$

is expanded into a Taylor series about the unit circle $r=1$. This mode of expansion assures the solution a wider range than the expansion about L_4 adopted by Deprit et al[2]. With the aid of the formulas

$$r = G^2 - \xi + m + \ldots, \qquad \Theta = \lambda + 2\eta + \ldots \quad .$$

R. L. Duncombe (ed.), Dynamics of the Solar System, 251–256.

of elliptic motion, the expression (1) is put into the form

$$F = \frac{1}{2}G^{-2} + G - \frac{1}{2}(\xi^2 + \eta^2) + \frac{3}{2}\rho(\xi^2 + \eta^2) - \frac{1}{2}m + mf_o$$
$$+ m[-2f_1\rho + (1 + f_1)\xi + f_2\eta] - m^2(f_1 + \frac{1}{2} + \psi) + \ldots \tag{2}$$

where the various symbols are defined below:

$$\begin{aligned} \rho &= G - 1, \\ f_o &= \frac{1}{2s} + 2s^2 - \frac{3}{2} + m\psi, \\ f_1 &= \frac{1}{4s} - 2s^2, \qquad\qquad s \equiv \sin(\lambda/2), \\ f_2 &= 4c(s - \frac{1}{8s^2}). \qquad\qquad c \equiv \cos(\lambda/2). \end{aligned} \tag{3}$$

Here the function $\Psi(\lambda)$ is the so-called "regularizing function" to be determined later so as to remove the Poincaré singularity in the solution.

The intermediate Hamiltonian F_o is chosen as

$$F_o = \frac{1}{2}G^{-2} + G + mf_o(\lambda) - \frac{1}{2}\omega_2(\xi^2 + \eta^2) - \frac{1}{2}m, \tag{4}$$

where ω_2 is the frequency associated with the short period. Clearly, the system splits into two autonomous subsystems of one degree of freedom each. The first subsystem, constituted by the first three terms of F_o, is identified with the previously formulated[4] (1976) Ideal Resonance Problem; the second subsystem, constituted by the last two terms of F_o, is the Simple Harmonic Oscillator.

As shown in 5), the equations of the intermediate orbit can be written down immediately as

$$\begin{aligned} G^o &= 1 - \frac{1}{3}\sqrt{6m(\alpha^2 - f_o)} + \frac{4}{9}m(\alpha^2 - f_o) + \ldots, \\ z^o &= \sqrt{2\Gamma}\ \exp i(\omega_2 t + \phi), \\ \int_{\lambda_1}^{\lambda}(\alpha^2 - f_o)^{-1/2}\, d\lambda &= \sqrt{6m}\,[t - t_1 + \frac{4}{9}(\lambda - \lambda_1)] + \ldots \end{aligned} \tag{5}$$

The time-dependence $t(\lambda)$ is furnished by the hyperelliptic integral in the last equation of (5), where λ_1 is the lower bound of the libration

$$\lambda_1 \leqslant \lambda \leqslant \lambda_2.$$

The function $\lambda(t)$ is obtained by inversion. Clearly, $\lambda(t)$ and $G(\lambda)$ are periodic in t, of long period T_1 given by

$$T_1 = 2\pi/\omega_1 = (6m)^{-1/2}\oint(\alpha^2 - f_o(\lambda))^{-1/2}\, d\lambda. \tag{6}$$

In contrast, $z(t)$ is short-periodic, inasmuch as $\omega_2 = 0(1)$.

The disturbing Hamiltonian is given by

$$\delta F = F - F_o = F_1 + F_2 + \dots , \tag{7}$$

and the perturbations of $0(m)$, arising from F_1 are calculated by the method of Lie-series in the version of Hori[8]. The result is of the form

$$\begin{aligned} \delta G &= \frac{2}{3} m f_1 + 0(\sqrt{\Gamma}) + \dots \\ \delta z &= m\,(1 + f + g + 0(\sqrt{\Gamma})) + \dots \\ \delta\lambda &= 0 \end{aligned} \tag{8}$$

Here the complex variable $f(\lambda)$ is defined by

$$f \equiv f_1 + i f_2,$$

while g is the <u>resonant term</u> of the form

$$\begin{aligned} g &= c_\kappa \exp(ik\omega_1 t)/D, \\ D &\equiv \omega_2 - k\omega_1 \end{aligned} \tag{9}$$

Here k is the integer nearest the ratio ω_2/ω_1, and D is the <u>critical divisor</u>.

From F_2 of (7), the regularizing function ψ is calculated[5] as

$$\psi = \frac{1}{6} f_1^2 + \frac{1}{2} f_2^2 . \tag{10}$$

Since the short-periodic terms in (5) and (8) carry $\sqrt{\Gamma}$ as a factor, such terms can be removed from the solution by a choice of initial conditions corresponding to $\Gamma=0$. The result is a one-parameter family of long-periodic orbits,

$$\begin{aligned} G &= G^\circ(\lambda) + \frac{2}{3} m f_1(\lambda), \\ z &= m[1 + f(\lambda) + g(t)], \\ t &= t(\lambda) \end{aligned} \tag{11}$$

The family-parameter α^2 is related to the Jacobi constant C by[5]

$$C = 2m\alpha^2 + 3.$$

Instead of α^2, it is convenient to use the <u>normalized Jacobi constant</u> α_o^2 defined by

$$\alpha_o^2 = \alpha^2 - m\psi(\lambda_2).$$

Then the range $0 \leq \alpha_o^2 \leq 1$ corresponds to the family of the tadpole-shaped orbits librating about L_4, while $\alpha_o^2 > 1$ includes the "horseshoes" encircling both L_4 and L_5.

The problem illustrates double resonance. For, in addition to the 1:1 resonance between the asteroid and Jupiter, there is also a k:1 resonance between the long and the short periods of the asteroid. Because of the apparently irremovable critical divisor D, the solution is local, rather than global. For its domain is restricted by the inequality of the form $|D| > \varepsilon$, or

$$|m - m_k| > \delta,$$

interpreted as avoidance of the set $\{m_k(\alpha)\}$ of the critical mass ratios corresponding to the exact commensurability of ω_1 and ω_2.

The mean value of the rapidly oscillating resonant term g(t), averaged over the "short" period T_1/k, is

$$\bar{g} = 0.$$

Accordingly, we define the mean orbit by writing

$$z = \bar{z} = m(1 + f)$$

in the second equation of (11). Clearly, the resonant term imparts to the mean orbit an epicyclic character. As m varies, D varies accordingly, and the epicycle develops cusps and loops, in qualitative accord with the results of numerical integration by Deprit and Henrard[3].

The presence of the resonant term g also serves to refute the Brown conjecture (1911) that the family of tadpoles terminates at L_3, thus confirming the earlier finding of Deprit and Henrard[3]. However, the conjecture is valid for the mean orbits and for the (G,λ)-projection.

§2 CONTRIBUTIONS OF PART II

The principal contributions of Part II of "Theory of the Trojan Asteroids", now in press, are itemized below:

1) The solution is carried from $O(m)$ to $O(m^{3/2})$, and the feasibility of a recursive algorithm to generate a solution to any order is shown.

2) The expression (6) for the period T_1 then becomes

$$T_1(\alpha_o^2,m) = (6m)^{-1/2} \oint (1 + mq)\ (\alpha^2 - f_o)^{-1/2}\ d\lambda,$$

$$q \equiv \frac{1}{36}(3s^{-3} - 4s^{-1} + 20\alpha^2 + 78 - 160s^2) \qquad (12)$$

For small oscillations about L_4, we calculate

$$\omega_1(0,m) = \sqrt{\frac{27}{4}m}\ (1 + \frac{23}{8}m + \ldots),$$

in agreement with the classical theory[9], which provides a further check of the regularizing function $\psi(\lambda)$, entering (12) through f_o of (3).

3) The regularizing function is extended to higher orders in m, and the result is used to prove the periodicity of the solution to any order.

§3 CONTRIBUTIONS OF PART III

Part III of the paper, now in progress, deals with the Hagihara integral:

1) The long period $\tau(\alpha_o^2,m)$, normalized so that $\tau(0,m) = 1$, is a <u>weak</u> function of m. Thus, it can be approximated by the Hagihara integral[7],

$$\tau(\alpha_o^2,0) = \frac{3}{4\pi} \oint [z/(1-z^2)(z-z_1)(z_2-z)(z-z_3)]^{1/2}dz$$

which is a relatively simple hyperelliptic integral of <u>class two</u>.

2) The latter is expanded into a convergent series,

$$\tau(\alpha_o^2,0) = \frac{2A}{\pi} \sum_{0}^{\infty} c_i c^i C_i,$$

where A, c_i, and c are known functions of α_o^2, and C_i are generated recursively in terms of the standard elliptic integral κ and E.

3) Asymptotic approximations to the Hagihara integral for the cases $\alpha_o \sim 0$ and $\alpha_o \sim 1$ are obtained in the form

$$\tau = 1 + \frac{1}{6}\alpha_o^2 + \ldots \qquad (\alpha_o \sim 0)$$

$$= \frac{3}{\pi\sqrt{14}} \log[28(3 - \sqrt{2})/(1 - \alpha_o^2) + \ldots (\alpha_o \sim 1)$$

The behavior of τ near $\alpha_o = 0$ had been studied by Deprit and Delie[1], while the logarithmic singularity at $\alpha_o = 1$ has been noted by Hagihara[6]. Incidentally, this singularity in the period serves to confirm the Strömgren Termination Principle when it is applied to the family of the "tadpoles".

4) The small correction ε to the Hagihara integral has been expanded into a Taylor series,

$$\tau(\alpha_o^2, m) = \tau(\alpha_o^2, 0)(1 + m\,\varepsilon_1 + m^2\varepsilon_2 + \ldots),$$

and the functions $\varepsilon_i(\alpha)$ have been calculated.

REFERENCES

1. Deprit, A. and Delie, A., 1965, Icarus 4, 242.
2. Deprit, A., Henrard, J., and Rom, A., 1967, Icarus 6, 381.
3. Deprit, A. and Henrard, J. 1970, "Periodic Orbits, Stability and Resonance", Ed. G.E.O. Giacaglia, p.1 (Reidel).
4. Garfinkel, B. 1976, Cel. Mech. 13, 229.
5. Garfinkel, B. 1977, Astron. Jr. 82, 368.
6. Hagihara, Y. 1944, Japanese Jr. Astron. Geophys. 21, 29.
7. Hagihara, Y. 1972, "Celestial Mechanics" vol.II, 299-307(MIT Press).
8. Hori, G. 1966, Publ.Astron. Soc. Japan, 18, 287.
9. Szebehely, V. 1967, "Theory of Orbits", 250 (Academic Press).

DISCUSSION

Message: You call the orbits obtained with $\Gamma=0$ "periodic solutions of the premiere sorte." However, they still contain long-period oscillations, while those solutions of the restricted three-body problem to which Poincaré gave that designation have short-period terms only.

Garfinkel: It is true that Poincaré's examples, as well as Hill's variation orbit, are short-periodic; however, one may define a "premiere sorte" orbit as a periodic orbit that reduces to a circle for m=0. My long-periodic orbit has this property.

Hori: The choice of a Keplerian intermediary, followed by an elimination of short-period terms, leads to simple resonance theories constructed by the previous investigators. It is necessary to include in the intermediary a part of the disturbing function of the same order of magnitude, and this inclusion accounts for the double resonance. Is this true?

Garfinkel: Yes. Indeed, the Ideal Resonance Problem, which I chose as my intermediary, incorporates the dominant term of the external 1:1 resonance.

A REVIEW OF DYNAMICAL EVIDENCE CONCERNING A FORMER ASTEROIDAL PLANET

T. C. Van Flandern
U. S. Naval Observatory
Washington, D. C. 20390, U.S.A.

This paper is a brief review of results presented elsewhere (Van Flandern 1977, 1978). The conclusion of these results is that at least the comets, and probably also the asteroids and meteorites, originated in the breakup of a major planet in the present location of the asteroid belt, at an epoch of just 5×10^6 years ago. Although there are many "well-known facts" about the solar system which seem to contradict this conclusion, these contradictory "facts", upon closer examination, are often not so convincing as we have been inclined to assume; in each instance so far suggested there is a plausible alternative interpretation of the data which is supportive of the breakup hypothesis. A compelling contradictory argument has not yet surfaced. In view of this, and in consideration of the strength of the arguments favoring the hypothesis, it will be necessary to judge the conclusion on the merits of the case, without the intervention of the *a priori* decision that it cannot be true.

Even with that granted, it must be admitted that two aspects of the hypothesis are very surprising. First, specific mechanisms for disrupting a large planet are unknown, and there is little in the evidence to suggest what the mechanism may have been. However, this ignorance is not a very good counterargument. Ramsey (1950) has discussed core instability conditions which could result in the explosion of terrestrial-sized planets. Stellar explosions are accepted by all, though not yet well understood theoretically. Recent work by Oort (1976) shows that even entire galaxies are observed to be exploding. It would therefore seem fair to point out that explosion is a fact of life in the observable universe, despite our discomfort with the contemplation of its applicability to the type of celestial body on which we reside.

R. L. Duncombe (ed.), Dynamics of the Solar System, 257–262.

The second puzzling aspect of the hypothesis is the recentness of the event. Since Copernicus we have learned not to place ourselves in a special location, or a special time, in the universe. However, it is possible that the solar system has had a rich and varied history. What event robbed Venus of its rotational angular momentum? How did Saturn's rings form? What tilted Uranus on its side? What event disrupted the satellite system of Neptune? When were the Martian satellites captured? There may have been a great many individually unique events in the solar system's history, of which the breakup of an asteroidal planet may have been only the most recent, and therefore the most easily reconstructed.

What, then, is the evidence? Because of planetary perturbations, principally by Jupiter, only two types of orbits can survive several million years after an explosion: low eccentricity orbits, as for the asteroids; and nearly parabolic once-around orbits, as for the very long-period comets. Oort (1950) discovered the remarkable property that many comets are in orbits having periods of revolution of 5×10^6 years, yet are known to be making their first return to the inner solar system! In a breakup origin, although comets with all periods are produced, only those with periods just equal to the time interval from the breakup to the present can be making first returns today. Hence the epoch of the event: $(5.5 \pm 0.6) \times 10^6$ years ago. These comets exhibit a strong bias in the directions of their perihelia. (Throughout this section, refer to Van Flandern 1977 and 1978 for specifics). While such a perihelion direction bias is inevitable in an explosion because of the "Sun-selecting influence", it is difficult to explain in any other uncontrived way. In particular, perihelion directions are very difficult to perturb, either by passing stars or galactic tidal forces. The comet perihelion distance distribution is likewise in agreement with a breakup origin, but is not reconcilable with the usual theory that this distribution must have been produced by passing stars. When the effects of galactic tidal forces and, statistically, passing stars are removed from the comet orbits in a five-million-year backwards integration, the tendency for the orbits to cluster (i.e. intersect at a point at that epoch) is truly remarkable. In every case where a measure is possible, the comet orbits have the distribution predicted by a breakup event.

Asteroid orbital elements display "explosion signatures", similar to artificial satellites which exploded in Earth orbit. Moreover, Bensusen and Van Flandern (1978) have discussed a statistically significant tendency for the orbits of the two largest asteroid families to intersect in a common point at an epoch of about 5×10^6 years ago. At no other epoch is the tendency to cluster statistically significant.

Meteoritic evidence consists of the very young cosmic ray exposures ages, such as for H-group chondrites, which cluster near 4×10^6 years old; chemical differentiation; isotopic anomalies; nondispersal of meteor shower radiants; the high frequency of meteorities with high geocentric velocities, and a great deal more. The physical and chemical evidence will not be reviewed extensively here.

In summary, there exists a great deal of dynamical, as well as physical and chemical, evidence that a large planet broke up in the asteroid belt 5×10^6 years ago. Although this hypothesis is inconsistent with the interpretation of the comet, asteroid, and meteorite data, there do not seem to be any irreconcilable conflicts with the data itself. Indeed the hypothesis provides fresh insights into many solar system anomalies, some of which have been most awkward to explain by the conventional hypotheses.

REFERENCES

Bensusen, S. J. and Van Flandern, T. C.: 1978, in "Abstracts of Papers of the 144th National Meeting of the AAAS; 12-17 February 1978", Washington, p. 105.

Oort, J. H.: 1950, "Bull. Astron. Inst. Neth." 11, pp. 91-110.

Oort, J. H.: 1976, in "The Galaxy and the Local Group", R. J. Dickens and J. E. Perry (eds.), R.G.O. Bull. #182, Herstmonceux, pp. 31-56.

Ramsey, W. H.: 1950, "Mon. Not. Roy. Astr. Soc." 110, pp. 325-338.

Van Flandern, T. C.: 1977, in "Comets, Asteroids, Meteorites", A. H. Delsemme (ed.), Toledo, pp. 475-481.

Van Flandern, T. C.: 1978, "Icarus" 36, (in press).

DISCUSSION

Scholl: 1) A good test for your theory would be a numerical calculation backwards of all the observed minor planets.
2) Why do we observe so few asteroids in the Kirkwood gaps and in the region between the Hecuba gap and the Hilda group?

Van Flandern: Thank you for your suggestion of an integration; I think it is a very good idea, although somewhat expensive.

The problem of the origin of the Kirkwood gaps is often still considered as unsolved. The planetary break-up model adds one new, interesting idea which has the potential to lead to an explanation; namely, that the initial asteroid orbits for any given semi-major axis had initially a relationship between eccentricity and longitude of perihelion (because initially all orbits intersected in a point). The dynamical consequences of such a relation have not yet been explored.

Kozai: I am afraid that even Williams' theory is not sufficient to compute positions of orbital planes of asteroids several million years ago.

Van Flandern: In general you are certainly correct. We confined our study to orbital planes of asteroids in the Eos and Koronis families, which have very similar elements within each family. We then looked only for a clustering of family members relative to one another. The result could be an artifact, but I think it is interesting that the only statistically significant clustering in each family occurred at $(4.8 \pm 0.1) \times 10^6$ years ago.

Kresak: I do not see how the distribution of meteor orbits can lend support to your hypothesis. The observed distribution pattern is due to (1) meteor showers, which are definitely of recent origins and mostly associated with existing short-period comets; (2) a strong prevalence of direct motions among the sporadic meteors; (3) a strong selection effect favoring the observation of meteors with high geocentric velocities; and (4) the fact that all optical observations are made at nighttime, and most of them are from the northern hemisphere. Even the best velocity measurements are unable to distinguish the intermediate-period meteors, like the Perseids or the Orionids, from the long-period ones. The relative contribution of long-period meteors is very small anyway. This is conceivable because solar radiation pressure would sweep out from the solar system all meteor-sized particles released by a "new" comet, so that they never return and cannot be observed outside the comet's tail.

Van Flandern: Sporadic meteors have several peaks in their semi-major axis distributions, one in the asteroid belt, and one at "large" (but not well-determined) semi-major axes somewhere between 20 AU and infinity. If these semi-major axes are really as large as for the long-period comets, this would favor the breakup model. Again, the distribution of radiants of shower meteors is not as dispersed as is expected, suggesting a relatively recent origin for their parent bodies (less than 10^6 years).

Weissman: I would like to offer one dynamical and one physical argument why your hypothesis is incorrect. On the distribution of perihelion directions and distances, I would say that there is no good statistical evidence that any preferred perihelion direction exists. The observational selection effects are quite complicated and no work I have seen has ever adequately accounted for them all. Moreover, in dealing with a statistical process like the perturbation of comets in the Oort cloud by passing stars there is no reason why we should not expect large fluctuations in the flux and directions of new comets when their periods are on the order of four million years and we only have good orbital data covering perhaps a 250-year span.

Concerning the perihelion distances of the comets, there is a shortage among small q comets but that shortage comes from a loss of "old" comets evolving to small semi-major axes and not the "new" comets from the Oort cloud. I will explain the reason for the loss of the older small q comets in my paper tomorrow. There is, however, no evidence that is statistically significant for a shortage of Oort cloud comets at small perihelia. Even if your hypothesis were correct the stellar perturbations on just a single five million year period orbit would destroy any record of the initial perihelion distribution of the material. The perturbations are sufficiently great to spread the comets over the complete range of observable perihelia, and in fact most comets would be perturbed out of the planetary region altogether.

On the physical side there is the evidence from the study of meteorites concerning their histories. The vast bulk of meteorites, the ordinary and carbonaceous chrondrites, are samples of primitive solar system materials. They were formed 4.5 billion years ago and since that time have not been heated or shocked to any significant degree. They have never been part of any planetary sized body and could not have undergone the catastrophic event you claim for your hypothetical planet. It is not possible for such undifferentiated, high in volatile content material to have survived for most of the history of the solar system as part of any major sized body.

By spectral reflectance measurements it has been shown that most asteroids are at least covered with material very similar to the ordinary and carbonaceous chondrites. Most of the asteroid belt in fact appears to be carbonaceous chondrites, the most primitive of the meteorite classes. The comets form a natural extension of the meteorite groups as having even more volatiles and thus being yet more representative of solar nebula material prior to the accretion of the planets. To hypothesize that all the meteorites, asteroids and comets have a 90 earth mass planet as a common parent body is totally inconsistent with all the physical evidence which has been accumulated on the meteorites.

Van Flandern: Although we cannot exactly remove the effects of observational selection on the perihelion directions of new comets, we can at least know that they operate in the sense that, when the true distribution is reconstructed, the asymmetry is greater (not less). Its statistical significance has been confirmed by many authors. No one has suggested a mechanism whereby such a strong asymmetry could have been produced by stellar encounters.

On the perihelion distance distribution, it is true that these may be changed by one, or even several, astronomical units. Therefore, I do not insist that these cut off at 2.8 AU, or vanish at the Sun; only that the general trend to diminish in number as one approaches the Sun should be preserved, as observed. Near the Earth's orbit, and after using Everhart's criteria to correct for observational selection, the observed distribution of perihelion distance of new comets is inconsistent with the uniform distribution required by the "Oort cloud" model at the 95% significance level. Many of Weissman's remarks about meteorites, concerning which neither of us is an expert, disagree with my reading of the literature. Authorities such as Brown, Patterson, Heezen, Glass, and many others have suggested a planetary origin for meteorites, particularly chondrites. The argument that they could not have been part of a much larger body is based on cooling rates, and assumes no active thermal processes other than the Sun's. These cooling rates are in agreement with a possible origin inside a planet large enough to have an internal heat source.

I have used the argument about the compositional similarities of comets, asteroids, and meteorites as one which favors a common origin for these bodies. Weissman's last statement should be modified to read, ". . . totally inconsistent with the existing interpretation of all the physical evidence." As I have pointed out, much of this evidence can be reinterpreted in a way which is at least consistent with, and often quite supportive of, the breakup hypothesis.

PART VI

COMETS

EMPIRICAL DATA ON THE ORIGIN OF "NEW" COMETS

A. H. Delsemme
Department of Physics and Astronomy
The University of Toledo
Toledo, Ohio 43606

INTRODUCTION

The different hypotheses proposed so far on the origin of comets have been reviewed in detail last year (Delsemme 1977). In particular, I mentioned why the orbital statistics seem to be consistent with only one hypothesis, namely that comets were accreted at unknown but moderate distances (10 to 1000 A.U.) within the protosolar nebula, and ejected later into a sphere whose radius is some 50,000 A.U., usually called the Oort's cloud. Safronov (1977) and Cameron (1977) have different scenarios to do just that. Even with improvements like that of Dermott and Gold (1978) the choice between these scenarios will remain impossible until we have a convincing model describing the gravitational collapse of an interstellar cloud into a planetary system.

The important point is that the primary source of the long- and short-period comets is unique and cannot be really doubted very much any more: It is supplied by those "new" comets coming straight from the Oort's cloud. In the recent list of original orbits (Marsden _et al_, 1978), the accumulation of 80 original values of 1/a below 100 (in 10^{-6} AU^{-1} units) leaves little doubt that most of these comets are new. I would however avoid to call all and every one of them "new" (as Marsden _et al_, do in their Table IV) because of Oort's definition: new comets are those that have never been through the solar system before.

This semantic distinction is important here, because it is unavoidable that a small fraction of the 80 very-long period comets (1/a original < 100 means P > 10^6 years) have already been once or several times through the solar system before, but have come back almost exactly to their previous value of 1/a. This is because of the random nature of the changes in their binding energies introduced by planetary perturbations. Hence these "young" comets cannot be distinguished from the "new" comets by orbital considerations. However, the mere accumulation of orbits below 1/a = 100 demonstrates that a steady state has not been reached by orbital diffusion, which implies that comets decay fast. The physical decay of new comets must be so fast that we can hope (although it has never been done) to separate "new" and "young" comets by using their gross physical properties.

R. L. Duncombe (ed.), Dynamics of the Solar System, 265–271.

PHYSICAL DECAY OF "NEW" COMETS

Oort (1950) had already mentioned that the observed number of "new" comets is approximately five times higher than that which would be expected from the number of other long-period comets. His interpretation was that "they must have a greater capacity for developing gaseous envelopes." Whatever the mechanism, the point is that they decay considerably in not much more than one passage, since some 80% of them seem to have disappeared from the statistics when they should have come back for their second or third passages.

A mere decay of the vaporization rate of the very volatile constituents may be a partial answer, but is unlikely to totally explain the dimming by 3 to 5 magnitudes that is needed to explain the total effect.

In a study of 13 split comets, Stefanik (1966) has shown that they were predominantly "new" comets in the Oort's sense, that ten split without the action of any tidal forces, and that some actually split on their way to their first perihelion passage (Whipple and Stefanik 1966). Repeated splitting is a low-energy fragmentation process from larger bodies that yields an exponent -2/3 in the final mass distribution of the splitted bodies. Most of the grinding processes known in nature also yield this type of mass distribution for the final grains. In particular, collision splitting also yields such a dependence, which is typical of the asteroids. Hartmann (1972,1975) finds that the exponent grows from -2/3 for smaller energies, to -1 for larger fragmentation energies. The existence of a mass distribution law with the exponent -2/3 could therefore be predicted if all comets are derived from "new" comets by a fragmentation process that uses a low-energy, like splitting. However, this does not imply that "new" comets should follow the same distribution, because their size distribution is likely to come from another mechanism, like accretion in the protosolar nebula.

It is therefore submitted here that:

a) the size distribution of "new" comets is likely to be very different from the size distribution of old long-period comets.

b) the size of a comet can be assessed statistically by assuming that its absolute brightness is in proportion to the surface area of the vaporizing nucleus (this is the basic assumption of the vaporization theory of comets, Delsemme and Miller 1971).

We can therefore predict that the distribution of the absolute brightnesses of the quasi-parabolic comets should be bimodal, being the mixture of some 80% (from Oort's remark) of "new" comets, with some 20% of much fainter fragments, that we have called "young" comets (Oort's ratio can of course be considerably biased by the faintness of the "young" comets).

BRIGHTNESS DISTRIBUTION OF "NEW" COMETS

Using the sample of Marsden _et al_ (1978), we exclude first from the 80 orbits those 5 that seem the least reliable (comets 1975 q, 1955 V, 1940 III, 1968 VI and 1959 III), because their osculating orbits have mean errors larger than 100 (in 10^{-6} AU^{-1}). Among the 75 orbits left, we exclude those whose perihelion is too far away ($q > 4$ AU) to extrapolate an absolute magnitude with any significance (comets 1974 XII, 1975 II,

TABLE I

H_0	Number	Individual Comets
1.1 to 2	1	1914 V.
2.1 to 3	0	
3.1 to 4	5	1889 I, 1890 II, 1905 IV, 1915 II, 1951 I.
4.1 to 5	10	1853 III, 1863 VI, 1886 IX, 1898 VII, 1903 II, 1914 III, 1919 V, 1947 VIII, 1955 VI, 1966 V.
5.1 to 6	17	1886 I, 1892 VI, 1895 IV, 1898 VIII, 1899 I, 1902 III, 1911 IV, 1922 II, 1925 I, 1925 VII, 1942 IV, 1947 I, 1948 V, 1957 III, 1971 V, 1973 XII, 1975 VIII.
6.1 to 7	12	1897 I, 1903 IV, 1904 II, 1907 I, 1912 II, 1917 III, 1921 II, 1941 I, 1946 I, 1948 I, 1954 X, 1975 V.
7.1 to 8	5	1849 II, 1900 I, 1941 VIII, 1944 IV, 1948 II.
8.1 to 9	0	
9.1 to 10	8	1932 VI, 1932 VII, 1952 VI, 1953 II, 1954 XII, 1967 II, 1972 VIII, 1975 XI.
10.1 to 11	1	1937 II.
11.1 to 12	3	1946 V, 1954 V, 1976 XIII.

1954 VIII, 1972 IX, 1956 I, 1925 VI, 1957 VI and 1936 I). Absolute magnitudes H_0 were determined for 62 of the 67 comets left, mainly from observations published in IAU circulars when available. The actual absolute magnitude deduced from the mean light curve reconstructed from observations when the comet crosses r = 1 A.U. before perihelion was preferred when possible. When the dependence on distance was not known, for comets with q > 1 A.U., the approximation known as H_{10} was used, in particular that given for the 19th century and early 20th century comets by Vsekhvyatsky (1964). When H_{10} varied during the cometary visibility, the value before perihelion was preferred. The results appear in Table I.

DISCUSSION

The expected bimodal distribution is present. However, a first bias is apparent. It is clear that all comets in group 2 ($9.1 < H_o < 12$) were found telescopically during the last 45 years, whereas group 1 ($1.1 < H_o < 8$) corresponds to a span of 130 years, roughly three times longer. The number of comets of the second group in an unbiased sample should probably be larger by a factor of 3. Second, the incompleteness grows fast in magnitudes 10 to 12, and the numbers of objects are too small to make any deduction on the shape of the distribution tail.

The interesting point is of course the gap from 8.1 to 9 already announced by the steady decline from the fifth to the eighth magnitude. If the gap were introduced by a systematic error in the absolute magnitudes of telescopic comets, the shift needed to suppress the bimodal distribution is about 4.0 magnitude, which seems too large to be possible. However, the major argument against a systematic error of that size is that the gap does not exist in the statistics for all long-period comets (Vsekhsvyatsky 1964). Since our very-long period comets are undistinguishable at discovery from other long-period comets, it would be difficult to explain the gap as an artifact coming from observational selection.

In particular, using Vsekhsvyatsky's incompleteness model, since 12 very-long period comets have been observed since the 1930'ies between absolute magnitudes 9 and 12, then at least 12 should have been discovered (instead of zero) between absolute magnitudes 8 and 9; 16 between 7 and 8 (instead of 3); 24 between 6 and 7 (instead of 5). This is a total of 44 missing comets that should have been discovered since the 1930'ies between absolute magnitudes 8 and 9, to fill up the gap with a constant distribution; many more would be needed for a unimodal distribution. The only alternate explanation is that all magnitudes of telescopic comets are biased by 3.5 to 4.0 magnitudes, and that not enough comets of this type are included in Vsekhsvyatsky's statistics to show the bias. It seems easier to believe that the gap is real. We can therefore tentatively identify those 50 brighter comets between magnitudes 1 and 8, as pristine "new" comets, whereas the 12 fainter comets, between magnitudes 9 and 12, would possibly be the fragments of those comets that have split or dimmed during their first passages, and whose orbits have come back by chance in the same energy range as that of "new" comets.

The distribution of the brighter objects seems to be rather narrow, corresponding to objects with a radius of 3 ± 2 km, (that is with a mass between 10^{16} and 10^{18} grams) The shape of their distribution is not known with accuracy; however, contrarily to that of the long-period comets, no stretch of the imagination can fit it in with a constant slope on a log-log diagram. It must therefore be explained by a formation mechanism different from a fragmentation process, since in particular it seems to possess a cutoff near magnitude +8. The distribution of the fainter objects is not known at all. Incompleteness could easily hide a constant slope for those faint magnitudes, therefore they might indeed result from the fragmentation of the brighter bodies. It is remarkable that Goldreich and Ward (1973) predict $R = 5\ \xi^{2/3}$ km for the size of the planetesimals accreted from gravitational instabilities in the solar nebula (ξ is smaller than, but near unity). The present results suggest therefore that new comets could be identified with pristine planetesimals, and that they decay very fast, probably by splitting when they come within the inner solar system. Grants of NSF (AST 78-08038) and NASA (NSG-7301, planetary atmosphere program) are gratefully acknowledged.

REFERENCES

Cameron, A. G. W. 1977, NATO's Newcastle Meeting on "The Origin of the Solar System," edit. Dermott (in press).

Delsemme, A. H. 1977 a, pp. 453-467, in "Comets, Asteroids, Meteorites," edit. A. H. Delsemme, publ. The Univ. of Toledo, Toledo, Ohio.

Delsemme, A. H. 1977 b, pp. 3-13 in "Comets, Asteroids, Meteorites," edit. A. H. Delsemme; publ. The Univ. of Toledo, Toledo, Ohio.

Delsemme, A. H., and Miller, D. C. 1971, Planet. SpaceSci., *19*, 1229.

Dermott, S. F., and Gold, T. 1978, Astronom. J., *83*, 449.

Goldreich, P., and Ward, W. R. 1973, Astrophys. J., *183*, 1051.

Hartmann, W. K. 1972, "Moons and Planets," Wadsworth Publ.

Hartmann, W. K. 1975, pp. 111-123, in "The Solar System," Freeman & Co., Publ.

Marsden, B. G. 1975, Catalogue of Cometary Orbits, Cambridge, Mass.

Marsden, B. G., Sekanina, Z., and Everhart, E. 1978, Astronom. J., *83*, 64.

Oort, J. H. 1950, Bull. Astronom. Inst. Netherlands, *11*, 91.

Safronov, V. S. 1977, p. 483 in "Comets, Asteroids, Meteorites, edit. A. H. Delsemme, publ. The Univ. of Toledo, Toledo, Ohio.

Stefanik, R. P. 1966, Colloq. Internat. Liege, *37*, 29.

Vanysek, V. 1977, pp. 499-503 in "Comets, Asteroids, Meteorites," A. H. Delsemme; publ. The Univ. of Toledo, Toledo, Ohio.

Vsekhsvyatsky, S. K. 1964, Physical Characteristics of Comets, NASA-TT-F80 OTS 62-11031, Washington, D.C.

Whipple, F. L., and Stefanik, R. P. 1966, Colloq. Internat. Liege, *37*, 33.

DISCUSSION

Van Flandern: I do not object to your interpretation of the data, but offer an alternative one, consistent with the planetary break-up model of cometary origin. In this model, much material is released suddenly into the vacuum of space, and the volatiles are suddenly frozen by the release of temperature and pressure. Such frozen volatiles will adhere only to masses large enough to have self-gravitation, which may explain the lower mass limit you have found. All except a few which are propelled close to the Sun will be somewhat protected from solar radiation by the optical depth of the debris cloud from the break-up. This may explain how they can have been continuously frozen since their origin. Would you care to comment on this alternative explanation of your data?

Delsemme: Your alternative explanation of my data maily demonstrates that you have a bright imagination; I have no quarrel with it. The real problem is to see how your planetary break-up model, which is at variance with the paradigm of recent work on the solar system and its origin, will stand the criticism of the scientific community when it is published. Since you reinterpret many of the observational data in a different way, only the future will tell us whether you have achieved a new paradigm without cracks, able to compete with the classical interpretation.

Kresák: I understand that the bimodality of your magnitude distribution can only be removed by an abrupt change of the discovery probability near H = 7-8^m. Maybe that such a discontinuity is not so unnatural as it appears at first glance. We have a rather complete record of the comets which become brighter than about 10th absolute magnitude, and are discovered in systematic visual searches; and we have a very poor coverage of the fainter comets, which are only detected by chance on plates taken for different purposes. Moreover, it appears that the current total absolute magnitudes of those comets which do not become bright enough for smaller instruments, are systematically underestimated by 2-3^m. It would be interesting to see an analogous distribution with the maximum absolute brightness as the parameter, and to look at the circumstances of discovery of the 12 comets forming the low-brightness peak.

Delsemme: If the bimodal distribution of the absolute magnitudes observed for new comets were an effect of observational selection, as proposed by Kresák, it would also show on the larger statistics of long-period comets. It does not. The remarkable fact is that, in spite of a selection effect that must be the same for both groups of comets, the distribution of "new" comets is entirely different from that of the long-period comets. It must mean something: My interpretation is that we see a distribution of pristine planetisimals, mixed up with 20% of broken fragments.

Weissman: I think this is a very important result and should be pursued to the fullest. I see two problems with it, however. First, the observed splitting rates for comets is about 10% for Ooor cloud comets and 4% for older long-period comets. Thus not all comets split on each return. The distribution of magnitudes should thus be a mix of the intrinsic distribution and the fragmentation distribution. Perhaps we simply do not have sufficient statistics to derive this structure in the distribution.

Secondly, Whipple has recently shown that there is no evidence for the existence of cometary groups, with the exception of the sun-grazing comets and a few additional pairs. Thus we do not see many families of fragmentation products. Again the problem may be only observational.

Delsemme: I think Weissman is right when he says that these problems are observational. Not only do we not have enough comets in the faint peak (fragments) of my bimodal distribution, but also this peak is considerably influenced by observational bias. If my interpretation is correct, we should see, when the observational bias is removed, a straight line with almost a -1 slope on a log-log diagram, because these comets must come from a fragmentation process, undistinguishable from that of the long-period comets.

THE SHORTAGE OF LONG-PERIOD COMETS IN ELLIPTICAL ORBITS

Edgar Everhart
Physics Department, University of Denver

Based on the number of 'new' comets seen on near-parabolic orbits, one can predict the number of comets that should be found on definitely elliptical orbits on their subsequent returns. We show here that about three out of four of these returning comets are not observed.

In this study a Monte Carlo model, which follows hypothetical comets, yields distributions that can be compared with those for real comets. The model takes hundreds of these hypothetical random comets entering the solar system with original 1/a-values, here called u-values, equal to 50×10^{-6} AU^{-1} . This is a typical value for near-parabolic new comets. Jupiter and Saturn perturb the total energies of the comets. Those that gain sufficient total energy are lost to infinity on hyperbolic orbits, but others lose total energy (i.e. their u-values become more positive) and return again. Sometimes comets return many times during their random walk on the u-axis, but eventually u becomes negative and all are lost to infinity.

The model also includes perturbations by passing stars.Comets with very small positive u-values have enormous aphelia. For these the gravitational impulses caused by passing stars can change their perihelion distances, sometimes so drastically that the comets do not then come near enough to the sun to be visible.

For the class of comets with perihelia between 0 and 1 AU we form a cumulative distribution N(u) vs u, where N(u) is the number of comets with u-values equal to or less than u. Figure 1 compares N(u) from the Monte Carlo experiment with the same distribution for real comets, normalizing both curves to 1000 new comets.

The data shown for real comets are found from the 1/a (original) column of Table III in Marsden, Sekanina, and Everhart (1978). Of the 82 comets in that table with perihelia less than unity we take 28 to be new, based on their original u-values. The observed N(u) curve is far lower than the predicted curve. This paper discusses the discrepancy.

R. L. Duncombe (ed.), Dynamics of the Solar System, 273–275.

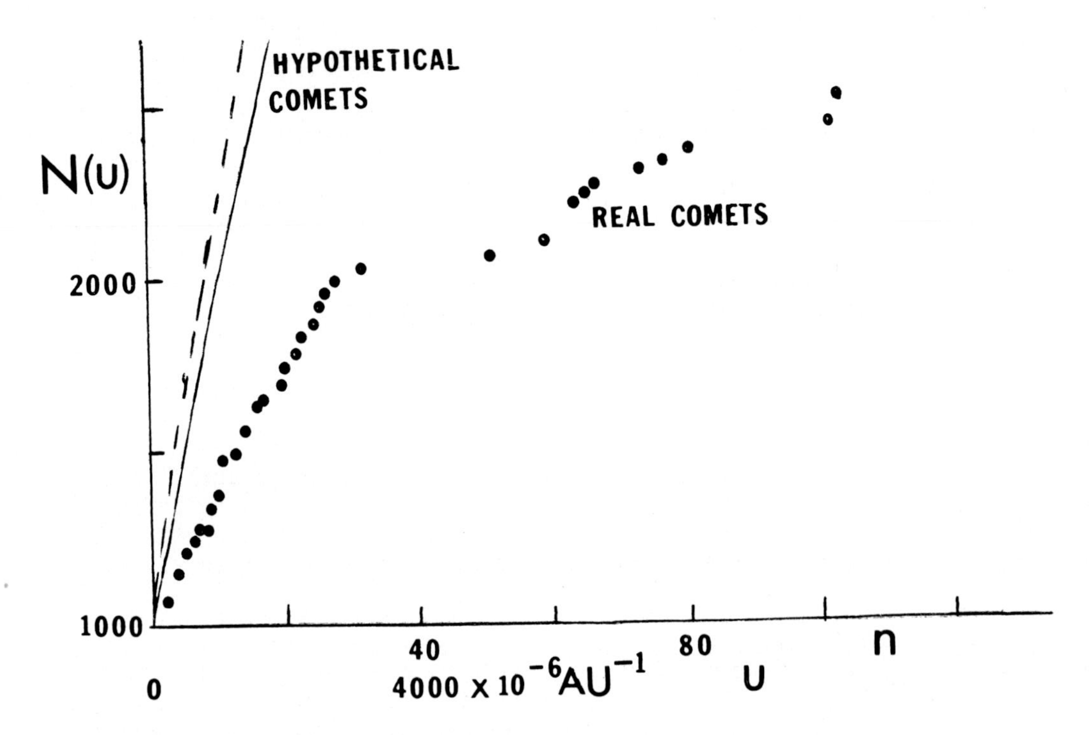

Fig. 1. Cumulative Distributions for Comets with $0<q<1$.

The dashed line in the figure shows the effect of not including stellar perturbations in the model. Allowing for them does lower the predicted curve, but not nearly enough to agree with the data.

Let us look at the Monte Carlo results in more detail. Let m be the return number of a hypothetical comet. Thus when m = 3 it is on its third return, not counting its initial appearance. Let n be the index specifying its u-value, as in the scale in Fig. 1. Define S_{mn} as the number of events in the (m,n) category, normalizing to 1000 entering comets. One can form the table of S_{mn} values shown below. We see that on their 3rd return there was 41 comets with a u-index of 2.This is only

n \ m	1	2	3	4	...
1	42	59	65	50	...
2	30	39	41	30	...
3	17	23	26	22	...
4	...	...	...	...	...

Table I. The S_{mn} Matrix

the upper left corner of a 100 by 200 matrix, since the experiment followed hypothetical comets up to 100 returns, and there were 200 u-indices. Summing S_{mn} of m for each row and then forming a cumulative sum up to index n yields the predicted N(n)-values of Figure 1.

Now let w_m be the probability that a comet is observed on its mth return. One might expect this to be considerably less than unity because of dissipation and observational selection. Then

$$w_1S_{1n} + w_2S_{2n} + w_3S_{3n} + \ldots = \Delta N_0/\Delta n, \tag{1}$$

where the right side is the slope of the data curve for real comets in Fig. 1 per unit change in n.

There are 200 such equations, each with 100 terms. Such a large matrix is not tractable. Accordingly, we group the m and n indices so that there are only 22 equations with 22 w-values to solve for.Although the 22 by 22 system is readily solved,the results turn out to be useless. The w-values, being probabilities, should be in the range of 0 to 1, but the solution has them wildly oscillating. Values of -94 and +80 are seen. Reasonable results are achieved only after restricting the w-variation to 2 parameters and solving for these in a least-squares sense.

For the set of comets where $0 < q < 1$ we require w to change only linearly as m increases with the grouped data. We then find w to be about 0.20 for the first return and about 0.23 for the group of returns numbered 90 to 100. This means that we see only about 1/5 to 1/4 of the returning comets in this range of perihelia. For $1 < q < 2$ we find w to be 0.28 for the first return and 0.13 for returns 90 to 100. For the range $2 < q < 3$ these probabilities are 0.33 and 0.04.

An improved model would show that the actual probabilities are lower than the above numbers. Because of observational selection not all real comets are actually observed on their initial appearance when they are 'new'. If we saw only 50% of these, this would change the normalization in Fig. 1 so as to double the discrepancy. Then the probabilities given above would be cut in half.

It is well known that comets with $q > 3$ AU are rarely seen, except as new comets. We see here nearly the same thing for comets of smaller perihelia. The model is now being improved, allowing for distributions in absolute magnitude of comets and observational selection. With the better model we ought to be able to make a quantitative estimate of dissipation, which is (at this writing) the only effect we know of which could account for so many returning comets being missed.

The support of the National Science Foundation is appreciated.

Reference:

Marsden, B.G., Sekanina, Z., and Everhart, E. "Astron. J." 83, pp.64-71.

PHYSICAL AND DYNAMICAL EVOLUTION OF LONG-PERIOD COMETS

Paul R. Weissman
Department of Earth and Space Sciences
University of California, Los Angeles, CA 90024
and
Jet Propulsion Laboratory
Pasadena, CA 91103

Oort (1950) first suggested that the source of the long-period comets is a large spherical cloud of comets surrounding the solar system and extending roughly halfway to the nearest stars. The observational evidence for this is the distribution of original inverse semi-major axes of the long-period comets which shows a large spike of comets at very small positive values of $1/a_o$, less than 10^{-4} AU^{-1}. Attempts to model the evolution of these comets by Oort in his original paper, by Kendall (1961), Shteins (1961), and Whipple (1962) were successful in recreating the general shape of the $1/a_o$ distribution. However in each case the authors were unable to match the observed ratio of new comets from the Oort cloud versus older comets evolving under the influence of planetary perturbations.

The work described here is a further study of the problem using a new method, a Monte Carlo simulation of the evolution of the long-period comets under the influence of a combination of physical and dynamical processes. Models were derived for the perturbation of cometary orbits by the major planets, by non-gravitational forces, and by random passing stars. Physical loss of comets due to planetary collision, random disruption (splitting), and loss of all volatiles was also modeled. These processes were combined in a computer simulation program which followed large numbers of hypothetical comets as they evolved from the Oort cloud to their eventual end-states. By changing input parameters to the computer program it was possible to vary the relative effect of each of the different processes.

In the case of planetary perturbations 5×10^4 hypothetical comets were integrated through a model solar system consisting of Jupiter and Saturn. The results were tabulated and used to derive the probability density function for the total change in 1/a per perihelion passage as a function of the perihelion distance and inclination of each comet. The density function was fit with a Gaussian distribution whose standard deviation was given by a second order polynomial expansion in q and cos i. For comets with perihelia uniformly distributed between 0.01 and 4 AU and with random inclinations, the standard deviation was

R. L. Duncombe (ed.), Dynamics of the Solar System, 277–282.

721×10^{-6} AU^{-1}.

Non-gravitational forces were studied using the model of Marsden et al. (1973). The change in 1/a was found for hypothetical comets passing through the solar system on parabolic orbits with non-gravitational accelerations similar to those measured for some real comets. It was found that non-gravitational changes in 1/a could exceed that due to planetary perturbations, particularly for comets with $q < 1$ AU. Also, if it was assumed that the orientation of the rotation axes of the cometary nuclei did not change from one perihelion passage to the next, then the non-gravitational forces could cause a regular stepping of the comets in 1/a, rather than the randomly positive or negative steps as a result of planetary perturbations.

For stellar perturbations a model similar to that by Wyatt and Faintich (1971) was used. Stellar perturbations were treated as a random perturbation in the aphelion velocity vector of each comet's orbit, dependent on the period of the orbit, and were only really important for comets with large aphelion distances, greater than about 10^4 AU.

A simple model for planetary collision showed that the probability of a random long-period comet passing within the Roche limit of a planet or colliding with an asteroid was 1.35×10^{-7} per perihelion passage. This was negligible compared to other loss mechanisms and was not included in the final version of the simulation program. A study of the statistics of observed splits of long-period comets yielded a disruption rate of 10% per perihelion passage for Oort cloud comets and 4% for older comets making subsequent returns. Except for four tidal splittings the disruption events appear to be random with respect to perihelion distance, inclination, time of splitting, etc. The lifetime of one kilometer radius water ice spheres in near-parabolic orbits was found as a function of perihelion distance in terms of the number of perihelion passages made, using a modification of the method described by Lebofsky (1975). This served as the basis for a model of cometary lifetime against loss of all volatiles.

The processes described above were combined in a Monte Carlo simulation program. In addition to being able to vary the effect of each of the processes, the program also incorporated several different initial perihelion distributions. Weissman (1977) has shown that the expected perihelion distances of Oort cloud comets are uniformly distributed with respect to q in the planetary region. However in running the simulation program it was found that a better fit to the observed results was obtained by using a perihelion distribution heavily skewed towards small perihelia orbits, much like the observed perihelion distribution for all long-period comets. This tended to compensate for observational selection effects in the data.

The program also allowed the user to vary the location of the source of the long-period comets. It was clearly demonstrated that the only way to achieve the observed $1/a_o$ distribution was to place the source at an aphelion distance of 2 x 10^4 AU or more, yet still gravitationally bound to the solar system. Thus the basic correctness of the Oort hypothesis was confirmed.

Initial runs with the simulation program using the nominal models for each of the processes described above gave a significantly better fit to the observed $1/a_o$ distribution than any of the previous studies. There were several reasons for this. First, due to the work of Everhart and Raghavan (1970) and Marsden et al. (1978) the catalog of observed original orbits has been significantly improved over past catalogs and a prior bias towards Oort cloud comets has been largely removed. Secondly, the random disruption rates used in the model, based on 24 observed disruption events, are significantly higher than those used in previous studies, in particular the 10% disruption rate for Oort cloud comets. Lastly the inclusion of the dependence of the planetary perturbations on perihelion distance and inclination, combined with the skewed initial perihelion distribution, is a better physical representation of the evolution of the long-period comets which are actually observed.

A comparison of the observed and computer generated $1/a_o$ distributions is shown in table 1. The observed distribution is based on the orbits of 184 selected comets and has been smoothed over varying intervals of $1/a_o$ and normalized so that the height of the Oort cloud spike is unity. The one standard deviation statistical error associated with each interval is also given. Two computer generated distributions are shown, the first for the nominal model that has already been described and the second for an improved model discussed below.

Table 1. Observed and modeled $1/a_o$ distributions.

$1/a_o$ range - 10^{-6} AU^{-1}	Nominal model	Improved model	Observed
0 - 100	1.0	1.0	1.0
100 - 500	.108	.098	.104 ± .025
500 - 1000	.103	.088	.067 ± .017
1000 - 2000	.077	.058	.037 ± .009
2000 - 4000	.043	.027	.023 ± .007
4000 - 6000	.022	.014	.013 ± .004
6000 - 8000	.013	.010	.012 ± .004

Although the fit to the $1/a_o$ distribution was substantially improved using the nominal model, the perihelion distribution resulting from it did not agree with the observed orbits. The simulation program was unable to account for the observed rapid disappearance from the solar system of small q comets with a < 400 AU. At the same time the model did not allow a sufficient number of intermediate q comets to evolve to orbits with small semi-major axes, as is observed. To study this various modifications to the nominal models were proposed and explored with the Monte Carlo program. Three major changes came out of this exercise. First, it was shown that a process similar to loss of volatiles but acting much more rapidly could account for the loss of small q comets. Whipple (1977) has suggested that as ice sublimates from a comet's surface a crust of silicate materials is left behind which halts further sublimation and renders the comet unobservable. Tests with the simulation program showed that a crust forming after 50 to 100 meters of water ice had been removed could account for the disappearance of small q comets at the required rate.

Second, it was demonstrated that the disruption probability can not be the same for all comets but must vary with some comets being highly susceptible to disruption and others relatively immune. As a first order improvement tothe nominal model the comets were split randomly into two groups: the first with a constant disruption probability and the second with zero disruption probability. The best fit to the observed $1/a_o$ and q distributions was obtained with 15% of all comets having zero disruption probability and the remainder having a probability of splitting of about 12% per perihelion passage. An additional change in the model to fit the observed statistics was that the difference in disruption probability between Oort cloud and older comets tended to disappear, nearly equal disruption probabilities being required for each group.

Finally, the effect of non-gravitational perturbations was greatly reduced. It would appear that this process is more random in nature than the model derived for it would indicate.

In addition to the improvement in the perihelion distribution of the comets, the improved model yielded a better fit to the observed $1/a_o$ distribution. This can be seen in table 1. The $1/a_o$ distribution for the improved model (unsmoothed) is shown in figure 1, based on 10^5 hypothetical comets followed for a maximum of 10^3 returns each with the Monte Carlo simulation program. Each interval in the histogram has a width of 10^{-4} AU^{-1}.

The primary end-states found for the long-period comets with the improved model were: ejection on hyperbolic orbit, 65.2%; random disruption, 27.6%; and formation of silicate crusts, 7.1%. Other significant end-states were capture to short-period orbit, 0.04%; and perturbation to a perihelion distance less than the solar Roche limit, 0.02%. The average comet made 4.4 perihelion passages with a mean time of 5.9×10^5 years between its first perihelion passage from the Oort cloud and the one which determined its particular end-state. The mean hyperbolic excess velocity for the ejected comets was 0.60 km/s.

This work has attempted to explore the various hypotheses concerning the origin of the long-period comets and to assess the relative role played by various physical and dynamical processes in the evolution of the comets. Though the solution found is not necessarily unique, it

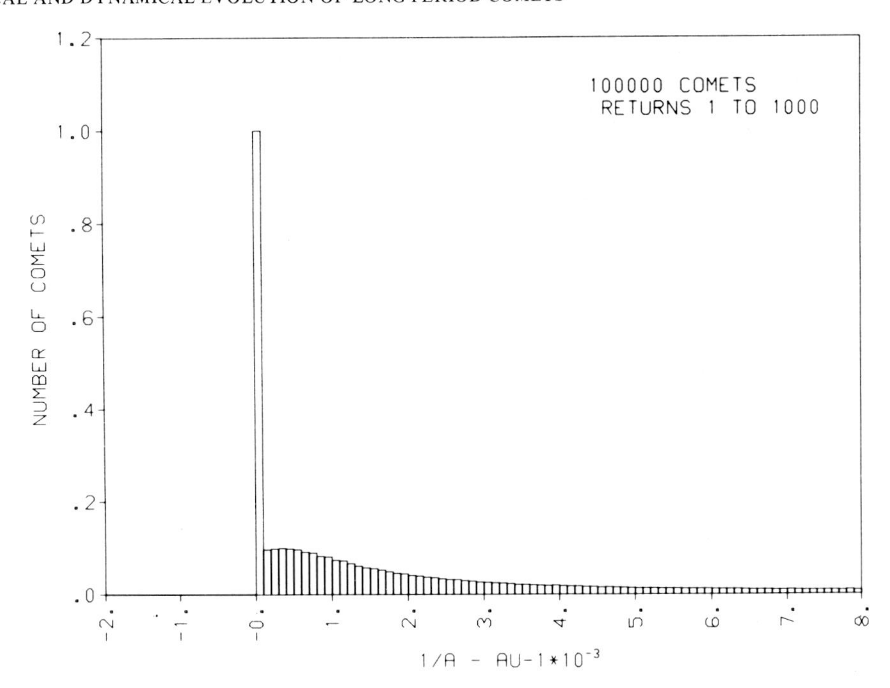

Figure 1. Computer generated $1/a_o$ distribution based on 10^5 hypothetical comets.

does identify the major processes which control the long-period comets and describes their histories and likely end-states.

The work described here is part of the author's doctoral dissertation in planetary and space physics at the University of California. The author wishes to thank his advisors, George Wetherill and William Kaula, for their guidance and their support during the course of this study. This work was supported in part by NASA grant NGL 05-007-002.

Everhart, E., and Raghavan, N.: 1970, "Astron.J." 75, 258.
Kendall, D.G.: 1961, "Fourth Berkeley Symposium on Mathematical Probability and Statistics" Univ. Calif. Press, 3, 99.
Lebofsky, L.A.: 1975, "Icarus" 25, 205.
Marsden, B.G., Sekanina, Z., and Yeomans, D.K.: 1973, "Astron.J." 78, 211.
Marsden, B.G., Sekanina, Z., and Everhart, E.: 1978, "Astron.J." 83, 64.
Oort, J.H.: 1950, "Bull.Astron.Inst.Neth." 11, 91.
Shteins, K.A.: 1961, "Soviet Astron.J." 5, 228.
Weissman, P.R.: 1977, "Comets, Asteroids, Meteorites: Interrelations, Evolution and Origins" Univ.Toledo Press, 87.
Whipple, F.L.: 1962, "Astron.J." 67, 1.

Whipple, F.L.: 1977, "Center for Astrophysics preprint" No. 814.
Wyatt, S.P., and Faintich, M.B.: 1971, "Bull.Am.Astron.Soc." 3, 368.

DISCUSSION

Kiang: I would like to point out that, at least for comets like P/Halley, the effect of persistent, nongravitational force is in most cases absorbed into a hyper period through the stabilization of the system by the Jovian perturbation. Detail in Paper VI.8.

Weissman: Marsden, Sekanina, and Yeomans have shown that the nongravitational perturbations seen in some long-period comets are an order of magnitude or more greater than those seen in short-period comets. These perturbations would in fact dominate the motion of the small perihelion comets if they were regular, greatly exceeding the effects of the random planetary perturbations in most cases.

Yabushita: Your calculated 1/a distribution: does it refer to a fixed value of N (perihelion passage) or to a fixed interval of time after cometary formation?

Weissman: The calculation assumes a continuous flux of comets from the Oort cloud and the resulting distribution is the steady state $1/a_o$ distribution. It is essentially the sum of $1/a_o$ distributions for all values of N from 1 to 1000.

ON SOME CHARACTERISTICS OF THE DISTRIBUTION OF PERIHELIA OF LONG-PERIOD COMETS

S. Yabushita
Department of Applied Mathematics and Physics,
Kyoto University, Kyoto, JAPAN

1. Introduction

Although the problem of the origin of comets has not yet been settled, there are only two possibilities; one is to regard that comets formed within the solar system when or soon after planets formed, and the second possibility is that comets have been captured from interstellar space by planetary perturbations or by perturbations of the galactic nucleus. To decide upon the possibility, one has to take into account statistical features of cometary orbits.

In this paper, we examine if there are statistical data which support interstellar origin of comets, and in doing so, we will take into account long-period comets (comets with periods exceeding 200 years) and disregards short-period comets. There are two reasons why only long-period comets are considered. Firstly, of nearly 600 comets observed so far, some 500 are long-periodic. Secondly, short period-comets are generally regarded as long-period comets which have been perturbed into short periodic orbits by planets (mainly Jupiter and Saturn).

If it is found that statistical data of the properties of cometary orbits are such that they support interstellar origin of comets, one should proceed to investigating physical processes which lead to cometary formation in interstellar space.

2. Perihelion points and solar motion apex

A. Direction of strong concentration.-In the cometary catalogue of Marsden (1972), 503 parabolic and nearly parabolic comets are contained. We will label each of them by suffix i. Take a system of rectangular coordinates (x,y,z), such that x axis coincides with L=0°, y axis with L=90°, and z axis with B=90°, where L and B are ecliptic longitude and latitude, respectively (equinox 1950). Let (Ω, i, ω) be the geometrical orbital elements of a comet. From observed values of (Ω, i, ω), the direction cosines of perihelion point (P_i) can be readily calculated, which will be denoted by (ℓ_i, m_i, n_i). Further, let $\overline{\ell}$, $\overline{m}$, $\overline{n}$ be direction cosines of a point Q which is to be determined. The angle

R. L. Duncombe (ed.), Dynamics of the Solar System, 283–287.

(α_i) between the two points P_i and Q is given by

$$\cos \alpha_i = \ell_i \bar{\ell} + m_i \bar{m} + n_i \bar{n} \ . \quad (1)$$

We consider the sum of $\cos \alpha_i$,

$$S \equiv \sum_{i=1}^{N} \cos \alpha_i \ . \quad (2)$$

We now require that the point Q be chosen such that S takes on a minimum value. This is achieved when $\bar{\ell}$, $\bar{m}$, $\bar{n}$ are so chosen that

$$\frac{\Sigma \ell_i}{\bar{\ell}} = \frac{\Sigma m_i}{\bar{m}} = \frac{\Sigma n_i}{\bar{n}} \ ,$$

or

$$\bar{\ell} = \frac{1}{N} \Sigma \ell_i , \quad \bar{m} = \frac{1}{N} \Sigma m_i , \quad \bar{n} = \frac{1}{N} \Sigma n_i \ . \quad (3)$$

Note that if a criterion other than minimizing S were adopted, the direction ($\bar{\ell}, \bar{m}, \bar{n}$) would be different from the one given by equation (3). Numerical values of L and B given by equation (3) were calculated using orbital elements of 503 comets contained in Marsden's catalogue. They are given by

$$L = 259\overset{\circ}{.}7 \ , \quad B = 66\overset{\circ}{.}0 \quad (4)$$

The direction of the solar motion referred to the neighbouring stars is characterized by

$$L_s = 260\overset{\circ}{.}6 \ , \quad B_s = 43\overset{\circ}{.}2 \quad (5)$$

which differs from the direction (4) by some 20° only.

B. The closeness of the two directions (4) and (5) suggests the possibility that they are not independent but are correlated. Therefore the next problem is to examine whether the closeness is merely a chance coincidence or otherwise. We therefore assume (null hypothesis) that the perihelion points are uniformly randomly distributed in the celestial sphere and investigate if the statistical data support the hypothesis. Under the hypothesis, the average values of (ℓ_i, m_i, n_i) will all be zero, while their dispersions are all equal to 1/3. Now, the direction of the solar apex is known *a priori* so that the cosine of the angle θ_i between the apex and P_i is such that the mean value is zero, while the standard deviation is 1/3. Now, according to the central limit theorem of probability, the distribution of sum of $\cos \theta_i$ approaches, as N is increased, a normal distribution such that the mean is zero and the dispersion is N/3. Since N is large (N=503), the above theorem should hold with sufficient approximation. For a normal variate, it should lie in the interval (-2.57σ, 2.57σ) with the probability of 99%, where σ is the standard deviation. On the null hypothesis, the standard deviation, σ of the normal distribution is $\sigma=(167.7)^{1/2}=12.9$, so that the above interval is (-33.15, 33.15). The actual value of $\Sigma\cos\theta_i$ is found to be 70.65, well outside the interval. Hence one may conclude that at 99% level of significance, the distribution of perihelion points are not unrelated to the solar motion apex.

3. Tamanov's law

Tamanov (1976) pointed out that if perihelia are concentrated at one point (L,B), then a relation of the form

$$\tan i = \tan B/\sin(L-\Omega) \qquad (6)$$

should exist between inclination (i) and node (Ω). We insert the values $L=L_s=260°.6$, $B=B_s=43°.2$, and obtain the curves shown in Fig. 1.

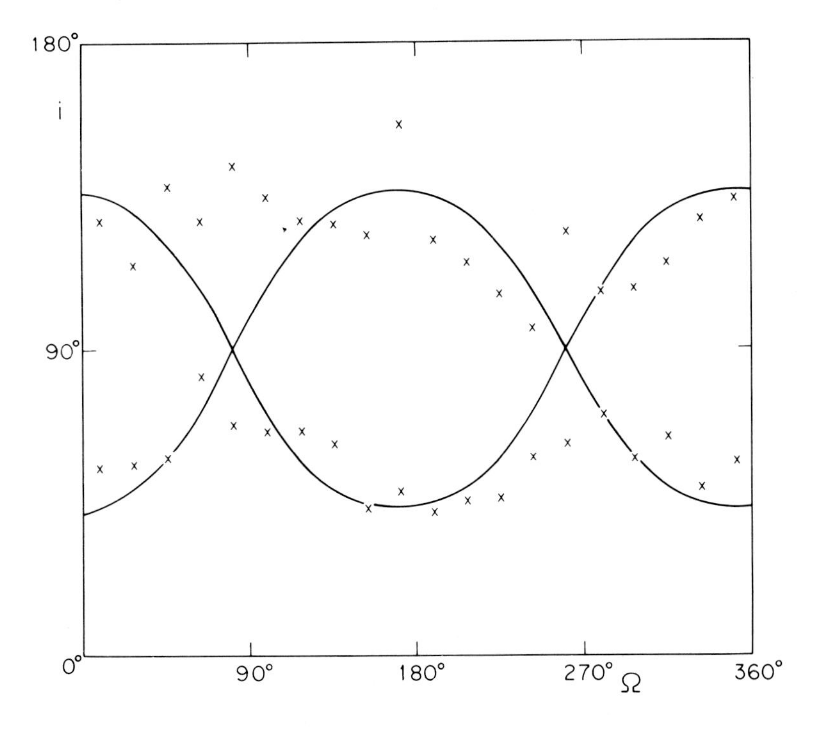

Figure 1. The curve gives the relation between i and Ω given by equation (6), while a cross denotes the average of observed i in each interval of Ω.

In order to see whether or not the observed comets satisfy the above relation, we once again used Marsden's catalogue and computed the average of i in each 18° interval of Ω. Those with i>90° and those with i<90° were averaged separately. Again, only those comets whose galactic longitudes lie in (25°, 155°) and in (205°, 335°) have been taken into account. For i<90°, there are 176 such comets and for i>90°, there are 192 such comets.

From the figure, the agreement between the relation (6) and observational data appears fair, if not excellent.

4. Discussions and Conclusions

We have followed the works of Oppenheim (1922), Tyror (1957) and Hasegawa (1976) regarding perihelion distribution by making use of Marsden's catalogue, and confirmed that perihelion points are concentrated in the direction of solar motion.

As to the relation (6) which should hold true if all perihelia were concentrated at a point, the agreement with observational data appears fair, though not excellent. When all of the observed comets are included, the agreement gets worse. Therefore, it appears that the gravitational field of the galaxy perturbs cometary orbits to some extent, especially those with large values of semi-major axes.

The two results arrived at in the present work indicate that cometary orbits are strongly related to the solar motion. This might be an indication that comets have interstellar origin. It has been pointed out (Yabushita & Hasegawa 1978) that some of comets with negative original values of 1/a calculated by Marsden *et al* (1978) may represent those comets which originated in interstellar gas clouds.

References

Hasegawa,I., 1976. Publ. Astr. Soc. Japan, 28, 259.
Oppenheim,S., 1922. Astr. Nachr., 216, 47.
Marsden,B.G., 1972. *Catalogue of cometary orbits*, Smithonian Astrophysical Observatory, Cambridge, Massachusetts.
Marsden,B.G., Sekanina,Z. and Everhart,E., 1978. Astron. J., 83, 64.
Tamanov,V.P., 1976. Soviet A. J., 19, 794.
Tyror,J.G., 1957. Mon. Not. R. astr. Soc., 117, 370.
Yabushita,S. and Hasegawa,I., 1978. Mon. Not. R. astr. Soc. (in the press).

DISCUSSION

Kiang: You have exaggerated the improbability of the observed closeness of your mean perihelion point to the solar apex, on the null hypothesis. Since for any distribution of the perihelion points, you will always find a mean point, what is remarkable is then that the mean point should be within $\theta(\sim 25°)$ of some special point. On the null hypothesis, the mean point should be uniformly distributed on the sphere. Hence the probability of the observed event on the null hypothesis is $(1-\cos\theta)/2 \sim .05$.

Van Flandern: Observational selection effects have operated to greatly reduce the number of observed cometary perihelia south of the equator. When these are allowed for, the center of the observed perihelion distribution moves far (perhaps 60°) away from the solar apex. It also moves to ecliptic longitude 222°. Therefore, the proximity you have discussed must be a coincidence. [Everhart and Weissman concurred with this remark.]

Yabushita: Dr. Hasegawa told me that those comets observed in the southern hemisphere have the same tendency as those found in the northern hemisphere. See Publ.Astr.Soc.Japan 28,1976.

Marsden: We find original hyperbolic orbits only for comets of small perihelion distance. We attribute this to the fact that comets of small perihelion distance are influenced by nongravitational forces, which do not affect large-perihelion distance comets as much. How do you explain the fact that the hyperbolic orbits are restricted to the comets of small perihelion distances?

Yabushita: Your comment is very interesting but does not refer directly to what I have spoken about. Your comment refers to a paper by Hasegawa and myself to be published shortly. Hasegawa and I will closely examine whether a correlation exists between q and 1/a values in the manner you have suggested.

DYNAMICAL PROBLEMS IN THE PREPARATION OF THE CATALOGUE OF ONE-APPARITION COMETS

G. Sitarski
Polish Academy of Sciences, Space Research Center, Warsaw

Subtle effects influencing the results of orbit determination have to be taken into consideration when recomputing the cometary orbits for the Catalogue. It is very important to apply the numerical integration method, taking into account the specific conditions of the comet's motion, which almost eliminates a cumulation of numerical errors of integration. In this point of view the recurrent power series integration is adapted for cometary orbital computations. The optimal value of step-size of integration according to the required accuracy is calculated at every integration step. The method gives excellent results for "normal" comets as well as in the cases of close approaches of a comet to the Sun and to the planets.

The idea of recomputing all the orbits of long-period comets arose about ten years ago. The work was undertaken at the Warsaw Astronomical Observatory with the cooperation of Slovakian astronomers at the Astronomical Institute in Bratislava and Tatranska Lomnica. To make reasonable the idea of the new catalogue, the following problems have been taken into consideration when recomputing the cometary orbits:

1. to collect all the observations of one-apparition comets and to reduce them to the one system of star catalogues;
2. to define precisely various types of observations and to include in them the corrections of precession, aberration, etc.;
3. to determine the mathematical criteria for eliminating and weighting the observations; and
4. to take into account nongravitational effects in the comet's motion.

Some of these problems were solved or mathematically interpreted by Dr. M. Bielicki (1970), Bielicki and Ziolkowski (1976), who have a great deal of experience in orbital computations having worked more than forty years on the orbits of the periodic comets Wolf 1 and Kopff.

R. L. Duncombe (ed.), Dynamics of the Solar System, 289–292.

There is also a problem which we cannot avoid in the cometary investigations: numerical integration of equations of motion of a comet. It is important to take this problem into consideration carefully since the construction of the catalogue including many subtle effects in the comet's motion could be destroyed by the simple cumulation of numerical errors when integrating the equations of motion.

In the last few years a number of papers were published in Celestial Mechanics on the method of recurrence power series expansion applied to the integration of equations of motion (Schanzle, 1971)(Broucke, 1971) (Black, 1973). It was shown that by using this method we can avoid a cumulation of truncation errors in the process of numerical integration. The method was successfully applied to the orbit computation of artificial satellites, in the restricted three-body problem, in the general n-body problem and in the planetary heliocentric motion. I adapted this method to the practical computations in cometary investigations and found it to be very convenient.

The equations describing the heliocentric motion of a comet as perturbed by the planets, p, are:

$$\frac{d^2x}{dt^2} = - k^2 \frac{x}{r^3} - k^2 \sum_p m_p \left(\frac{x-X_p}{\rho_p^3} + \frac{X_p}{R_p^3}\right)$$

and similarly for y and z, where k is Gaussian gravitation constant, m_p is mass of a planet in terms of the solar mass, x,y,z are the rectangular coordinates of a comet and X_p, Y_p, Z_p those of a planet. r, ρ_p, R_p are the distances of a comet to the Sun, of a comet to the planet, and of a planet to the Sun, respectively.

Suppose that the dependence of time of a function f=f(t) is given in the form of power series

$$f = f_o + \sum_{n=1}^{N} f_n(t-t_o)^n$$

and that just in such a form we want to obtain the solution of equations of motion, i.e., $x=x_o + \sum_{n=1}^{N} x_n(t-t_o)^n$ and similarly for y and z.

Let us define

$$\xi_p = x-X_p, \quad \eta_p = y-Y_p, \quad \zeta_p = z-Z_p$$

and

$$s =-k^2/r^3, \quad \sigma_p = -k^2/\rho_p^3, \quad S_p = -k^2/R_p^3$$

where

$$r^2 = x^2 + y^2 + z^2 , \quad \rho_p^2 = \xi_p^2 + \eta_p^2 + \zeta_p^2 , \quad R = X_p^2 + Y_p^2 + Z_p^2 .$$

Then the equations of motion can be written in the form:

$$\ddot{x} = sx + \sum_p m_p \sigma_p \xi_p + \sum_p m_p S_p X_p$$
$$\ddot{y} = sy + \sum_p m_p \sigma_p \eta_p + \sum_p m_p S_p Y_p$$
$$\ddot{z} = sz + \sum_p m_p \sigma_p \zeta_p + \sum_p m_p S_p Z_p$$

If we complete these equations by the relations

$$r\dot{r} = x\dot{x} + y\dot{y} + z\dot{z}, \quad r\dot{s} = -3s\dot{r}$$
$$\rho_p\dot{\rho}_p = \xi_p\dot{\xi}_p + \eta_p\dot{\eta}_p + \zeta_p\dot{\zeta}_p , \quad \rho_p\dot{\sigma}_p = -3\sigma_p\dot{\rho}_p$$
$$R_p\dot{R}_p = X_p\dot{X}_p + Y_p\dot{Y}_p + Z_p\dot{Z}_p , \quad R_p\dot{S}_p = -3S_p\dot{R}_p$$

we are able to derive the recurrence relations which allow us to compute consecutively the values of coefficients x_n, y_n, z_n starting from the initial values of x_o, y_o, z_o and x_1, y_1, z_1 as given for $t = t_o$. The recurrence relations for the two-body problem, given by Schanzle (1971), may serve as an example.

The recurrence relations for a cometary motion were derived on the assumption that for each moment of integration we have ready the power expansions for the planetary coordinates since the motion of the planets we know very well. Thus we can regard the cometary problem as a special case of the n-body problem when the equations of motion of disturbing bodies do not have to be integrated.

The power series integration method allows us to determine an optimal value of integration step as dependent on the desired accuracy of results of computations. Let us take the function

$$A(t) = A_o + \sum_{n=1}^{N} A_n (t-t_o)^n$$

where $A_n = |x_n| + |y_n| + |z_n|$ for $n = 0,1,\ldots,N$. Suppose that for $h = t_h - t_o$ we want to obtain $A(t_h) = A_o + \sum_{n=1}^{N} A_n h^n$ with the accuracy ε. The value of h may be determined as follows: Taking $A_{N-1}h^{N-1} = \varepsilon$ we have $h_o = (\varepsilon/A_{N-1})^{1/(N-1)}$; h_o is the preliminary value for the size of the integration step. Let us assume now that

$$\frac{A_{N+1}}{A_N} = \frac{A_N}{A_{N-1}}$$

and that $(A_{N-1} + A_N h + A_{N+1} h^2) h^{N-1} = \varepsilon$. Hence we have

$$h = \left[\frac{\varepsilon}{A_{N-1} + A_N h_o (1 + \frac{A_N}{A_{N-1}} h_o)}\right]^{1/N-1}$$

Thus for every moment of integration we compute the values of coefficients of the N terms of power expansion and then we calculate the optimal value of step-size according to the defined function A(t).

This method was applied to the integration of equations of various types of cometary motion, including, e.g., the close approach of the comet to the Sun (the orbit with a very small value of the perihelion distance, q), the close approach of the comet to Jupiter, the long integration interval (several revolutions of a minor planet). In all the cases the method gave excellent results.

It is worth noticing that the computed values of coefficients of power series expansion can be readily used in an interpolation formula to compute cometary coordinates for the moments of observations; thus it is convenient in the orbit improvement process.

REFERENCES

Bielicki, M. (1970): Leningrad IAU Symposium No. 45.
Bielicki, M., and Ziolkowski, K. (1976): Acta. Astron.
Black, W. (1973): Celestial Mech. 8, 357.
Broucke, R. (1971): Celestial Mech. 4, 110.
Schanzle, A.F. (1971): Celestial Mech. 4, 287.

RECENT DYNAMICAL HISTORY OF THE SIX SHORT-PERIOD COMETS DISCOVERED IN 1975

Hans Rickman
Observatoire de Nice, France

In 1975 a remarkably large number of short-period comets were discovered - only the most recent number from 1977 is comparable. While the average discovery rate has been 0.8-1.0 new short-period comets per year (Kresák 1974), in 1975 there were six discoveries. In five of the cases IAU Circulars soon afterwards contained indications that close encounters with Jupiter had recently taken place (Marsden 1975, Kastel' 1975). For two of the comets, P/Kohoutek and P/Smirnova-Chernykh, also pre-encounter orbital elements were outlined, suggesting that substantial reductions of the perihelion distances had occurred.

There is at present a general consensus that, at least within the framework of the capture hypothesis, the immediate origin of the Jupiter family is a population of short- or intermediate-period comets (orbital periods $\lesssim$ 100 years) with low inclinations and perihelia near Jupiter's orbit (see e.g. Everhart 1972, Delsemme 1973, Vaghi 1973, Rickman and Vaghi 1976). Backward integrations of the motions of observed Jupiter family comets show a general agreement with this picture (e.g. Kazimirchak-Polonskaya 1972). The present sample of six newly discovered comets offers good material for such an investigation. Long-term integrations could not be performed with any confidence for these one-apparition comets due to the uncertainties of their observationally determined "starting" orbits, but indeed a very short time interval (20 years) is sufficient in order to find major changes of the orbits in five of the six cases.

COMPUTATIONS

The bulk of the calculations to be reported were carried out by N. Carlborg and the author at the Stockholm Observatory using a Cowell N-body integration program written by Carlborg and earlier used in other investigations (e.g. Danielsson and Ip 1972). Initial data for the planetary system were adopted from Oesterwinter and Cohen (1972), and starting orbital elements for the comets were kindly communicated by B.G. Marsden. For a few comets varied orbits were treated - these include an

R. L. Duncombe (ed.), Dynamics of the Solar Systems, 293–298.

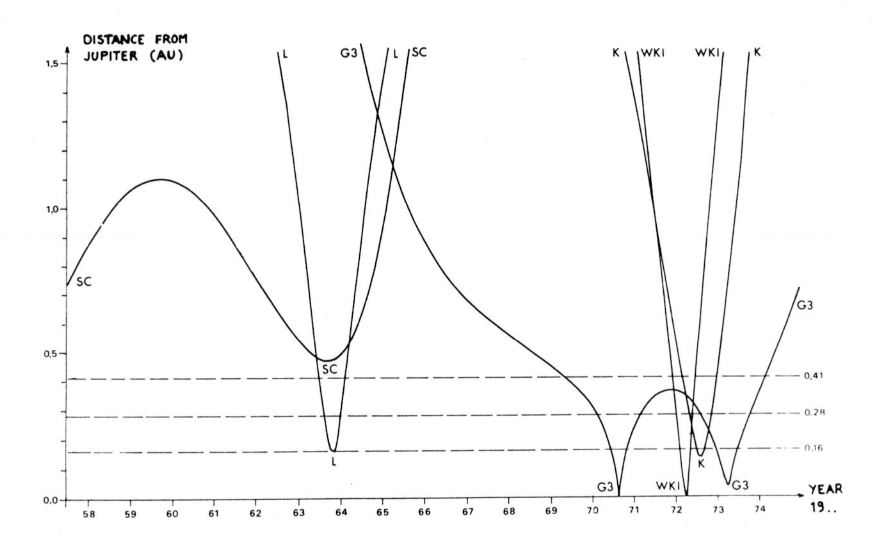

Figure 1. Distances of comets from Jupiter.

orbit computed by N.A. Belyaev for P/Smirnova-Chernykh. In the backward integration Mercury was thrown into the Sun after all the comets had been picked up at their respective osculation dates. The step-length of the integration program was automatically halved whenever necessary to keep the desired accuracy, and doubled whenever the current value was smaller than necessary. The maximum value was four days.

In Figure 1 the distances from Jupiter are represented as functions of time for the close approaches during 1957-75. The minimum extensions of three "spheres of action" around Jupiter are shown for comparison. Extremely close encounters were found in two cases. P/West-Kohoutek-Ikemura approached Jupiter to a minimum distance of 0.011 au on March 23, 1972, whereby the step-length was reduced to a minimum of $0^d.0625$. Comet P/Gehrels 3 is an even more extreme example. On August 15, 1970, it approached Jupiter to 0.0014 au (1.9 Jupiter radii from the planetary surface) whereby the step-length was reduced to $0^d.007813 \simeq 11$ minutes. Additional reversed integrations have been performed for the time intervals around the closest encounters, and the results show agreement with the original values to within a relative error of 10^{-10} in the positions and 10^{-9} in the velocities.

RESULTS

Table 1 contains a summary of initial and final (pre- and post-encounter) orbital elements for all the six comets. Detailed pre-encounter orbits will be published elsewhere (Carlborg and Rickman, in prep.). Here we turn instead to an account of particular results for the different comets.

Comet	Perihelion distance		Aphelion distance		Orbital period		Incli- nation	
	1957	1975	1957	1975	1957	1975	1957	1975
P/Boethin	1.08	1.09	8.76	8.83	10.9	11.0	6.0	5.9
P/We-Ko-Ik	4.80	1.40	16.19	5.29	34.0	6.1	20.1	30.1
P/Kohoutek	2.52	1.57	5.84	5.21	8.6	6.2	4.4	5.4
P/Smi-Che	5.24	3.57	5.88	4.78	13.1	8.5	6.0	6.6
P/Longmore	3.02	2.40	4.79	4.90	7.7	7.0	26.1	24.4
P/Gehrels3	5.71	3.42	8.18	4.65	18.3	8.1	3.1	1.1

Table 1. Orbital elements in 1957 and 1975.

Comet P/Boethin, 1975a.

A remarkable fact concerning this comet is that it is the only one under consideration, which has not recently experienced any significant reduction of the perihelion distance, although it was the brightest short-period comet observed in 1975. During the interval covered by our investigation it has not penetrated to less than 2.9 au from Jupiter or to less than 6.5 au from Saturn. The 1975 apparition of comet P/Boethin did not occur at the most favourable geometric circumstances (opposition perihelion passage). The closeness of its orbital period to eleven years indicates that it may also have escaped such configurations for a long time before discovery. Our investigation has yielded an ephemeris for comet P/Boethin at its perihelion passage in 1964, and this shows that the observational circumstances were then slightly less favourable than in 1975.

Comet P/West-Kohoutek-Ikemura, 1975b.

This comet encountered Jupiter rapidly and very closely in March 1972. The Laplacean approximation of matched conic sections would have yielded a fairly accurate representation of this encounter: the heliocentric orbital elements were essentially unperturbed at distances $\Delta >$ 0.28 au from Jupiter, while the jovicentric elements were very stable at $\Delta < 0.28$ au. The asymptotic deflection angle of the jovicentric hyperbola was 77° (mean eccentricity = 1.60). While the velocity with respect to Jupiter at the beginning of the encounter pointed less than 90° from Jupiter's heliocentric velocity, at the end of the encounter these two velocities were broadly antiparallel. As a result the heliocentric motion was severely braked, causing a decrease of the orbital period and of the perihelion distance. Comet P/West-Kohoutek-Ikemura is presently close to 2/1 resonance with Jupiter, indicating the possibility of repeated close encounters with the planet - particularly in early 1984 - in spite of the relatively high inclination of this comet's orbit.

Comet P/Kohoutek, 1975c.

This comet is at present moving in an orbit rather similar to the one of comet P/West-Kohoutek-Ikemura. The two comets encountered Jupiter at roughly the same time (see Figure 1), but the encounters were different as well as the pre-encounter orbits. Comet P/Kohoutek approached Jupiter to a minimum distance of 0.14 au in late July 1972. As may be expected from a comet which barely enters the zone of instability of heliocentric orbits ($\Delta < 0.16$ au), the orbital elements of this comet do not exhibit any drastic changes. Nevertheless the encounter resulted in appreciable decreases of the perihelion and aphelion distances, and in particular the reduction of the perihelion distance greatly enhanced the probability of discovery of the comet. In this sense the orbital evolution of P/Kohoutek between 1957 and 1975 may be viewed as a capture into the Jupiter family, as well as the more obvious case of the evolution of P/West-Kohoutek-Ikemura.

Comet P/Smirnova-Chernykh, 1975e.

This comet has the third largest perihelion distance (3.6 au) known among short-period comets. Because of the low eccentricity (0.15) of its orbit the comet may be observable even at aphelion oppositions. Twenty years ago, however, it would not have been observable even at perihelion. Figure 1 shows that the encounter with Jupiter started before 1957 for P/Smirnova-Chernykh, and the osculating elements in Table 1 for this comet are hence insignificant. An extended integration has therefore been performed back to January 1941. The pre-encounter orbital elements found in this way are: $q = 5.7$ au, $Q = 12.3$ au, $P = 27$ years, $i = 5.7°$.

A major transformation of the cometary orbit occurred in this case without any extremely close encounter with Jupiter ($\Delta_{min} = 0.20$ au in November 1955). This is consistent with the extremely long duration of the approach ($\Delta < 1.5$ au from August 1953 to August 1965). Comet P/Smirnova-Chernykh experienced a double encounter with Jupiter, a second approach to $\Delta = 0.47$ au having occurred in September 1963. While the first approach mainly decreased the aphelion distance, the second one implied the decrease of perihelion distance which was necessary for discovery of the comet. Between the two encounters the comet's heliocentric orbit had an extremely small eccentricity ($e_{min} = 0.015$ in early 1959).

Comet P/Longmore, 1975g.

This comet encountered Jupiter to a minimum distance of 0.16 au in October 1963. The evolutions of the heliocentric orbital elements are similar to the ones for comet P/Kohoutek in a broad sense - no extremely rapid transformations occur, and the resulting changes of orbital elements are comparable. For comet P/Kohoutek we used the term "capture" to signify a reduction of the perihelion distance which greatly enhances the probability of discovery. In this sense one may use the word also for the calculated evolution of comet P/Longmore.

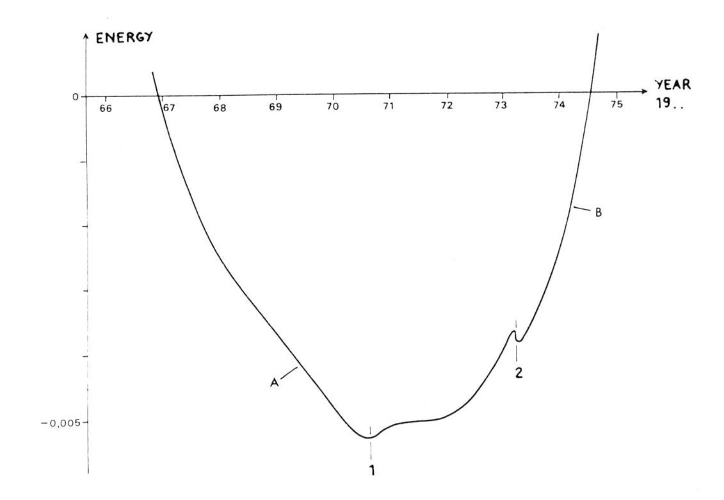

Figure 2. Jovicentric orbital energy for P/Gehrels 3. A and B denote the passages of 0.41 au jovicentric distance, 1 and 2 are the perijove passages.

Comet P/Gehrels 3, 1975o.

Like P/Smirnova-Chernykh this comet experienced a double encounter with Jupiter (see Figure 1). Both the encounters of comet P/Gehrels 3 were extremely close, however. As already mentioned, the first one implied a near-collision with the planet, and the second one took place only 2.6 years later to a minimum distance of 0.04 au on March 25, 1973. The comet remained within the stability zone of jovicentric motion ($\Delta < 0.41$ au) for the whole interval between the encounters.

Figure 2 shows the jovicentric orbital energy of comet P/Gehrels 3 as a function of time. It is evident that the jovicentric orbit was elliptic for 7.6 years, from December 1966 to July 1974, until 15 months before discovery. Together with the extended sojourn in the $\Delta < 0.41$ au sphere (April 1969 to February 1974) this makes the term "quasi-satellite" appropriate for the comet. Satellite captures by giant planets have been found before, but only concerning hypothetical "comets" (Everhart 1973, Kazimirchak-Polonskaya 1972). Numerical integration errors are of no importance in the present case. The inaccuracy of our calculation depends mainly on the fact that we took no account of Jupiter's oblateness or the perturbations by the Galilean satellites. Four different starting orbits were treated, the divergence of which throughout the whole double encounter is found to be remarkably small.

Comet P/Gehrels 3 first approached Jupiter along an extremely elongated, retrograde elliptic trajectory, the perijove distance being 0.0014 au and the apojove distance gradually decreasing to 0.36 au. After passing perijove on August 15, 1970, the comet went into its first and only apojove passage. Solar perturbations placed the actual apojove at 0.37 au from the planet, and the sense of motion was changed to prograde. The perijove distance was raised to 0.04 au while the apojove distance continued to increase beyond 0.41 au. After the second perijove passage the comet was carried away by the Sun's gravitation to enter into its present heliocentric orbit.

DISCUSSION

For comet P/Kohoutek our pre-encounter orbit is in good agreement with the preliminary estimate by Marsden (1975), while for comet P/Smirnova-Chernykh our results do not confirm those of Kastel' (1975), namely, $q = 5.7$ au, $Q = 6.7$ au before the encounter in 1963. Such a discrepancy is to be expected, since the starting orbit used by Kastel' is based upon a 38 days observational arc only and differs substantially from the ones we used.

A few general conclusions are apparent from Table 1. For all comets except P/Boethin the perihelion distances in 1957 were so large that the comets could hardly have been observed. On the other hand the orbital periods were still relatively short. Hence the above-mentioned picture of the immediate origin of the Jupiter family is supported by the present investigation. We find a high rate of transformations of cometary orbits (i.e. a high "capture rate") by Jupiter, in accordance with the recent high discovery rate of Jupiter family comets.

REFERENCES

Danielsson, L., Ip, W.-H.: 1972, *Science* 176, pp. 906-907.
Delsemme, A.H.: 1973, *Astron.&Astrophys.* 29, pp. 377-381.
Everhart, E.: 1972, *Astrophys. Letters* 10, pp. 131-135.
Everhart, E.: 1973, *Astron. J.* 78, pp. 316-328.
Kastel', G.R.: 1975, *IAU Circular* 2772.
Kazimirchak-Polonskaya, E.I.: 1972, *IAU Symp.* 45, pp. 373-397.
Kresák, L.: 1974, *IAU Coll.* 22, pp. 193-203.
Marsden, B.G.: 1975, *IAU Circulars* 2756, 2766, 2812, 2865.
Oesterwinter, C., Cohen, C.J.: 1972, *Cel. Mech.* 5, pp. 317-395.
Rickman, H., Vaghi, S.: 1976, *Astron.& Astrophys.* 51, pp. 327-342.
Vaghi, S.: 1973, *Astron.& Astrophys.* 24, pp. 107-110.

CHARACTERISTICS OF SINGLE ENCOUNTERS OF LONG-PERIODIC COMETS WITH JUPITER

Tsuko Nakamura
Dodaira Station of the Tokyo Astronomical Observatory, Tokigawa
Saitama 355-05, Japan

Original nearly parabolic orbits of comets are known to be evolved toward short-periodic elliptic orbits as statistical results of hundreds of encounters with Jupiter. There seems to be two methods to handle the process, namely, the method by exact numerical integrations for each orbit (Everhart, 1972) and random walk approach by using probability distributions of perturbations after single encounters(Lyttleton and Hammersley, 1963; Shteins, 1972). Since both methods need a great number of input parabolic comets to have only a few tens of short-periodic ones, the second method may save time compared with the first one, which is in turn more accurate. The purpose of this paper is to clarify the characteristics of single-encounter effects, in order to develope the second method more elaborately and extensively.

The second method often has been done by adopting simple, empirical and/or assumed Gaussian forms for the distributions of perturbations of the barycentric total energy ΔE. On the other hand Everhart(1968) gives one of the most detailed distributions. His results show that forms of the distributions are very sensitive to the variation of the adopted parameters q(perihelion distance) and i(inclination). This means that if q and i change considerably in the course of the evolution, the distributions of their perturbations must be calculated as well. As is said later, when nearly parabolic orbits evolve to short periodic ones, there is some possibility that q and i change greatly, and in fact Everhart(1972) showed such an example by using the first method.

Instead of the perturbations of q and i I compute those of barycentric total amount and z-component of the angular momentum, ΔG and ΔH, besides ΔE and describe the 3-dimensional distributions. The computations are carried out by integrating the equations of motion for the restricted three body problem. The equations of motion are regularized near the Sun and Jupiter. For one set of the parameters, a(semi-major axis), q and i, more than one thousand orbits are integrated by changing Ω(node) and ω(perihelion argument) randomly. The parameters are varied in the intervals of $1/500 \geq -1/a \geq -1$, $0.1 \leq q \leq 1.2$ and $0° \leq i \leq 180°$, where the unit of length is Jupiter's orbital radius. Thanks to the existence of the Jacobian inte-

R. L. Duncombe (ed.), Dynamics of the Solar System, 299–301.

gral, ΔE is always approximately equal to ΔH so that the distributions actually reduce to the two-dimensional ones.

The marginal distributions of ΔG is found to be almost always symmetric with respect to zero and can be well represented by a single- or double-peaked Gaussian form with extended skirts. The situation is very similar to the distributions of ΔE(Everhart, 1968). Then let us see the two-dimensional distribution of points on the ΔE- ΔG plane. One might expect such simple bi-variate Gaussian forms as the distributions of ΔE. However, every pattern is far from such simple forms. Some characteristics of the distributions are as follows:(1)Patterns are symmetric with respect to the two straight lines which are orthogonal to each other, regardless the values of the parameters. (2) The distributions generally consist of a dense core and a sparse envelope. (3) The distributions are almost insensitive to the variation of a, especially for small q. This fact seems to suggest a possibility that we might extrapolate the random-walk method to the orbits with small semi-major axis, say $a \leq 1.0$, which, exactly speaking, should be treated by numerical integrations or the theory of secular perturbations. (4) For small q the dense core appears as a rectangle whose sides are very sharp. Those sides or its inside may correspond to some quasi-integral of motion. (5) For large q the distributions show approximately elliptical forms though they have complex structure inside. (6) For direct motions ΔE and ΔG have the same sign, while for retrograde motions the sign changes. However, for large q, ΔE and ΔG have the same sign even for a great part of retrograde orbits. (7) ΔE and ΔG usually are of the same order of magnitude.

According to the last two items there comes a possibility that q and i can change considerably when the original nearly parabolic orbits of comets evolve to short-periodic ones. So the assumption that the perihelion distances are invariable under random-walk process is not accurate although it is often assumed. Consulting with the distributions I derived I can infer local pathes of the evolution to some extent. If, as the second step, I express these all distributions by appropriate functions, I will be able to treat the random-walk process rather in details. As the concrete forms and the values of the distribution functions are somewhat arbitrary, they are not presented here.

REFERENCES

Everhart, E.: 1968, "Astron. J.". 73, p.1039.
Everhart, E.: 1972, "Astrophys. Letters", 10, p. 131.
Lyttleton, R.A. and Hammersley, J.M.: "Mon.Not. R. Astron. Soc.",127,p.257.
Shteins, K. A.: 1972, in G.A.Chebotarev et al(ed.), "The Motion, Evolution of Orbits, and Origin of Comets", p.347.

DISCUSSION

Yabushita: I have two comments. The assumption of constant q (perihelion distance) is good when you consider the diffusion of comet with $a=10^4$ AU to $a=20$ AU or so. The relation $(\Delta E)=(\Delta G)$: is that not a consequence of Jacobi integral?

Nakamura: As for the first point, I quite agree with you; however, since we want to handle uniformly the diffusion of comets to, say, $a \simeq 5$ AU, we must take into account the variation of perihelion distance and inclination. In fact, we have found that our method can treat the diffusion process to within the region of a, mentioned above. The relations $0(\Delta E)=0(\Delta H)$ and sign $(\Delta E)=$ I sign (ΔH) are not a consequence of Jacobi integral but come from another quasi-integral.

LONG PROPAGATION PERIODS OF RESIDUALS IN THE MOTION OF A COMET

T. Kiang
Dunsink Observatory, Castleknock, County Dublin, Ireland.

P/Halley In a previous paper (Kiang 1973), I derived a differential equation which governs the behaviour of the residual O − C in the time of perihelion passage of a comet. The work was stimulated by the discovery (Brady 1972) that a long periodicity of about 600 years seemed to be present in the residual for P/Halley. The derivation was based on the following assumptions:

1. The system consists of only the Sun, Jupiter and P/Halley, and Jupiter is assumed to move in a circular orbit with a period $P' = 11.8614$ yr.

2. P/Halley is assumed to move in a *fixed* Keplerian orbit with a period $P = (13/2)\,P'$ exactly, so that the whole system is strictly recurrent with a period $P_0 = 13\ P' = 154.198$ yr.

A fixed Keplerian orbit is, of course, not a rigorous solution of the restricted problem of three bodies, but it is a simplifying assumption that made the derivation of the differential equation possible. The result was a second-order differential equation with periodic coefficients of period P_0. It was readily reduced to the standard form of a Hill's equation (Hill 1886), the solution of which consists essentially in the evaluation of Hill's exponent c.

In the practical solution of the equation, the following should be noted:

1. The coefficients of the equation are ultimately based on the values of the partial derivatives, with respect to Jupiter's longitude λ', of the rate of change of the mean motion n and of the mean anomaly M due to the action of Jupiter, through a whole recurrence period. Because these values can change violently at certain phases, we should not use their instantaneous values, but, rather, their average values, averaged over the adopted time interval (usually about 1 or 0.5 yr).

2. Hill himself dealt with a rather special case: his "intermediary orbit" had a two-fold symmetry, hence his formulae (copied in all text books) must be modified for the general case of an "orbit" with no symmetry. The necessary modifications were given in my paper, which contains, however, the following error and misprints: the error was that a factor of 2 was left out of the right side of Expression (40) of my paper and the misprints were (i), in (39) and (40), the argument of

R. L. Duncombe (ed.), Dynamics of the Solar System, 303–306.

cot should read $\pi\sqrt{\theta_0}$ and (ii) the denominator in (39) under the summation should read $k^2 - 4\theta_0$.

The system contained one free parameter λ_0', the longitude of Jupiter at the start of each recurrence period, which I took to correspond to a perihelion passage of the comet. It was found that c was real for some values of λ_0' and imaginary for others. A real c means that the behaviour of the residual is dominated by a long-period sinusoidal variation with period $P^* = P_0/c$ (the *hyperperiod*), while an imaginary c means that a secular component is present which will eventually dominate the behaviour.

I have now made calculations for more values of λ_0' and with the error in Expression (40) removed. The new results are shown in TABLE 1.

TABLE 1 Results for P/Halley (Fixed input parameters: eccentricity e = 0.96814, inclination of orbital planes $I = 161^\circ.25$, longitude of ascending node of Jupiter reckoned from the direction of perihelion $\Omega = 113^\circ.8$, $P_0 = 2P = 13P' = 154.198$ yr)

λ_0' =	0°	12°	14°	16°	18°	20°	22°	30°
c =	$-.51i$	$-.67i$	$-.14i$	$-.072$	$-.128$	.170	.187	.235
P^*(yr)=	...	...	...	2160.	1203.	910.	825.	655.

λ_0' =	60°	90°	120°	150°	154°	158°	160°
c =	.220	.201	.228	.307	.231	$-.024$	$-.17i$
P^*(yr)=	702.	766.	676.	503.	668.	6400.	...

Thus it appears that just over one-sixth of possible values of λ_0' give imaginary values of c; these correspond to those configurations in which the comet have close encounters with Jupiter. The rest five-sixths give real values of c, and values of hyperperiod mostly between 500 and 800 years, except near either end of the range, when the hyperperiod becomes indefintiely large.

A Qualitative Re-statement of Result Solution of Hill's equation proceeds in the frequency domain; the result obtained are mathematically precise but not easily graspable. I shall now try to re-state qualitatively the results obtained in terms of ordinary space and time. Let us start first with just the Sun and the comet. The latter then persues a Keplerian orbit. Now let us introduce a small, constant, non-gravitational force at every perihelion passage. Then the residual defined above will simply grow and grow; in other words, the system is unstable against persistent impulses of the same sign. Now introduce the massive Jupiter whose perturbation is many hundreds times greater than the non-gravitational impulse, but it can be of either sign and is of a higher frequency. Then, apparently, provided a certain condition is satisfied, the residual due to the non-gravitational force will not grow indefinitely, or the high-frequency perturbation by Jupiter has stabilised the system against persistent impulses. The condition is that there should be no close encounters between the comet and Jupiter. It is easy to see that this condition is necessary, for if there are close encounters, then even a very small change in the circumstance of encounter, corresponding to a very small residual, will result in very large effects.

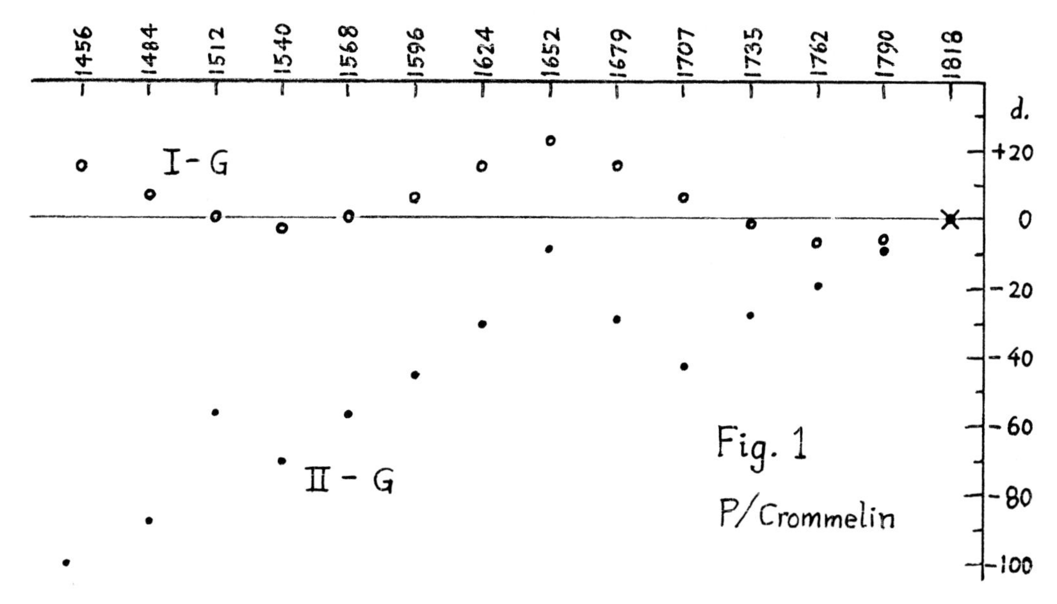

Fig. 1
P/Crommelin

P/Crommelin Dr. B. G. Marsden (1973) kindly supplied me 3 sets of perihelion passage times of this comet,[1] calculated for a purely gravitational model (G) and 2 non-gravitaional models (I: $A_1 = +1$, $A_2 = +0.1$; II: $A_1 = +1$, $A_2 = -0.1$). Fig. 1 shows the two runs of residuals I - G and II - G. There is a hint of a period of about 230 yr in the first while the second appears to be non-periodic. After having idealised the comet to have a period of $P = (7/3)\,P'$ exactly, the results in TABLE 2 were obtained.

TABLE 2 Results for P/Crommelin ($e = 0.919$, $I = 30°$, $\Omega = 335°.3$, $P_8 = 3P = 7P' = 83.030$ yr)

λ_0' =	0°	15°	30°	45°	60°	75°	90°	105°
c =	$-.50\,i$	.1654	.1086	.1458	.1610	.1932	.0675	$-.23\,i$
P^*(yr) =	...	501.	764.	570.	515.	429.	1229.	...

It appears that no value of λ_0' can give a hyperperiod as low as 230 yr, while about one-fourth of the initial conditions will give a non-periodic variation. This suggests that the non-gravitational model II may be closer to truth than model I.

Leonids and P/Tempel-Tuttle In a compilation of Chinese historical records of shower meteors, recently become available in English, Zhuang Tian-shan (1977) pointed out that there may be a 300-yr period in the visibility of the Leonid Meteor Stream. This stream is known to be associated with the comet P/Tempel-Tuttle, which has a period of 33yr. I therefore made calculations of c and P^* for an ideal comet with $P = (11/4)\,P' = 32.6189$ yr and the same eccentricity and orientation as P/Tempel-Tuttle. The results are shown in TABLE 3.

TABLE 3 Results for P/Tempel-Tuttle ($e = 0.9044$, $I = 163.^\circ6$, $\Omega = 169.^\circ3$
$P_0 = 4P = 11P' = 130.475$ yr)

λ_0' =	0°	15°	30°	45°	60°	75°
c =	$-.29\,i$	$-.09\,i$	.2210	.2426	.2406	$-.17\,i$
P^*(yr) =	...	...	590.	538.	541.	...

It appears that the "observed" period of 300 yr must be due to factors other than a residual in the time of perihelion passage. The visibility of metoers is, of course, critically dependent on the position of the node in relation to Earth, and the present theory does not apply to any oscillation in the node.

REFERENCES

Brady J.L., (1972) *Pub. astr. Soc. Pacific*, 84, 314-322.
Hill G.W., (1886) *Acta Math.*, 8, 1-36.
Kiang T., (1973) *Mon. Not. astr. Soc.*, 162, 271-287.
Marsden B.G., (1973) Private correspondence.
Zhuang Tian-shan (1977) *Chinese Astronomy* 1, 197-220; first published in Chinese in *Acta Astronomica Sinica* 14 (1966) 37-58.

DISCUSSION

Comment by Schubart: May I call your attention to the big display of Leonids seen in Europe in 1366. There are records from Portugal and from a monestary in Bohemia; compare Alexander von Humboldt's collection of references in "Cosmos." This shower was perhaps not visible in China.

ON THE ORBIT MECHANICS OF COMETARY DUST PARTICLES

H. Kimura
Purple Mountain Observatory, Academia Sinica, Nanking, China
and
Department of Astronomy, University of Tokyo, Tokyo, Japan

It is generally believed that the structure of dust tails of comets can be described in terms of the mechanical theory. The basic parameters of this theory are the repulsive force and initial emission velocity of particles. However, the effect of initial velocity does not seem to have been considered seriously in the discussion of the overall structure of dust tails. Finson and Probstein (1968) have recently proposed an elaborated method to analyse the tail brightness profiles by introducing the distribution function of particles. The initial velocity effect was, however, taken into account only in an approximate way. They assumed that a group of particles are emitted isotropically with a single speed, and afterwards forms a spherical shell expanding uniformly in time. The isotropic emission is suggested by the fluid-dynamical considerations on the gas-dust interaction in the inner head region, but the subsequent spherical expansion implies merely a simplification for the calculation to obtain the line-of-sight integrals analytically. Due to this simplification, the validity of the Finson-Probstein method is seriously limited, particularly when a tail after perihelion passage is concerned.

In order to proceed without such an approximation, we have reconsidered the orbit mechanics of cometary particles. Our problem is not an N-body problem, but a synthesis of a large number of two-body problems. The mathematics involved is rather elementary, but the three-dimensional treatment is required because the numerous orbital planes to be concerned are inclined to each other. If there is an effective formulation available to follow, and to synthesize, the orbit mechanics of many particles, then the surface density integrals can be estimated, directly and more rigorously, by counting of sample particles.

The underlying physics of the dust emission process suggested by Finson and Probstein (1968) will be retained. According to this picture, the dust emission process can be generally idealized as an isotropic one from a point source. A sample particle is labelled by the following parameters; the effective force parameter, μ_j, the time of emission, t_i, and the relative velocity of emission, $\vec{v}_k$; hence, the initial conditions for its Keplerian motion can be taken as $\vec{r}_0=\vec{r}_c(t_i)$, and $\vec{V}_0=\vec{V}_c(t_i)+\vec{v}_k$; where

R. L. Duncombe (ed.), Dynamics of the Solar System, 307–310.

the suffix c indicates the quantity referred to the comet nucleus. The orbital motion of the nucleus will be assumed to be parabolic. This is the case so far as dust-rich comets are concerned. We also note that the ratio (v_k/V_c) is very small ($\lesssim 0.01$) in most cases.

The equation of the Keplerian motion is

$$d^2\vec{r}/dt^2 + (\mu k^2/r^2)(\vec{r}/r) = 0 \quad , \quad \text{where} \quad \mu = 1-\gamma \quad . \tag{1}$$

As is well-known, this equation has two vector integrals; the one is the angular momentum vector, $\vec{h}=\vec{r}\times\vec{V}$ ($\vec{V}=(d\vec{r}/dt)$), and the other is given by $\vec{b}=(1/k^2)(\vec{V}\times\vec{h})-\mu(\vec{r}/r)$, which may be termed the eccentric vector (the terminology after Fushimi, 1954).

Employing these two constant vectors of motion, we obtain the equation of orbit in a vectorial form,

$$\vec{r} = q\,\{(1-z^2)\vec{P} + 2z\vec{Q}\} \,/\, (1-Sz^2) \quad , \tag{2}$$

where $\vec{P}=(\vec{b}/b)$ and $\vec{Q}=(\vec{h}/h)\times\vec{P}$ are unit vectors, $q=(h^2/k^2)/(b+\mu)$ denotes the perihelion distance, and the parameter $S=(b-\mu)/(b+\mu)$ is a modification or a generalization of the eccentricity. The variable $z=\tan(\theta/2)$, tangent of half the true anomaly, is a function of time, and the (z-t) relation is given by the integral of motion,

$$t - T = t_S \, J_S\,(z) \quad , \tag{3}$$

where $t_S=(2q^2/h)$ is a characteristic time of the motion, $T=t_0-t_S J_S(z_0)$ the epoch of perihelion passage, and $J_S(z)=\int_0^z (1+z^2)/(1-Sz^2)^2 dz$. The initial value of z required in the evaluation of T, is determined by the relation $z_0=\{(r_0-q)/(Sr_0+q)\}^{1/2}\text{sign}(\vec{r}_0\cdot\vec{V}_0)$.

Equations (2) and (3) are convenient and sufficient for our purpose to follow orbital motions of many particles. The generalization with respect to μ is also completed in these equations.

It is convenient to introduce the concepts of elementary space distributions or structural elements of the tail. First comes an ijk particle. Next, we have synchronic and syndynamic lines, which are familiar examples if constructed on assuming zero-emission velocity. Taking account of the velocity dispersion of emitted particles, we find that synchronic particles extend a three-dimensional volume, which may be termed a "synchronic tube". Similarly, one can define a "syndynamic tube". A "tube element" is also defined as a crossing of two tubes of different kind.

In order to see how these structural elements change their sizes and shapes with time, it is useful to consider the vertical motion of particles relative to the comet orbit plane. The height above the comet orbit plane is given by

$$\zeta = v_{0\zeta} t_S (z-z_0)(1+zz_0)/(1-Sz_0^2)(1-Sz^2) \quad (z \geq z_0) \quad . \tag{4}$$

Combining this with Eq. (3), one can obtain ζ as a function of time. It is readily seen that the linear approximation $\zeta = v_{0\zeta}(t-t_0)$ applies only for particles of small $\tau(=t-t_0)$ or very small μ. We also note that some particles recross the comet orbit plane when the condition $z=-(1/z_0)$ is satisfied. This elementary fact leads to an interesting result, that is, the appearance of a vertically depressed part in the tail, which may be termed the "neckline structure".

For the understanding of such a structural feature, it will prove convenient to define a "neckline" as the locus of particles originally emitted into vertical directions and just recrossing the comet orbit plane. For a vertically emitted particle, it is rather easy to determine the position (r_e, θ_e) and time (t^*) of its recrossing to occur. The results are expressed symbolically as, $t^*=t^*(t_i, \mu_j)$, $r_e=r_e(t_i, \mu_j)$, and $\theta_e=\theta_e(t_i)$. Based on these relationships, one can define several curve families on the comet orbit plane, as follows; $f_1(\theta_e)=c_1(t_i)$, $f_2(r_e, \theta_e)=c_2(\mu_j)$, $f_3(r_e, \theta_e)=c_3(t^*)$. The last one represents the family of necklines.

If further the distribution function of particles, $F(t, \gamma, \vec{v})$ is introduced, one can discuss the tail brightness profiles. With a modification by cross-sections for light-scattering, F can be resolved into three functions, that is, $C_{sca}(\gamma)F(t, \gamma, \vec{v}) = \langle C_{sca} \rangle \dot{N}_d(t) f(\gamma; t) \psi(\vec{v}; \gamma, t)$, where $\dot{N}_d(t)$ is the dust emission rate, $f(\gamma; t)$ the modified size-distribution function, and $\psi(\vec{v}; \gamma, t)$ the velocity distribution. $\dot{N}_d(t)$ will be assumed to follow the inverse square law of $r_c(t)$ as suggested by the nucleus model of Delsemme and Miller (1971). The velocity distribution is an isotropic one with a single speed v_0, and the time- and γ-dependency of v_0 is assumed to follow Probstein's formulae. The size-distribution $f(\gamma)$ is left to be determined through fitting procedure of the calculation to observed profiles. One can now construct "syndynamic brightness profiles" for different sample values of γ, by means of the counting method. A suitably weighted superposition of them is expected to reproduce the observed feature. If an appropriate weighting function is found, it gives the information on $f(\gamma)$.

We have applied this method for Comet Arend-Roland, and found that both of the main and anti-tails can be reproduced in a unified manner, namely, no temporal anomalies in the source functions need not be introduced. In other words, the anomalous tail of this comet can be well understood in terms of the neckline structure. The nature of the derived and assumed source functions will not be discussed here. The point of the present argument is that the neckline structure is a common feature in the dust tails of comets after perihelion passage, and so it seems responsible for some peculiar appearance of dust tails. The anomalous sunward tail of Comet Arend-Roland, consisted of a sharp spike and a faint extended halo around it, was a striking example.

This presentation is a part of the work by C.P. Liu and the present author (published in Acta Astron. Sinica, Vol. 16, 138, 1975, in Chinese)

REFERENCES

Delsemme, A.H. and Miller, D.C. : 1971, Planet. Space Sci., 19, 1229.
Finson, M.L. and Probstein, R.F. : 1968, Astrophys. J., 154, 327 and 353.
Fushimi, K. : 1954, "Mechanics", (in Japanese), Iwanami, Tokyo.

DISCUSSION

Comments by Kiang: 1. The mathematics used is most elegant. The parameter S orders the various forms of the conic sections much more neatly than the usual elements a and e. 2. The neckline structure predicted by the author is a string-like concentration of material starting at the nucleus. As such it should sharpen considerably any interpretation of cometary tails. 3. A complete English translation of the original Chinese paper is now available ("Chinese Astronomy" Vol 1, No 2, December 1977).

THE SPLIT COMETS: GRAVITATIONAL INTERACTION BETWEEN THE FRAGMENTS

Z. Sekanina
Harvard-Smithsonian Center for Astrophysics,
Cambridge, Massachusetts, U.S.A.

1. THE APPROACH

The n-body computer program by Schubart and Stumpff (1966) has been slightly modified to study the gravitational interaction between two fragments of a split comet nucleus in the sun's gravitational field. All calculations refer to the orbit of Comet West (1976 VI), the velocity of separation of the fragments is assumed to be equal in magnitude to the velocity of escape from the parent nucleus, and the numerical integration of the relative motion of one fragment (called the companion) with respect to the other (principal fragment) is carried over the period of 200 days from separation. The motion of the companion is modeled as a function of the following circumstances at splitting:

(1) Position of the comet in heliocentric orbit at the time of splitting. Considered are three locations, whose true anomalies are $-90°$, $0°$, and $+90°$, corresponding, respectively, to heliocentric distances of 0.393 A.U. prior to perihelion, 0.197 A.U. (perihelion), and 0.393 A.U. subsequent to perihelion.

(2) The mass of the principal fragment. Two cases are considered: 10^{17} and 10^{15} grams.

(3) The mass ratio of the principal fragment to the companion. Calculations are made for ratios 1:1 and 100:1.

(4) The average bulk density of the fragments. Densities used are 1 and 0.1 g cm^{-3}, same for both fragments.

(5) Direction of separation of the companion from the principal. Considered are six cardinal directions: sunward; forward, i.e., perpendicular to the sunward direction in the orbit plane in the sense of the comet's orbital motion; northward, i.e., in the direction of that orbital pole from which the comet is seen to orbit the sun counterclockwise; and the respective opposite directions: anti-sunward; backward; and southward.

R. L. Duncombe (ed.), Dynamics of the Solar System, 311–314.

(6) Area of separation of the companion on the surface of the parent nucleus. Six basic locations are investigated: subsolar area; leading-side area, oriented in the forward direction from the center of the nucleus; north-pole area; and the areas located opposite these: antisolar area; trailing-side area; and south-pole area.

All possible combinations of the six variables represent a total of 720 patterns. Fortunately, symmetries with respect to the orbit plane reduce the number of independent patterns by 216, and other symmetries that lead to virtually identical patterns reduce the number by additional 240 to leave a total of 264 patterns. Furthermore, the only measurable effect of the mass of the principal fragment is that on the dimensions of the companion's orbit, which scale with the cube root of the mass. This brings the number of patterns under consideration down to 132.

2. THE RESULTS

Computer generated plots of the variations with time of the separation between the fragments, of their relative velocity, and of the angle subtended by the separation and the velocity vectors have been examined for similarity in appearance, separately for each of the three positions in heliocentric orbit. It has been found that the differential gravitational effect of the sun becomes noticeable about 0.5 to 1 day after separation for the breakup taking place at perihelion, and about 1 to 2 days after separation for the breakup at heliocentric distance twice the perihelion distance. It has further been found that the visual aspect of the plots is not affected fundamentally by the bulk density and by the fragment mass ratio, although the outcome of the dynamical solution may so be affected. Consequently, it has been possible to categorize the patterns in terms of only the direction of separation and the area of separation by considering the following eleven standard patterns: separation sunward from (1) subsolar area, (2) leading-side area, (3) north-pole area, (4) trailing-side area; separation forward from (5) subsolar area, (6) leading-side area, (7) north-pole area; separation northward from (8) subsolar area, (9) leading-side area, (10) north-pole area; and (11) separation backward from subsolar area.

For the splitting at true anomaly $-90°$ the calculated patterns can be divided into seven categories. Category 1.1 encompasses separations from the leading-side area sunward, forward, and northward, and from the trailing-side area sunward. The separation distance increases monotonically. The companion escapes along a very strongly hyperbolic orbit; inclination to the comet's orbit plane is zero or very low; prograde motion. Category 1.2 includes separations from the north-pole area sunward and forward. Except for a shallow minimum shortly after perihelion the separation distance increases monotonically. The companion escapes along a high-inclination, strongly hyperbolic orbit; prograde motion. Category 1.3 is a separation forward from the subsolar area. The separation distance increases monotonically at rather a slow rate. The velocity surges at perihelion, then follows a profound decrease and a moderate upswing. The companion escapes along a slightly hyperbolic orbit; zero inclination; retrograde motion. Category 1.4 is a separation northward from the subsolar area. The separation curve has a complicated shape; the velocity curve, after a sudden upswing near perihelion, displays a downward trend. The companion gets gradually into a highly elongated, quasi-stable elliptical orbit with a period $\sim$240 days; high inclination; retrograde motion. Category 1.5, a separation sunward from the subsolar area, exhibits a separation curve that gradually levels off and a velocity curve with a very deep minimum about 30 days after splitting. The companion gets into a highly elongated, quasi-stable elliptical orbit with a period of 226 days; zero inclination; retrograde motion. Category 1.6, a separation backward from the subsolar area, displays very complicated separation and velocity variations. The companion pursues a retrograde orbit, whose eccentricity is slightly less than unity. The patterns with the bulk density 0.1 g cm^{-3} show the fragments crashing into each other 120 days after breakup for the mass ratio 1:1, 129 days for 100:1. Category 1.7, a separation northward from the north-pole area, results in a crash of the fragments 15 days after breakup regardless of the mass ratio and the density.

For the splitting at perihelion there are five categories of patterns. Category 2.1 is identical in contents with Category 1.1. The companion's motion displays similar but still more vigorous escape characteristics than were those for the Category 1.1 patterns; prograde motion. Category 2.2 encompasses the patterns of Categories 1.3, 1.4 and 1.5. The companion's motion differs from that in Category 2.1 by appreciably milder hyperbolic excess; prograde motion. Category 2.3 is identical in contents with Category 1.6. Escape is still more hesitant than in Category 2.2, and in the case of the high mass ratio combined with the low bulk density the motion becomes elliptical; zero inclination; prograde motion. Category 2.4, equal in contents to Category 1.2, shows the companion in a high-inclination, elliptical, quasi-stable orbit. The eccentricity and the revolution period depend strongly on the mass ratio and on the density; prograde motion. Category 2.5 contains the pattern of Category 1.7 and is again characterized by collision of the fragments, this time 16 days after breakup.

For the splitting at true anomaly +90° we have another five categories. Category 3.1, encompassing Categories 1.3, 1.4, 1.5 and 1.6, has the companion's behavior reminiscent of the one in Category 2.2; prograde motion. Category 3.2 is identical in contents with Category 1.1 except for a separation sunward from the trailing-side area which now makes up Category 3.3. The companion escapes almost exactly radially, the sense of motion being changed from retrograde early after separation to prograde later; zero or low inclination. In Category 3.3 the companion gets into a high-eccentricity elliptical orbit; zero inclination; retrograde motion. Category 3.4 is identical with Category 1.2. The companion's motion is elliptical, somewhat similar to the one in Category 2.4, but less dependent on the mass ratio and the density; high inclination; prograde motion for the separation sunward, retrograde for the forward. Category 3.5, identical with Category 1.7, shows again the two fragments crashing into each other, this time 90 days after separation.

3. FINAL REMARKS

A great variety of the companion's dynamical behavior results from the strong dependence on the circumstances at breakup, particularly on the position of the comet in orbit, on the direction of separation and on the area of separation on the parent nucleus. The calculated patterns vary from zero to very high inclinations, from a prograde to a retrograde sense of motion, and from escape of the companion along very strongly hyperbolic trajectories to its pursuance of periodic, quasi-stable orbits around the principal nucleus terminated in some cases by collision of the fragments. We can state plainly that the velocity of separation as derived from observations of the split comets (Sekanina, 1977, 1978) offers by itself no information on the mass of the comet. The interpretation of the empirically determined time of splitting and the separation velocity, the interaction between the gravitational and the nongravitational perturbations, and the possibility of the existence of binary and multiple comet nuclei are among the most interesting problems that remain to be explored.

This work has been supported by Grant NSG 7082 from the National Aeronautics and Space Administration.

REFERENCES

Schubart, J. and Stumpff, P.: 1966, *Veröffentl. Astron. Rechen-Inst. Heidelberg* No. 18, pp. 1-31.
Sekanina, Z.: 1977, *Astrophys. Letters* 18, pp. 55-59.
Sekanina, Z.: 1978, *Icarus* 33, pp. 64-71.

ON THE EFFECT OF BINARY ENCOUNTERS

Tsutomu Shimizu
Bukkyo University, Kyoto 603, Japan

1. INTRODUCTION

The effect of binary encounters is examined by restricting considerations to a sphere of finite radius R, instead of an infinite one assumed hitherto. The sphere is centered at an assigned test-particle with velocity $\mathbf{V}_0$. The corresponding modified assumption is that particles lying outside the sphere are scattered randomly with an average number density n and their velocity distribution is Maxwellian with an r.m.s. of $\sqrt{3}\sigma$ in the space velocity. As for the particles' masses the same M is presumed. Now the number of particles, with relative velocities $\mathbf{V}(V, \theta, \varphi)$ referred to the test-one, entering into a spherical band between the colatitudial angles χ and $\chi + d\chi$ per unit time, dN_1, is given by multiplying $nV\cos\theta \cdot 2\pi R^2 \times \sin\chi d\chi$ with the frequency distribution of $\mathbf{V}$ as follows (c.f. Fig. 1).

$$dN_1 = 2\sqrt{\frac{2}{\pi}} n R^2 \sigma \operatorname{Exp}[-Y_0^2 - Y^2 + 2Y_0 Y(\cos\theta\cos\chi + \sin\theta\sin\chi\cos\varphi)] Y^3 dY \times \sin\theta\cos\theta d\theta d\varphi \sin\chi d\chi, \qquad (0 \leq \theta \leq \pi/2), \tag{1}$$

where $Y = V/(\sqrt{2}\sigma)$ denotes the non-dimensional relative speed of a field-particle and $Y_0 = V_0/(\sqrt{2}\sigma)$ the non-dimensional speed of the test-one.

2. DENSITY DISTRIBUTION OF ENCOUNTING PARTICLES

Taking into account that the encounter phenomena are symmetric around

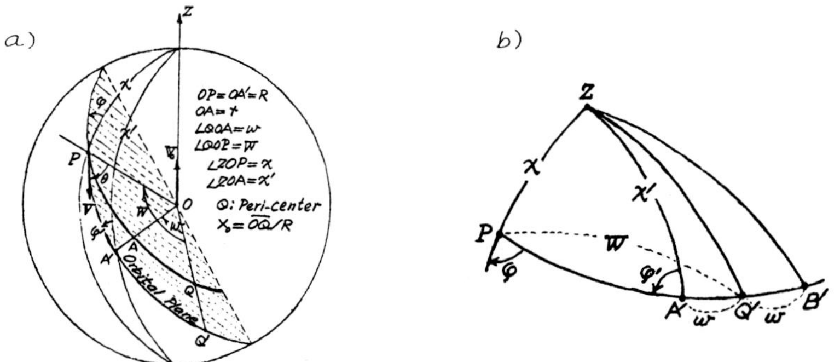

Fig. 1. Notations concerning an encounting particle's orbit.

R. L. Duncombe (ed.), Dynamics of the Solar System, 315–318.

the z-axis (c.f. Fig. 1) let us evaluate the number density per unit time of the entered particles at a point with radius r and colatitudial angle χ', $\nu(r,\chi')$. Since each of the entered particles moves supposedly along a Keplerian orbit, its position and its motion at some instant are expressed in terms of Y, Y_0, θ, φ, and χ. Hence, the change of variables from χ and φ to χ' and φ' in eq. (1) provides information on the number of encounting particles in a volume element $2\pi r^2 dr \sin\chi' d\chi'$. However, the time-interval for staying in a spherical shell between the radii r and r+dr, namely $dr/|\dot{r}|$ ($\dot{r}$: the radial velocity at r) depends on r, so that it must be taken into account. Then, letting X be r/R and since $\sin\chi d\chi d\varphi = \sin\chi' d\chi' d\varphi'$ we have

$$\nu(X,\chi')2\pi r^2 dr \sin\chi' d\chi' = 2\sqrt{\frac{2}{\pi}}\, nR^2\sigma \int_{\alpha_1}^{\infty} \mathrm{Exp}(-Y_0^2-Y^2)Y^3 dY \int_0^{\theta_1} \frac{dr}{|\dot{r}|}\sin\theta\cos\theta\, d\theta$$
$$\times \int_0^{2\pi} \mathrm{Exp}[2Y_0Y(\cos\theta\cos\chi+\sin\theta\sin\chi\cos\varphi)](\sin\chi' d\chi')\, d\varphi', \quad (2)$$

where the integral range of Y is taken in order to exclude the elliptic orbits, since the escape velocity at R corresponds to $\sqrt{2GM/(\sigma^2 R)} = \alpha_1$, while that of θ is conditioned so as to reject the imaginary radial velocity by putting $\theta_1 = \mathrm{Sin}^{-1}\sqrt{\alpha_1^2 X(1-X)/Y^2+X^2}$. After cancelling $dr\sin\chi' d\chi'$ on both sides and carrying out some calculations eq. (2) is reduced to

$$\nu(X,\chi') = \frac{2}{\sqrt{\pi}}\frac{n\alpha_1^3}{X^2}\int_1^{\infty} \mathrm{Exp}(-Y_0^2-\alpha_1^2y^2)y^3 dy \int_0^{\theta_1} \frac{1}{\rho}\sin\theta\cos\theta\, d\theta$$
$$\times\{\mathrm{Exp}[2\alpha_1Y_0y\cos(W-w+\theta)\cos\chi']\, I_0[2\alpha_1Y_0y\sin(W-w+\theta)\sin\chi']$$
$$+\,\mathrm{Exp}[2\alpha_1Y_0y\cos(W+w+\theta)\cos\chi']\, I_0[2\alpha_1Y_0y\sin(W+w+\theta)\sin\chi']\}, \quad (3)$$

where y denotes Y/α_1, $\rho = |\dot{r}|/(\sqrt{2}\sigma\alpha_1)$ the non-dimensional radial velocity, W the true anomaly at an incident point P on the sphere, w the true

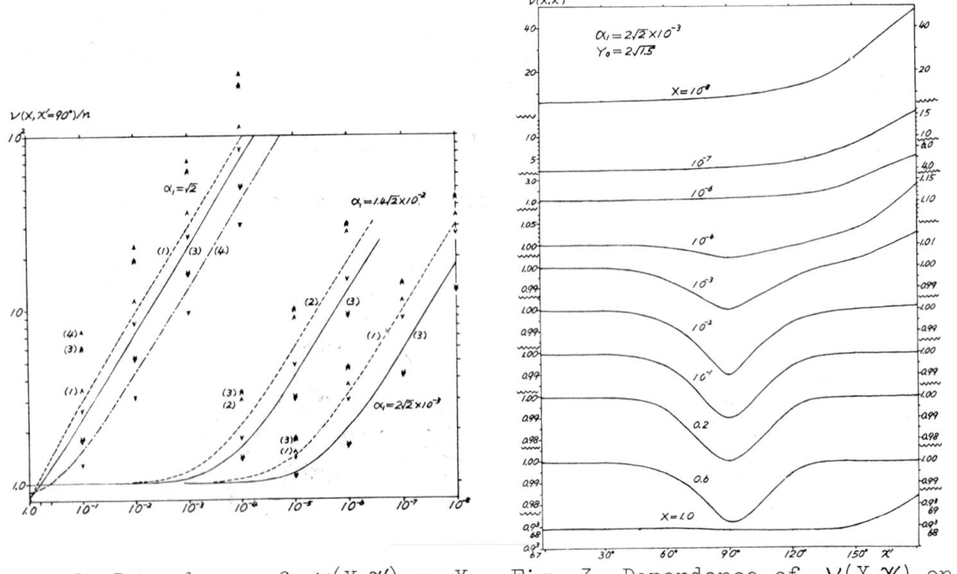

Fig. 2. Dependence of $\nu(X,\chi')$ on X. Fig. 3. Dependence of $\nu(X,\chi')$ on χ'.

anomaly at r, and $I_0(x)=\sum_{i=0}^{\infty}(x/2)^{2i}/(i!i!)$ the modified Bessel function of the zeroth order. It should be noticed that the second and the third line in eq. (3) correspond to the particles going into and those going out from a sphere with radius r respectively.
From eq. (3) the number of particles per unit time in a spherical shell between the radii r and r+dr, $dN(X)$, is derived as follows.

$$dN(X)=4\sqrt{\pi}nR^3\frac{\alpha_1^2}{Y_0}X^{3/2}dX\int_1^{\infty}\{\mathrm{Exp}[-(Y_0-\alpha_1 y)^2]-\mathrm{Exp}[-(Y_0+\alpha_1 y)^2]\}\sqrt{1+X(y^2-1)}\,dy. \quad (4)$$

This indicates that the proportionality of $dN(X)$ to X holds more closely as X becomes smaller. Accordingly a nearly dependence of $\nu(X,\chi')$ on $X^{-1/2}$ is expected for sufficiently small values of X.
Our numerical results substantiate it as exemplified in Fig. 2, where full and broken lines indicate runs of $\nu(X,\chi'=90°)/n$, while $\nu(X,\chi'=179°)/n$ and $\nu(X,\chi'=1°)/n$ are specified by upward marks (△, ∧, or ⩓) and downward ones (▽, ∨, or ⩔) respectively. In the vicinity of the test-particle, where such linearity holds approximately, the dependence of $\nu(X,\chi')$ on α_1 appears to be roughly linear for a certain set of Y_0 and χ'.
As for the relationship between $\nu(X,\chi')/n$ and χ' it is illustrated in Fig. 3, in which a dence crowding of particles is detectable in the stern of the test-particle. Such a trend is magnified as α_1 increases and Y_0 increases.

3. DYNAMICAL FRICTION

Now the gravitational force acting on a unit mass of the test-particle due to the encounting particles lying outside an assigned radius r, F(X), can be obtained by multiplying $(GM/r^2)\cos\chi'\cdot 2\pi r^2 dr \sin\chi' d\chi'$ with eq. (3) and integrating with respect to χ' and r. After lengthy calculations we have

$$F(X)=-\sqrt{\pi}\frac{n\mu^2}{\sigma^2 Y_0}\mathrm{Exp}(-Y_0^2)\int_1^{\infty}\mathrm{Exp}(-\alpha_1^2 y^2)\left[\cosh(2\alpha_1 Y_0 y)-\frac{\sinh(2\alpha_1 Y_0 y)}{2\alpha_1 Y_0 y}\right]\frac{y\,dy}{(y^2-1)}\left\{1-\frac{(2y^2-1)^2}{2y^2(y^2-1)}\ln(2y^2-1)\right.$$
$$\left.-\frac{1}{y}\sqrt{\frac{1+X(y^2-1)}{X}}+\frac{(1-X)^2}{2Xy^2}(y^2-1)\ln\left|\frac{\sqrt{1+X(y^2-1)}-y\sqrt{X}}{\sqrt{1+X(y^2-1)}+y\sqrt{X}}\right|-\frac{[y^2+X(y^2-1)]^2}{2Xy^2(y^2-1)}\ln\left|\frac{\sqrt{1+X(y^2-1)}-\sqrt{X}(y-1/y)}{\sqrt{1+X(y^2-1)}+\sqrt{X}(y-1/y)}\right|\right\}. \quad (5)$$

For $X=\varepsilon\ll 1$ all the terms in the second line of eq. (5) reduce to $4(y-1/y)\sqrt{\varepsilon}$. Thus eq. (5) without the second line represents F(0), namely the force due to all the particles in the whole sphere of the radius R. This F(0) is proved to be identical with the velocity change per unit time of the test-particle, $\Sigma\Delta v_{\parallel}$, derived under the same assumption as made here, so that F(X) is nothing but a part of the so-called dynamical friction.
Numerical integrations of eq. (5) for different values of X furnish information concerning which part of the hypothetical sphere contributes significantly to the encounter effect. Fig. 4 illustrates the dependence of F(X)/F(0) on X and the value of F(0) for each value of Y in the case of $\alpha_1=2\sqrt{2}\times 10^{-3}$. It is found therein that the encounter effect on a test-particle is mainly caused by its neighbouring particles crowding backwards, and the other ones lying in the outer part of the sphere contribute a little even though their number is much larger.
Our results show that as α_1 increases and Y decreases a smaller range of logX covers nearly the same curve of F(X)/F(0) as shown in Fig. 4.

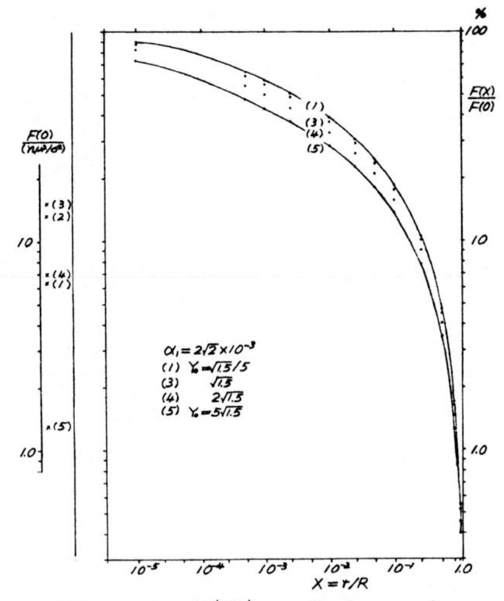

Fig. 4. F(0) and dependence of F(X)/F(0) on X.

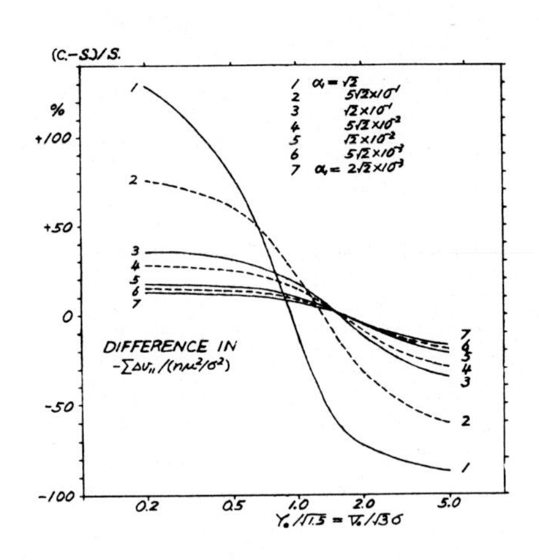

Fig. 5. Percentage difference in $\sum \Delta v_{\parallel}$ ((C-S)/S).

4. NUMERICAL DIFFERENCES BETWEEN CHANDRASEKHAR'S $\sum \Delta v_{\parallel}$ AND OURS.

Here we compare numerical values from Chandrasekhar's formula of $\sum \Delta v_{\parallel}$ with those from our F(0). Supposing D_0=R and denoting $\Phi(x)$ the error integral, we can write the former in our notations as

$$\sum \Delta v_{\parallel} = -\pi \frac{n\mu^2}{\sigma^2 Y_0^2} \ln\left(\frac{3}{\alpha_1^2}\right)\left[\Phi(Y_0) - Y_0 \Phi'(Y_0)\right]. \tag{6}$$

Numerical differences in $\sum \Delta v_{\parallel}/(n\mu^2/\sigma^2)$ between eq. (6) and F(0) from eq. (5) are shown graphically in Fig. 5 with different values of α_1 and Y . It is found there that deviations remain insignificant so long as α_1 is small enough as for the stellar field in the solar neighbourhood, but for a larger α_1 as in the cores of globular clusters and of clusters of galaxies the deviation becomes larger as Y_0 departs far from $\sqrt{1.5}$ corresponding to $V_0 = \sqrt{3}\sigma$. This may be anticipated since the initial assumption in deriving eq. (6) was $R \to \infty$, so $\alpha_1 \to 0$. On this connection it should be remarked that the impact parameter and the inclination of the orbital plane have usually been taken as the pair of independent variables, while in our treatment the colatitudial angle and the azimuthal one of the incident point on the sphere are adopted, consequently a modification sets in even when R tends to infinity.

Reference.

Chandrasekhar, S.: 1943, "Astrophys. J." 97, pp. 255-273.

SOME STATISTICS ON LONG-PERIOD COMETS

I. Hasegawa and T. Shimizu
2-3-11, Saidaijinogami, Nara 631, and
Bukkyo University, Kyoto 603, Japan

Distributions of the original reciprocal semimajor axes $1/a$ and the perihelion distances q of eighty "new" comets (Marsden et al, 1978) are discussed from the view point of a hypothesis of the interstellar origin (Hasegawa, 1976).

Consider a sphere of the radius R centered at the sun which moves with a velocity V_0 and assume that comets outside the sphere are distributed uniformly in space and their velocities obey the Maxwellian law, but if they happen to enter into the sphere they move along Keplerian orbits around the sun. Then, the frequency distribution of both the perihelion distance and the eccentricity of the comet inside the sphere is derived from equation (1) given in the preceding article by Shimizu as follows:

$$dN(e, X_0) = \frac{1}{4}\sqrt{\frac{\pi}{2}}\, n R^2 \sigma \frac{\alpha_1^3}{Y_0}\left\{ \mathrm{Exp}\left[-\left(Y_0 - \alpha_1\sqrt{1+\frac{e-1}{2X_0}}\right)^2 - \mathrm{Exp}\left[-\left(Y_0 + \alpha_1\sqrt{1+\frac{e-1}{2X_0}}\right)^2\right]\right\}\left(1+\frac{e-1}{2X_0}\right)^{-\frac{1}{2}} de^2\, d(\log X_0), \quad (1)$$

where X_0 and $\sqrt{3}\sigma$ denote q/R and the dispersion of comets' velocities outside the sphere respectively, while $Y_0 = V_0/(\sqrt{2}\sigma)$ and $\alpha_1^2 = GM/(\sigma^2 R)$ (G: gravitational constant, M: the sun's mass) are the parameters. The notations used here are identical with those of Shimizu. Rejecting the imaginary radial velocity at a point on the assumed sphere, the permissible values of X_0 should be in such a range as

$$\text{for } e \leqq 1,\ 1 \geqq X_0 \geqq \frac{1-e}{1+e} \geqq 0;$$

$$\text{for } e \geqq 1,\ 1 \geqq X_0 \geqq 0. \quad (2)$$

It is concluded therefore that for $e = \varepsilon\ (0 < \varepsilon \leqq 1)$ the value of X_0 is limited to a narrow range between $1 - 2\varepsilon$ and 1, while for $e > 1$ any value of X_0 in the whole range between 0 and 1 is admissible. It is noted here that when $e = 1$ it follows that $\sin^2\theta = X_0$, but for $e = 1 \mp \varepsilon$ we have $\sin^2\theta$

R. L. Duncombe (ed.), Dynamics of the Solar System, 319–322.

$= X_0(1 \mp \varepsilon/2)/(1+\varepsilon/2X_0)$, and for an assigned $X_0 \leqq 1$ such elliptic orbits as with the eccentricities less than $1-2X_0$ could not be found.

Then, with the aid of Shimizu's equation (1) above mentioned and the energy equation for the two body problem, the frequency distribution of R/2a with respect to small range of eccentricities from e to e'(>e), excluding e=1 in this interval, is derived as follows:

$$dN\left(\frac{R}{2a}\middle|e,e'\right)=\frac{1}{4}\sqrt{\frac{\pi}{2}}\,nR^2\sigma\alpha_1^3\frac{e'^2-e^2}{Y_0}\left\{\mathrm{Exp}\left[-\left(Y_0-\alpha_1\sqrt{1-\frac{R}{2a}}\right)^2\right]-\mathrm{Exp}\left[-\left(Y_0+\alpha_1\sqrt{1-\frac{R}{2a}}\right)^2\right]\right\}\left(1-\frac{R}{2a}\right)^{-\frac{1}{2}}\frac{d\left(\frac{R}{2a}\right)}{\frac{R}{2a}}. \quad (3)$$

With an assumed value of 1 km/s for both the mean relative velocity of the sun and the r.m.s. of comets' velocities, and some values for R between 10^3 and 10^5 AU, the theoretical distributions of 1/a and q are calculated. The observed data are taken from Table III of Marsden et al. (1978). Among both the first- and the second-class orbits in this table, we have eighty comets with the original 1/a values smaller than $+100 \times 10^{-6}$ AU^{-1}. These can be considered to be "new" comets which come near the sun probably for the first time.

For all of these eighty "new" comets and some additional ones, the frequency distribution of the original 1/a is obtained as shown by the shaded areas in Fig.1. The dotted lines in the figure represent the theoretical frequency distributions calculated from equation (3) with

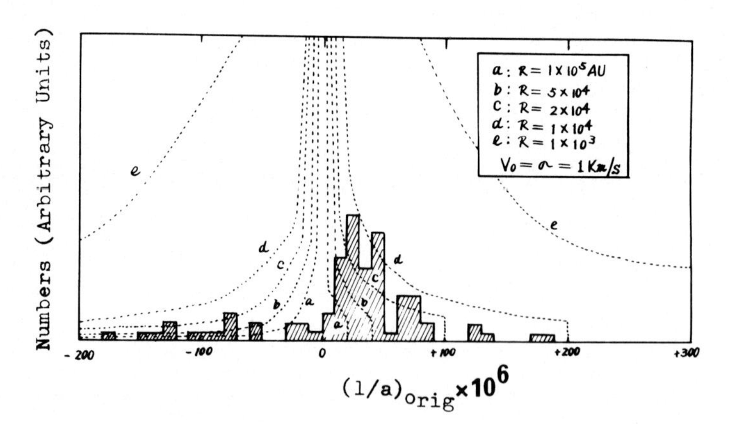

Fig.1 Distribution of 1/a

different parameters entered therein. These theoretical distributions seem to resemble closely the observed one in general trend, though a deviation is found around the zero value of 1/a.

In Fig.2 the frequency distribution of the perihelion distance q from the same data is shown together with the theoretical ones from equation (1) with different parameters written in. A general similarity is also discernible here, though it fails for small values of q.

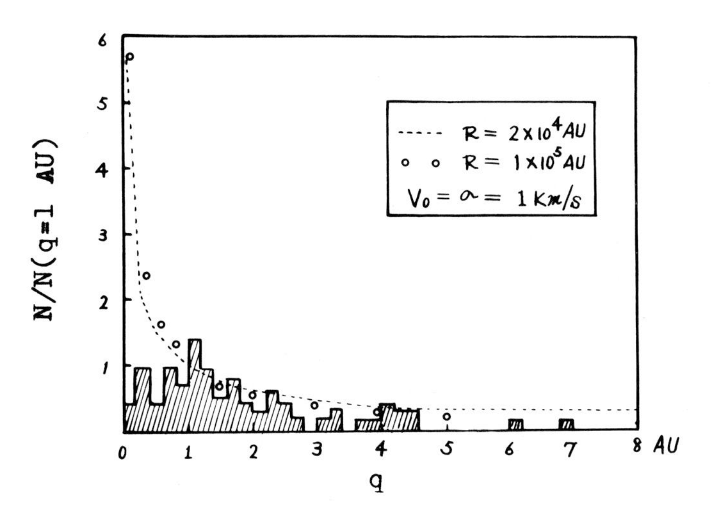

Fig.2 Distribuion of q

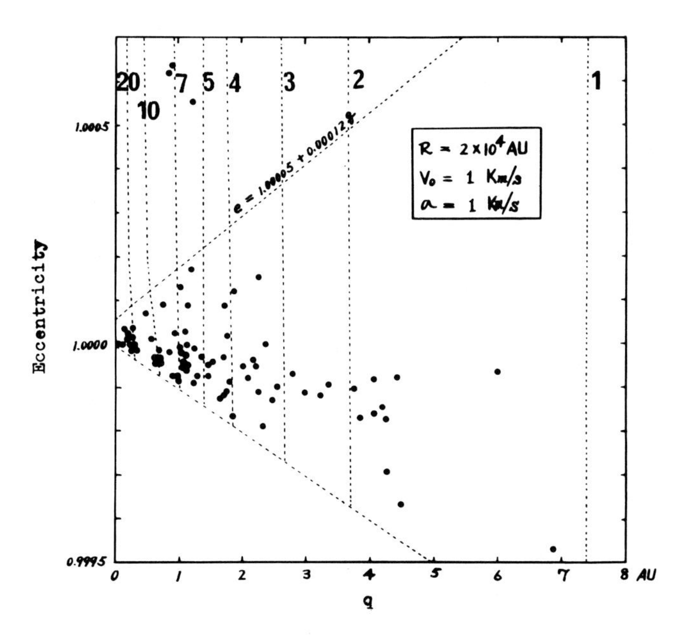

Fig.3 Correlations between e and q

On the other hand, Fig.3 is a correlation diagram between q and e (= 1-q/a), where the numbers entered between pairs of successive nearly vertical lines indicate the corresponding number of comets calculated from equation (1). It is seen there that almost all of the available data are included in a triangular region wedged between two dotted lines specified as follows: the lower one represents $e = 1-2X_0$, namely, the existence limit of comets with elliptic orbits and the upper one represents

$$e = 1.00005 + 0.00012q \text{ (AU)}. \qquad (4)$$

Assuming $R = 2 \times 10^4$ AU, the numerical values in equation (4) correspond to $\theta = 0^\circ\!.6$, V=0.50 km/s for q=1 AU, and $\theta = 1^\circ\!.5$, V=0.46 km/s for q=10 AU.

If the distribution of the comet's space velocity is not approximated well by the Maxwellian type assumed here, some deviations, as seen in Figs. 1 and 2, may be caused. This is because θ and V should vary in a narrow range such as above mentioned at the assumed distance R from the sun. Furthermore the selection effect sets in as for the observed data adopted here, especially for those with small perihelion distances, so that some deformations in the distributions of 1/a and q may be produced.

Therefore, our hypothesis that the comets originated in interstellar space seems to explain, at least in a general way, their observed distributions of 1/a and q by merely assuming a Maxwellian velocity distribution.

References

Hasegawa, I.; 1976, "Publ. Astron. Soc. Japan", 28, pp. 259-276.
Marsden, B. G., Sekanina, Z., and Everhart, E.; 1978, "Astron. J.", 83, pp. 64-71.

INDEX OF NAMES

INDEX OF SUBJECTS